Framework MATHS 8 E

TEACHER'S BOOK

David Capewell	Formerly Westfield School, Sheffield
Marguerite Comyns	Queen Mary's High School, Walsall
Gillian Flinton	All Saints Catholic High School, Sheffield
Paul Flinton	Chaucer School, Sheffield
Geoff Fowler	Maths Strategy Manager, Birmingham
Derek Huby	Mathematics Consultant, West Sussex
Peter Johnson	Wellfield High School, Leyland, Lancashire
Penny Jones	Waverley School, Birmingham
Jayne Kranat	Langley Park School for Girls, Bromley
Ian Molyneux	St. Bedes RC High School, Ormskirk
Peter Mullarkey	Netherhall School, Maryport, Cumbria
Nina Patel	Ifield Community College, West Sussex

OXFORD
UNIVERSITY PRESS

D0480988

OXFORD
UNIVERSITY PRESS

Great Clarendon Street, Oxford OX2 6DP

Oxford University Press is a department of the University of Oxford.
It furthers the University's objective of excellence in research,
scholarship, and education by publishing worldwide in

Oxford   New York

Auckland   Cape Town   Dar es Salaam   Hong Kong   Karachi
Kuala Lumpur   Madrid   Melbourne   Mexico City   Nairobi
New Delhi   Shanghai   Taipei   Toronto

With offices in

Argentina   Austria   Brazil   Chile   Czech Republic   France   Greece
Guatemala   Hungary   Italy   Japan   Poland   Portugal   Singapore
South Korea   Switzerland   Thailand   Turkey   Ukraine   Vietnam

Oxford is a registered trade mark of Oxford University Press
in the UK and in certain other countries

British Library Cataloguing in Publication Data

Data available

ISBN   978 0 19 914851 6

10   9

Typeset by Mathematical Composition Setters Ltd.

Printed and bound in Great Britain by Bell and Bain.

Acknowledgements
The photograph on the cover is reproduced courtesy of Graeme Peacock.

The publisher and authors would like to thank the following for permission
to use photographs and other copyright material:
Stone, page 1, Corbis, page 15 and 72, Empics, page 29, Science Photo Library,
page 57, Alamy Images, page 72 and 126, Rex Features, page 119, Ordnance
Survey, page 214.

Figurative artwork by Paul Daviz.

FSC
www.fsc.org

MIX
Paper from
responsible sources
FSC® C007785

About this book

This book has been written specifically to help you implement the Framework for Teaching Mathematics with higher ability students in Year 8. The content is based on the Year 9 teaching objectives from the Framework, and is targeted at students who have achieved Level 5 or above at the end of Year 7. To make the most of the material contained in this book it is strongly recommended that your students use the corresponding student book as shown on the back cover.

The authors are experienced teachers and maths consultants, who have been incorporating the Framework approaches into their teaching for many years and so are well qualified to help you successfully introduce the objectives in your classroom.

The book is made up of units based on the sample medium term plans that complement the Framework document, thus maintaining the required pitch, pace and progression.

The units are:

Each unit comprises a double page spread which should take a lesson to teach. These are shown on the full contents list.

References are made to resource material available on CD-ROM. There is more information about the Coursemaster CD-ROM on the back cover of this book.

This book is organised into double-page spreads that correspond to a 50–60 minute lesson. Each page shows the corresponding Student Book page, making it very easy to use.

The left-hand page gives suggestions for an engaging three-part lesson.

The **mental starter** is designed to be inclusive, that is all students should be able to participate most of the time. It usually provides a lead-in to the concepts of the main lesson.

An overview of the **teaching objectives** covered in each unit is provided on the first page of a unit so you can include references in your scheme of work.

Useful resources are listed here including CD-ROM references. These are also listed at the beginning of each unit so you can be fully prepared in advance.

The **introductory activity** will help you bring the associated student book to life as it provides engaging questions that will help students discuss the mathematical ideas.

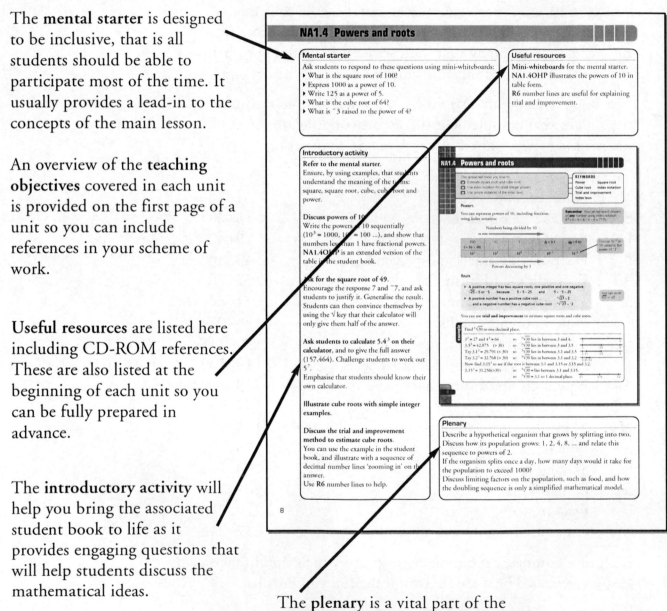

The **plenary** is a vital part of the three-part lesson, suggesting a way of rounding up the learning and helping to overcome any difficulties students may have faced.

The right-hand page of each spread corresponds to the Student Book exercise and will help you to make the most of the material provided.

Further activities are suggested that extend the questions provided, meaning you are unlikely to run out of work for students to do.

ICT resources are highlighted here.

The exercises contain three levels of **differentiation**: lead-in, focus and challenge. These three levels are highlighted to make it easier for you differentiate within ability groups and to manage the learning environment effectively.

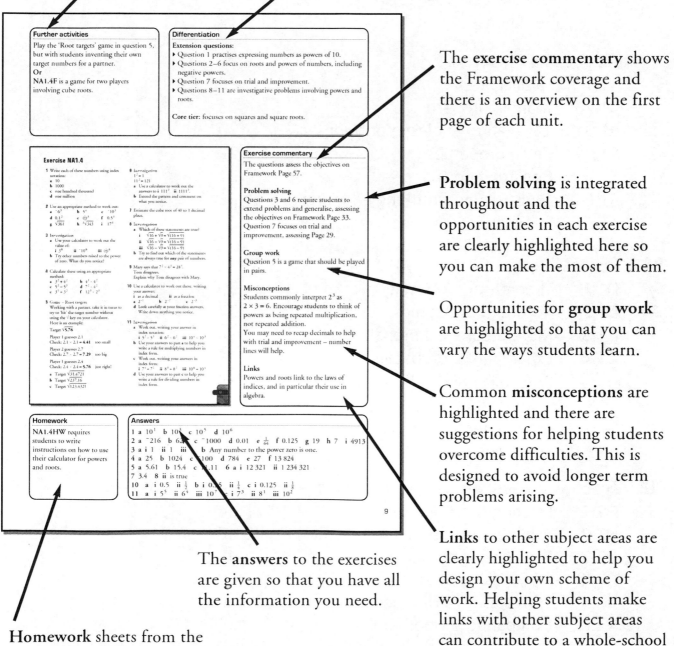

The **exercise commentary** shows the Framework coverage and there is an overview on the first page of each unit.

Problem solving is integrated throughout and the opportunities in each exercise are clearly highlighted here so you can make the most of them.

Opportunities for **group work** are highlighted so that you can vary the ways students learn.

Common **misconceptions** are highlighted and there are suggestions for helping students overcome difficulties. This is designed to avoid longer term problems arising.

Links to other subject areas are clearly highlighted to help you design your own scheme of work. Helping students make links with other subject areas can contribute to a whole-school numeracy policy.

The **answers** to the exercises are given so that you have all the information you need.

Homework sheets from the CD-ROM are described so you can quickly choose which ones to use.

Contents

Mental starters

Objectives covered in this unit:
▸ Order, add, subtract, multiply and divide integers.
▸ Calculate using knowledge of multiplication and division facts and place value.
▸ Know and use squares, cubes, roots and index notation.

Resources needed

* means class set needed
Essential:
NA1.2OHP shows the rules for multiplying and dividing directed numbers.
NA1.3OHP illustrates finding HCF and LCM using prime factors.
NA1.4OHP illustrates the powers of 10 in table form.
NA1.5OHP – position-to-term table for linear sequence.
NA1.6OHP – position-to-term table for quadratic sequence.
Useful:
Dice
Mini-whiteboards*
R1 – 0–9 digit cards*
Stopwatch
R6 – number lines*
NA1.4F – cube roots
NA1.6ICT – linear sequence

NA1 Numbers and sequences

This unit will show you how to:

▸▸ Extend mental methods of calculation, working with decimals, factors, powers and roots.
▸▸ Use standard column procedures to add and subtract integers and decimals of any size.
▸▸ Use the sign change key on a calculator, and function keys for powers and roots.
▸▸ Use the prime factor decomposition of a number.
▸▸ Use index notation for integer powers.
▸▸ Make and justify estimates and approximations of calculations.

▸▸ Generate terms of a sequence using term-to-term and position-to-term definitions.
▸▸ Write an expression to describe the nth term of an arithmetic sequence.
▸▸ Solve increasingly demanding problems and evaluate solutions.
▸▸ Solve substantial problems by breaking them into simpler tasks.
▸▸ Use trial and improvement where a more efficient method is not obvious.

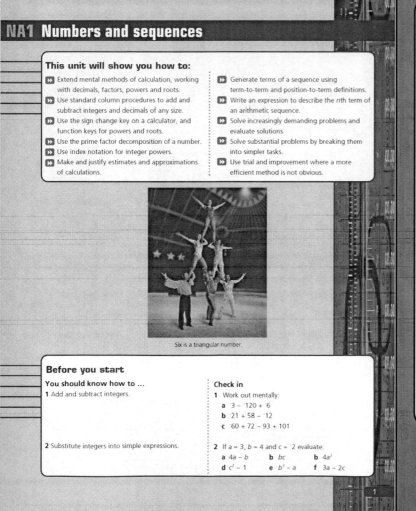

Six is a triangular number.

Before you start

You should know how to ...
1 Add and subtract integers.

2 Substitute integers into simple expressions.

Check in
1 Work out mentally:
a $^-3 - ^-120 + ^-6$
b $^-21 + 58 - ^-12$
c $^-60 + 72 - 93 + 101$

2 If $a = 3$, $b = 4$ and $c = ^-2$ evaluate:
a $4a - b$ b bc b $4a^2$
d $c^2 - 1$ e $b^3 - a$ f $3a - 2c$

Unit commentary

Aim of the unit

The first four sections develop skills in calculating with negative numbers, identifying multiples, factors and primes, including LCM and HCF, and powers and roots. The algebra sections look at linear and quadratic sequences, going on to develop position-to-term formulae for both types of sequence.

Introduction

Emphasise the importance of sequences in mathematics: this subject has its roots in spotting and developing patterns.

Framework references

This unit focuses on:
▸ Teaching objectives Framework Pages: 49, 51, 55, 57, 67, 91, 97, 105, 149, 151–153, 155–157
▸ Problem-solving objectives Framework Pages: 7, 29, 31, 33

Check in activity

The class stand in a circle. Give the first few terms of a sequence (5, 10, 20, ...) and, moving clockwise, the class must continue the sequence, one term each. The aim is to complete as many subsequent terms as possible. Discussion can bring out key vocabulary, for example was that sequence linear?

Differentiation

Core tier

Focuses on integer arithmetic, powers and roots, and sequences. The topics are covered parallel to the extension tier but at a more basic level.

NA1.1 Adding and subtracting negative numbers

Mental starter

On the line

Ask students to imagine a number line from ⁻1 to 1.
Ask questions about decimal numbers on the line:
▸ What is the largest 2-digit number?
▸ What is the smallest number with exactly one decimal place and no zeros?
▸ What is the mean of the largest and smallest numbers?
▸ Extend the activity by changing the number line from ⁻2 to 2.

Useful resources

R6 number lines may be useful for the mental starter and introductory activity.

Introductory activity

Discuss where negative numbers occur in real life.
Encourage students to demonstrate with examples (temperature, money, and height) how it helps to think of the context when comparing decimals.

Write five decimals on the board.
You could choose:
3.18, ⁻3.1, ⁻3.02, 3.3, ⁻3.25
Ask students to order the numbers.
Encourage students to explain and justify their answers. Demonstrate the use of a number line to help with solving the problem.

Ask students to add or subtract pairs of numbers mentally.
Start with a negative and a positive:
For example ⁻0.7 + 0.3
Then progress to two negatives:
⁻1.1 − ⁻1.2
Clarify the rules for adding and subtracting negative numbers. Emphasise the need to simplify a calculation:
⁻2.5 − ⁻3.6 = ⁻2.5 + 3.6

Ask students in pairs to think of two decimal problems: one that can be done using a mental method and one that will require a written method.
Encourage students to demonstrate their examples and solutions to the class.

NA1.1 Adding and subtracting negative numbers

This spread will show you how to:
▸▸ Order positive and negative integers and decimals.
▸▸ Add and subtract integers and decimals.

KEYWORDS

Integer	Positive
Negative	Inequality

You can use negative decimals to describe money that is owed.

example

Matt, Phoebe and Justine are students.
They are all overdrawn at the bank.
Their bank statements read ⁻£34.28, ⁻£7.43 and ⁻£12.89 respectively.

a Who owes i the most amount of money
            ii the least amount of money?
b How much do the three students owe the bank in total?

a i   Matt owes the most amount of money (£34.28).
   ii Phoebe owes the least amount of money (£7.43).
b 34.28 + 7.43 + 12.89 = 54.6 (using a written method).
   The students owe £54.60 in total.

Line up the decimal points:
34.28
 7.43
12.89
54.60

To compare decimal numbers, first check the sign and then compare the place values of each digit.

example

Complete these inequalities, using either of the symbols > or <.

a 0.0234   0.3
b ⁻32.4   4.3
c ⁻3.1246   ⁻3.1252

a Compare the first decimal place:
   0.0234 < 0.3
b Compare the sign:
   ⁻32.4 < ⁺4.3
c Compare the third decimal place:
   ⁻3.1246 < ⁻3.1252

You can add and subtract negative numbers either mentally or using a written method.
You need to remember these rules:

▸ Adding a negative number is the same as subtracting a positive number.
  3.4 + ⁻2.3 = 3.4 − 2.3 = 1.1

▸ Subtracting a negative number is the same as adding a positive number.
  3.4 − ⁻2.3 = 3.4 + 2.3 = 5.7

2

Plenary

Discuss the standard historical time line, with years measured as BC and AD.
Identify the fact that there was no year zero (you may discuss why this was the case – zero started as a medieval Arab concept), and pose problems such as:
▸ How many years were there between 58BC and 37AD?
▸ A man celebrates his 43rd birthday in 19AD. When was he born?

Further activities

Students can invent number pyramids and arithmagons for a partner to solve.

Students could investigate a link between the size of the arithmagon or pyramid, and the amount of numbers that you need at the start.

Differentiation

Extension questions:

▸ Question 1 practises adding and subtracting negative integers.
▸ Questions 2 to 8 focus on directed decimals.
▸ Question 9 is an investigative problem involving missing numbers.

Core tier: focuses on adding and subtracting negative integers and simple decimals in context.

Exercise NA1.1

1 Look at these integers.

    5   ⁻3   2   ⁻6   ⁻10

  a Which pair of integers gives:
   i the largest answer when added?
   ii the largest answer when subtracted?
   iii the smallest answer when added?
   iv the smallest answer when subtracted?
  b What is the sum of all the integers?

2 Calculate, using a mental or written method:
  a 4.3 + 3.69   **b** 4.3 + ⁻2.7
  c 1.037 + 3.063   **d** 12.65 − ⁻3.48
  e 4.005 + 5.095   **f** ⁻6.13 + ⁻2.8

3 Write all the decimal numbers with exactly two decimal places that lie between ⁻1.4 and ⁻1.3.

4 Puzzle
  1.551 is a palindromic number.
  It reads the same backwards and forwards.
  a How many other palindromic numbers can you find between ⁻2 and 2 that contain:
   i no zeros?
   ii exactly four digits
  b What is the total of all of these palindromic numbers?

5 a Add together each pair of numbers in the brackets. What do you notice?

   (3.43, ⁻3.2) (⁻6.7, 6.93) (⁻7.56, 7.79)

  b Subtract the larger number in each bracket from the smaller.

6 Calculate the following using a mental or written method, or, where appropriate, using a calculator.
  a ⁻4.3 + 2.7   **b** 37 − ⁻19 + 99
  c 42.1 + 36.7 + 12.4
  d ⁻37.5 + ⁻12.5   **e** 143 + 16.8 − ⁻4.9
  f 16.03 − 8.4 + 11.25 − ⁻3.75
  g ⁻³⁄₄ + ⁻²⁄₃   **h** ²⁄₅ − ⁻1¾

7 Puzzle

  Here are six numbers.
  You can add or subtract the numbers to make a target number, for example:
  12.23 − ⁻6.7 = 18.93.
  a Find two numbers to make ⁻18.5.
  b Find three numbers that you can use to make 12.
  c Find four numbers that you can use to make 16.48.

8 Look at the numbers in the cloud:

  a What is the total of the numbers in the cloud?
  b What is the mean of these numbers?
  c Which number is closest to zero?
  d What number lies exactly halfway between the largest and the smallest number?

9 Investigation
  In an arithmagon you add the numbers in the circles to produce the numbers in the squares.
  a Solve this arithmagon.

  b Describe and explain your method for solving this problem.
  c Invent an arithmagon problem of your own which involves **subtracting** the numbers in the circles to give the numbers in the squares.
  d Explain your method for solving subtraction arithmagons. Comment upon any problems in designing a subtraction arithmagon.

3

Exercise commentary

The questions assess the objectives on Framework Pages 49 and 105.

Problem solving
Questions 4, 7 and 9 are number problems and assess the objectives on Framework Page 7.

Group work
Question 9 can be attempted in pairs. Students can invent an arithmagon problem for their partner to solve.

Misconceptions
Students often have difficulty in ordering negative numbers, and may assume that ⁻2.3 is smaller than ⁻2.6.
Encourage students to think of a number line.

Links
Negative numbers link to other areas of the curriculum, in particular Science and Geography.

Homework

NA1.1HW provides practice in adding and subtracting directed numbers in context.

Answers

1 a 5 + 2   **b** 5 − ⁻10   **c** ⁻6 + −10   **d** ⁻10 − 5   **e** ⁻12
2 a 7.99   **b** 1.6   **c** 4.1   **d** 16.13   **e** 9.1   **f** ⁻8.93
3 ⁻1.39, ⁻1.38, ⁻1.37, ⁻1.36, ⁻1.35, ⁻1.34, ⁻1.33, ⁻1.32, ⁻1.31
4 a i ⁻1.991, ⁻1.881, ... ⁻1.111, 1.111, ... 1.991
    ii ⁻1.991, ⁻1.881, ... ⁻1.001, 1.001, 1.111, ... 1.991   **b** 0
5 a 0.23   **b** ⁻6.63, ⁻13.63, ⁻15.35
6 a ⁻1.6   **b** 155   **c** 91.2   **d** ⁻50   **e** 164.7   **f** 22.63   **g** ⁻1⁵⁄₁₂   **h** 2³⁄₂₀
7 a ⁻6.7 + ⁻11.8   **b** 5.02 + 0.28 − ⁻6.7   **c** 0.28 + ⁻2.3 − ⁻6.7 − ⁻11.8
8 a ⁻10.026 24   **b** ⁻1.114   **c** ⁻0.000 24   **d** ⁻2.333   **9 a** ⁻4.5, 15.1, ⁻0.1

3

Mental starter

Display this grid:

1	2	3	4	5	6
⁻5	⁻3	⁻6	0.4	7	1.5

Roll a dice twice and read out the corresponding numbers.
For example if you roll a 4 and a 2, you would read out 0.4 and ⁻3.
Students work out the product and write down their answer on a
mini-whiteboard.
Extend the activity to the product of three numbers.

Useful resources

Mini-whiteboards are useful for the mental starter.
NA1.2OHP shows the rules for multiplying and dividing directed numbers.
Dice

Introductory activity

Refer to the mental starter.
Discuss which products give negative answers. Encourage students to generalise.

Summarise the rules for multiplying and dividing directed numbers.
You could refer to the table given in the student book, or you could show **NA1.2OHP**, which gives the rules and examples.
Encourage students to memorise the rules.

Revise mental strategies for multiplying and dividing with decimals.
Give examples of:
▶ Using factors
▶ Using place value
Refer to the examples in the student book, and emphasise that the rules work the same for negative numbers.

Ask students to describe a mental method of working out 25 × 37.
Now ask for the answer to:
2.5 × 37, ⁻25 × ⁻37, ⁻0.25 × 37 …
Encourage students to link multiplication and division:
925 ÷ 25 = 37
Now ask for the answer to, for example,
925 ÷ ⁻2.5
Students can respond by whole-class discussion.

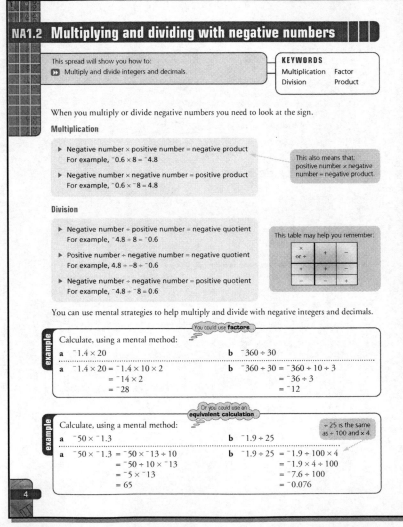

Plenary

Play **× and ÷ Countdown.**
▶ Split the class into small groups
▶ Write the numbers 1.5, ⁻2, 6, ⁻0.2, ⁻10, ⁻2.5
▶ Write a target number ⁻12.5
▶ Each group tries to reach the target number using any of the six numbers and the operations × and ÷
▶ The closest to the target number wins.
(One possible solution is ⁻10 × ⁻2.5 ÷ ⁻2)

Further activities

Students can invent multiplication arithmagons, like in question 8, for a partner to solve.
Arithmagons should include negative numbers and decimals.

Differentiation

Extension questions:

▸ Questions 1 and 2 practise multiplication of directed integers.
▸ Questions 3 to 7 focus on multiplying and dividing pairs of decimals.
▸ Question 8 is a problem-solving activity.

Core tier: focuses on multiplying and dividing directed integers.

Exercise NA1.2

1 Copy and complete this multiplication grid.

×	⁻4	⁻2	0	2	4
3					
2					
1					
0					
⁻1					
⁻2					
⁻3					

2 Here are four integers:

⁻3　⁻4　⁻5　10

Answer the following questions.
a Which pairs of numbers, when multiplied, will give a negative answer?
b Which pairs of numbers, when divided, will give a positive answer?
c Find a product of three numbers that will give an answer of 120.

3 Calculate these, using a mental or written method.
a ⁻0.6 × ⁻7　　**b** 240 ÷ ⁻48
c 1.7 × ⁻9　　**d** ⁻2.6 × ⁻40
e ⁻18 × 19　　**f** 43 × ⁻24
g ⁻60 × ⁻1.4　**h** ⁻644 ÷ ⁻28

4 Puzzle
Use two or more of the numbers in Box A, together with the operators × or ÷, to make the target number.
For example, for target number ⁻2.4:
⁻6 × 2 ÷ 5 = ⁻2.4
Target numbers
a ⁻4.2
b 4.43 (to 2 d.p.)
c 8.3
d 323 to the nearest whole number

Box A

⁻6	2
⁻10	⁻0.31
5	0.7

5 Investigation
Here is the ⁻1.45 times table!
1 × ⁻1.45 = ⁻1.45　　6 × ⁻1.45 =
2 × ⁻1.45 =　　　　　7 × ⁻1.45 =
3 × ⁻1.45 = ⁻4.35　　8 × ⁻1.45 =
4 × ⁻1.45 =　　　　　9 × ⁻1.45 =
5 × ⁻1.45 =　　　　　10 × ⁻1.45 = ⁻14.5

Write down strategies for calculating each of the missing products.
For example,
$$5 \times {}^-1.45 = \frac{10 \times {}^-1.45}{2} = \frac{{}^-14.5}{2} = {}^-7.25$$

6 Use the sign change key on your calculator to work these out.
In each case estimate your answer first.
a ⁻1.34 × 4.5
b 4.87 ÷ ⁻0.2
c ⁻3.1 + ⁻6.7
d 0.004 × ⁻3 000 000

7 Find the value of each of the expressions given, using these values:

$a = {}^-3$　　$b = {}^-4$　　$c = {}^-5$
　　　$d = 10$　　$e = {}^-0.5$

a $3a^2$　　　**b** $e - 2(\frac{d}{4})$
c $c^2 + 0.5b$　**d** $abc \div de$
In each case, show your working.

8 In these arithmagons the numbers in the squares are the products of the numbers in the circles on either side.
Solve these arithmagons.
a　　　　　　**b**

⁻182　154　　⁻399　437
　⁻143　　　　　⁻483

Exercise commentary

The questions assess the objectives on Framework Pages 51 and 97.

Problem solving
Questions 4, 5 and 8 assess the objectives on Framework Page 7.

Group work
In question 5, students can compare their strategies in small groups.

Misconceptions
Encourage students to realise that the mental strategies for multiplication and division apply equally for negative numbers.
Students should decide on the mental strategy to use, then think about the sign of the answer.

Links
The harder questions link to algebraic substitution: Framework Page 139.

Homework

NA1.2HW extends the link between number and algebra.

The questions require students to substitute numbers, including negatives and decimals, into expressions involving powers.

Answers

2 a ⁻3, ⁻4 or ⁻5 with 10　**b** Two negative　**c** ⁻3 × ⁻4 × 10
3 a 4.2　**b** ⁻5　**c** ⁻15.3　**d** 104　**e** ⁻342　**f** ⁻1032
　g 84　**h** 23
6 a ⁻6.03　**b** 4.67　**c** ⁻9.8　**d** ⁻12 000
7 a 27　**b** 0.1　**c** 23　**d** 12
8 a 14, 11, ⁻13　**b** 19, 23, ⁻21

Mental starter

Draw two digits at random from a set of 0–9 digit cards.
Students must make the two-digit number that has the greatest
number of divisors from this list: 2, 3, 4, 5, 6, 7, 8, and 9
For example, you draw out 3 and 6.
▸ 36 can be divided by 2, 3, 4, 6 and 9.
▸ 63 can be divided by 3, 7 and 9.
▸ 36 has the most divisors.

Useful resources

0–9 digit cards for the mental starter.
(You could use **R1**)
NA1.3OHP illustrates finding HCF and
LCM using prime factors.

Introductory activity

Refer to the mental starter.
Discuss strategies for dividing by single-
digit numbers.

Pose the question: 'If you know that 3 and
4 will divide into a number, does this also
mean that 12 will divide into it?'
Verify with examples, and extend to other
numbers: for example, will 21 go into any
number that both 3 and 7 will go into?

Recap definitions. What is a...
▸ Factor
▸ Prime number
▸ Prime factor?
Illustrate with the prime factors of 12
(2 and 3).
Invite a volunteer to show how 12 can be
written as a product using only its prime
factors: $2 \times 2 \times 3 = 12$

**Demonstrate the two common methods
for finding prime factors:**
▸ the factor tree method, and
▸ division by primes.

**Show how you can use prime factors to
find the LCM and HCF of two numbers.**
NA1.3OHP illustrates the techniques
using a Venn diagram.

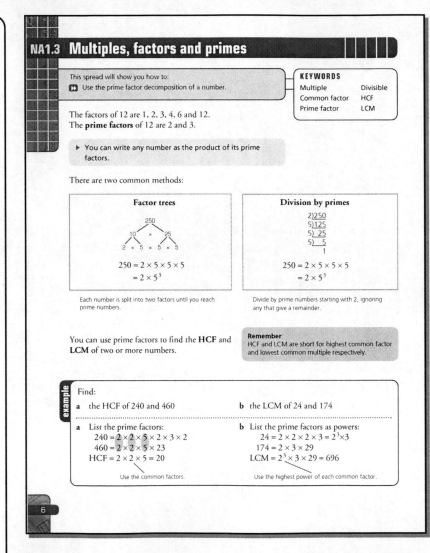

Plenary

Challenge students to calculate:
$\frac{12}{25} + \frac{14}{35}$
Discuss the use of prime factors to help find the lowest common
denominator.

Set this LCM problem:
Arjuna's Comet can be seen in the night sky every 72 years.
Nandi's Comet appears every 160 years. Given that they both appeared in the year 782AD, when should the two comets next appear in the same year?

(Answer 2222AD)

Differentiation

Extension questions:

▸ Questions 1 and 2 revise tests of divisibility, and listing prime factors in simple cases.
▸ Questions 3–6 focus on finding the HCF and LCM of a set of numbers, and using the LCM to add and subtract fractions.
▸ Questions 7–10 are puzzles relating to multiples and factors.

Core tier: focuses on finding the HCF and LCM of simpler numbers.

Exercise NA1.3

1 Find the HCF of each pair of numbers:
  a 25 and 46
  b 40 and 48
  c 24, 46 and 60
  d 15, 30 and 40

2 Find the LCMs of these sets of numbers:
  a 16 and 5
  b 12 and 20
  c the first three square numbers
  d the first three triangular numbers

3 Express each of these numbers as a product of its prime factors:
  a 34
  b 54
  c 485
  d 350
  e 83
  f 62
  g 815
  h 648

4 Find the HCF of each of these pairs of numbers.
  a 215 and 326
  b 532 and 359
  c 468 and 324
  d 238 and 430
  e 4950 and 4840
  f $6d$ and $24d$

5 Find the LCM of each of these sets of numbers:
  a 31 and 26
  b 34 and 39
  c 278 and 694
  d 648 and 132
  e 12, 18 and 30
  f 16, 9 and 12

6 Work out these problems involving fractions.
  (Hint: first find the LCM of the denominators.)
  a $\frac{23}{72} + \frac{12}{90}$
  b $\frac{13}{24} - \frac{60}{122}$
  c $\frac{24}{125} - \frac{24}{200}$
  d $\frac{43}{56} + \frac{67}{80}$

7 Puzzle
  a Which of these numbers is 2004 divisible by? Explain your method.
    i 15
    ii 12
    iii 22
    iv 25
    v 30
  b What is the smallest number that can be divided by 12, 15, 22, 25 and 30 without leaving a remainder?

8 Puzzle
  Find all the pairs of numbers that have an HCF of 30 and an LCM of 3150.

9 Puzzle
  Use the digits 0, 1, 5, 6, 7, and 8 to make two 3-digit numbers with an HCF of 45. Explain your method.

10 Puzzle
  a Write down a 2-digit number (for example, 47).
  b Reverse the digits (74).
  c Subtract the smaller number from the larger number (74 − 47 = 27).
  d Prove that the answer is always a multiple of 9.

Exercise commentary

The questions assess the objectives on Framework Pages 55, 67 and 91.

Problem solving
Questions 7–9 are puzzles requiring students to break the problem into simpler tasks, assessing the objectives on Framework Page 29.
Question 10 requires proof, assessing Page 31.

Group work
Questions 7–10 can be attempted and discussed in pairs.

Misconceptions
Students may incompletely decompose a number, for example writing $8 = 2 \times 4$. Encourage students to recognise the prime numbers up to 50.
Students often get LCM and HCF mixed up – using a Venn diagram as a visual aid may help.

Links
Finding the LCM links to adding and subtracting fractions: Framework Page 67.
Finding the HCF links to cancelling numerical and algebraic fractions: Pages 61 and 63.

Homework

NA1.3HW provides practice at finding and using prime factors, and reinforces rules for divisibility.

Answers

1 a 1 **b** 8 **c** 2 **d** 15 **2 a** 80 **b** 60 **c** 36 **d** 6
3 a 2×17 **b** 2×3^3 **c** 5×97 **d** $2 \times 5^2 \times 7$ **e** 83
  f 2×31 **g** 5×163 **h** $2^3 \times 3^4$
4 a 1 **b** 1 **c** 36 **d** 2 **e** 110 **f** $6d$
5 a 806 **b** 1326 **c** 96 466 **d** 7128 **e** 180 **f** 144
6 a $\frac{163}{360}$ **b** $\frac{73}{1464}$ **c** $\frac{9}{125}$ **d** $1\frac{399}{560}$ **7 a** 12 **b** 3300
8 30, 3150; 90, 1050; 150, 630; 210, 450 **9** 180, 675

NA1.4 Powers and roots

Mental starter

Ask students to respond to these questions using mini-whiteboards:

▸ What is the square root of 100?
▸ Express 1000 as a power of 10.
▸ Write 125 as a power of 5.
▸ What is the cube root of 64?
▸ What is ⁻3 raised to the power of 4?

Useful resources

Mini-whiteboards for the mental starter.
NA1.4OHP illustrates the powers of 10 in table form.
R6 number lines are useful for explaining trial and improvement.

Introductory activity

Refer to the mental starter.
Ensure, by using examples, that students understand the meaning of the terms: square, square root, cube, cube root and power.

Discuss powers of 10.
Write the powers of 10 sequentially ($10^3 = 1000$, $10^2 = 100$ ...), and show that numbers less than 1 have fractional powers. **NA1.4OHP** is an extended version of the table in the student book.

Ask for the square root of 49.
Encourage the response 7 and ⁻7, and ask students to justify it. Generalise the result. Students can then convince themselves by using the √ key that their calculator will only give them half of the answer.

Ask students to calculate 5.4^3 on their calculator, and to give the full answer (157.464). Challenge students to work out 5^7.
Emphasise that students should know their own calculator.

Illustrate cube roots with simple integer examples.

Discuss the trial and improvement method to estimate cube roots.
You can use the example in the student book, and illustrate with a sequence of decimal number lines 'zooming in' on the answer.
Use **R6** number lines to help.

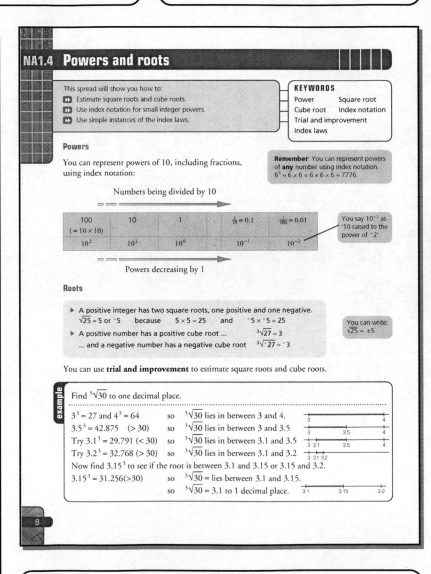

Plenary

Describe a hypothetical organism that grows by splitting into two.
Discuss how its population grows: 1, 2, 4, 8, ... and relate this sequence to powers of 2.
If the organism splits once a day, how many days would it take for the population to exceed 1000?
Discuss limiting factors on the population, such as food, and how the doubling sequence is only a simplified mathematical model.

Further activities

Play the 'Root targets' game in question 5, but with students inventing their own target numbers for a partner.
Or
NA1.4F is a game for two players involving cube roots.

Differentiation

Extension questions:
▶ Question 1 practises expressing numbers as powers of 10.
▶ Questions 2–6 focus on roots and powers of numbers, including negative powers.
▶ Question 7 focuses on trial and improvement.
▶ Questions 8–11 are investigative problems involving powers and roots.

Core tier: focuses on squares and square roots.

Exercise NA1.4

1 Write each of these numbers using index notation:
a 10
b 1000
c one hundred thousand
d one million

2 Use an appropriate method to work out:
a $^-6^3$ b 5^4 c $^-10^3$
d 0.1^2 e $(\frac{1}{4})^3$ f 0.5^3
g $\sqrt{361}$ h $^3\sqrt{343}$ i 17^3

3 Investigation
a Use your calculator to work out the value of:
  i 3^0 ii $^-10^0$ iii $(\frac{1}{4})^0$
b Try other numbers raised to the power of zero. What do you notice?

4 Calculate these using an appropriate method:
a $3^2 + 4^2$ b $4^3 \times 4^2$
c $5^2 - 5^3$ d $7^2 \times 4^2$
e $3^5 \div 3^2$ f $12^3 \times 2^3$

5 Game – Root targets
Working with a partner, take it in turns to try to 'hit' the target number without using the √ key on your calculator.
Here is an example:
Target $\sqrt{5.76}$

Player 1 guesses 2.1
Check: $2.1 \times 2.1 = \mathbf{4.41}$ too small

Player 2 guesses 2.7
Check: $2.7 \times 2.7 = \mathbf{7.29}$ too big

Player 1 guesses 2.4
Check: $2.4 \times 2.4 = \mathbf{5.76}$ just right!

a Target $\sqrt{31.4721}$
b Target $\sqrt{237.16}$
c Target $\sqrt{123.4321}$

6 Investigation
$1^2 = 1$
$11^2 = 121$
a Use a calculator to work out the answers to i 111^2 ii 1111^2.
b Extend the pattern and comment on what you notice.

7 Estimate the cube root of 40 to 1 decimal place.

8 Investigation
a Which of these statements are true?
  i $\sqrt{16} + \sqrt{9} = \sqrt{(16 + 9)}$
  ii $\sqrt{16} \div \sqrt{9} = \sqrt{(16 \div 9)}$
  iii $\sqrt{16} - \sqrt{9} = \sqrt{(16 - 9)}$
b Try to find out which of the statements are always true for **any** pair of numbers.

9 Mary says that $7^2 \times 4^3 = 28^5$.
Tom disagrees.
Explain why Tom disagrees with Mary.

10 Use a calculator to work out these, writing your answer:
  i as a decimal   ii as a fraction
a 2^{-1} b 2^{-2} c 2^{-3}
d Look carefully at your fraction answers. Write down anything you notice.

11 Investigation
a Work out, writing your answer in index notation:
  i $5^2 \times 5^3$ ii $6^2 \times 6^2$ iii $10^4 \times 10^3$
b Use your answers to part **a** to help you write a rule for multiplying numbers in index form.
c Work out, writing your answers in index form:
  i $7^4 \div 7^1$ ii $8^3 \div 8^2$ iii $10^6 \div 10^4$
d Use your answers to part **c** to help you write a rule for dividing numbers in index form.

9

Exercise commentary

The questions assess the objectives on Framework Page 57.

Problem solving
Questions 3 and 6 require students to extend problems and generalise, assessing the objectives on Framework Page 33. Question 7 focuses on trial and improvement, assessing Page 29. Questions 8 and 11 assess Page 31.

Group work
Question 5 is a game that should be played in pairs.

Misconceptions
Students commonly interpret 2^3 as $2 \times 3 = 6$. Encourage students to think of powers as being repeated multiplication, not repeated addition.
You may need to recap decimals to help with trial and improvement – number lines will help.

Links
Powers and roots link to the laws of indices, and in particular their use in algebra.

Homework

NA1.4HW requires students to write instructions on how to use their calculator for powers and roots.

Answers

1 a 10^1 b 10^3 c 10^5 d 10^6
2 a $^-216$ b 625 c $^-1000$ d 0.01 e $\frac{1}{64}$ f 0.125 g 19 h 7 i 4913
3 a i 1 ii 1 iii 1 b Any number to the power zero is one.
4 a 25 b 1024 c $^-100$ d 784 e 27 f $13\,824$
5 a 5.61 b 15.4 c 11.11 **6** a i $12\,321$ ii $1\,234\,321$
7 3.4 **8** ii is true
10 a i 0.5 ii $\frac{1}{2}$ b i 0.25 ii $\frac{1}{4}$ c i 0.125 ii $\frac{1}{8}$
11 a i 5^5 ii 6^4 iii 10^7 c i 7^3 ii 8^1 iii 10^2

9

Mental starter

Give the students a starting number, for example 30. Ask them to count on in their heads in jumps of, say, 11.

Time them for 1 minute, then see who 'wins' with the highest score.

Repeat with different starting numbers and rules, for example go down in even numbers.

Useful resources

Stopwatch
Mini-whiteboards may be useful for the plenary activity.
NA1.5OHP – position-to-term table for linear sequence.

Introductory activity

Discuss the mental starter activity.
When you start at 30 and count up in 11s, can you reach a total of 585? Encourage the response that all the possible answers must be 30 + a multiple of 11.
585 – 30 = 555, which is not a multiple of 11, so 585 is not a possible answer.

Emphasise that terms that go up or down by a constant amount each time are called linear sequences.
Ask for examples of linear sequences.

Are these sequences linear?

$$5, 20, 15, 20, 25$$
$$^-7, ^-14, ^-21, ^-28, ^-35$$
$$1, 4, 9, 16, 25$$

Discuss how students could win the mental starter game by cheating.
Discuss using differences (+11 each time) and the connection to multiples of 11. Derive the position-to-term sequence $T(n) = 11n + 30$. What is the 100th term of this sequence?

Compare the sequence $T(n) = 9n + 45$. What is the first term? What does the sequence go up in? Which of the two sequences has the largest 50th term?

Extend to sequences that go down by a fixed amount, for example $T(n) = ^-2n + 22$. **NA1.5OHP** contains a table illustrating the position-to-term rule for the sequence 8, 11, 14, 17, ... which is discussed in the student book.

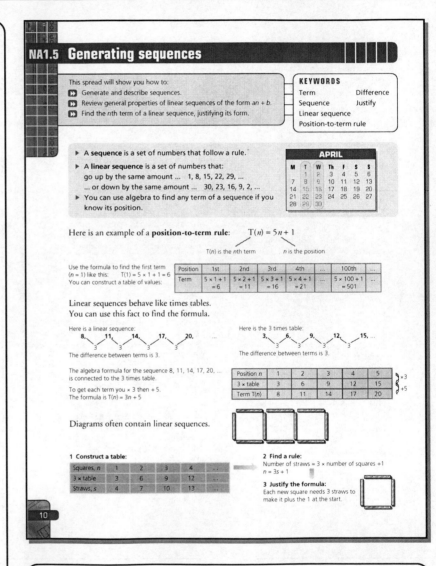

Plenary

Encourage students, as a class, to generate as many linear sequences as possible that include the term 58.

How many can they generate to include the terms 58 *and* 72?

Further activities

Students can use spreadsheets to generate linear sequences, extending to more complex sequences.

In **NA1.6ICT**, students generate terms of a linear sequence using a spreadsheet. The activity extends to estimating cube roots.

Differentiation

Extension questions:

▸ Questions 1–4 focus on completing terms in a sequence and using algebraic substitution to generate terms of a sequence.

▸ Questions 5–8 focus on finding a position-to-term formula to describe a sequence.

▸ Question 9 focuses on finding and justifying a formula for a sequence.

Core tier: focuses on generating sequences through using flowcharts.

Exercise NA1.5

1 Here is a page from a calendar.

Write down as many **linear** sequences as you can find.

2 Copy and complete these linear sequences:
a 5, 16, 27, 38, __, __, ...
b 36, 33, 30, 27, __, __, ...
c 4, __, 8, __, 12, ...
d 3, __, __, 33, __, ...
e 6, __, __, __, 26, ...
f 98, __, __, __, 78, ...

3 a Write down the first five terms of the sequences described by these position-to-term rules:

i $T(n) = 4n + 1$ ii $T(n) = 6n - 3$

iii $T(n) = 7n$ iv $T(n) = 18 - 3n$

b Name the sequence you have generated in **a** part **iii**.

4 a Generate the linear sequence that this flow chart describes.

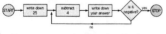

b Design your own flow chart for the sequence 4, 7, 10, 13, 17, 20, 23.

5 Find a formula for T(n), the general term of each sequence, in terms of n, the term number:
a 4, 9, 14, 19, 24, ...
b 3, 5, 7, 9, 11, ...
c 17, 30, 43, 56, 69, ...
d 20, 18, 16, 14, 12, ...
e multiples of 12
f counting down from 100 in 5s.

6 a Write down three different linear sequences with a third term of 10.
b Find a formula for T(n) in each case and give the 500th term.

7 The general formula for a linear sequence is $T(n) = an + b$, where a and b are constants (numbers).
What can you say about:
a a if the linear sequence is increasing in threes?
b b if the sequence is multiples of 10
c sequences where the value of a is 2
d a if the linear sequence is going down by a constant amount?

8 The 100th, 101st and 102nd terms of a linear sequence are 307, 310 and 313.
a Find a formula for T(n).
b Use it to generate the first five terms of the sequence.

9 Use this pattern to find a formula connecting the length of the middle square (L) and the number of white tiles (T). Justify why the formula works.

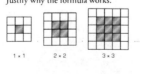

1 × 1   2 × 2   3 × 3

11

Exercise commentary

The questions assess the objectives on Framework Pages 149 and 155–157.

Problem solving

Question 7 assesses Framework Page 31. Question 9 provides an opportunity to assess Page 7, 31 and 33.

Group work

In question 6 students can compare their answers, aiming to generate more than three sequences.

The investigation in question 7 is best discussed in pairs or small groups.

Misconceptions

Students often confuse the position-to-term formula with the term-to-term rule. For example for 4, 7, 10, 13, they write $T(n) = n + 3$. The use of the difference idea needs to be reinforced repeatedly.

When justifying formulae, as in question 9, students may explain *how* a formula works, rather than *why* it works.

Links

Linear sequences are used in graph work, particularly to patterns noticed in line graphs, which are covered in A5.

Homework

NA1.5HW provides further opportunity to generalise in algebra and to justify formulae.

Answers

2 a 49, 60 b 24, 21 c 6, 10 d 13, 23, 43 e 11, 16, 21 f 93, 88, 83
3 a i 5, 9, 13, 17, 21 ii 3, 9, 15, 21, 27 iii 7, 14, 21, 28, 35 iv 15, 12, 9, 6, 3
   b Multiples of 7
4 a 25, 21, 17, 13, 9, 5, 1, ⁻3
5 a $5n - 1$ b $2n + 1$ c $13n + 4$ d $22 - 2n$ e $12n$ f $105 - 5n$
7 a $a = 3$ b $b = 0$ c Increase in 2s d Negative
8 a $3n + 7$ b 10, 13, 16, 19, 22 9 $T = 4L + 4$

Mental starter

Round the world game

Choose two students, one at either side of the room, to stand up.
Ask them a question on square numbers or roots (for example 8^2, $\sqrt{196}$, 20^2).
The first student to answer correctly remains standing. The other sits down and his/her neighbour stands up.
Continue round the room until everyone has had a question.
The last student standing is the winner.

Useful resources

NA1.6OHP – position-to-term table for quadratic sequence.

Introductory activity

Discuss how the Chinese used to find square roots (see students' book). This leads into the definition of a quadratic sequence as a sequence with constant second difference.

Encourage students to suggest formulae that will generate quadratic sequences (including the term n^2). Generate the first few terms of these sequences, reminding students to use BIDMAS.

Go through the method for finding the general term from the second difference, in the student book. Practise with other quadratic sequences, in each case using a table as in the example.

Ask students to suggest geometric representations of sequences:

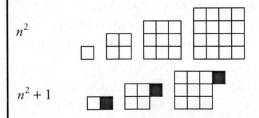

n^2

$n^2 + 1$

This links to work on justifying sequences, as in question 6.

NA1.6OHP illustrates the second example in the student book.

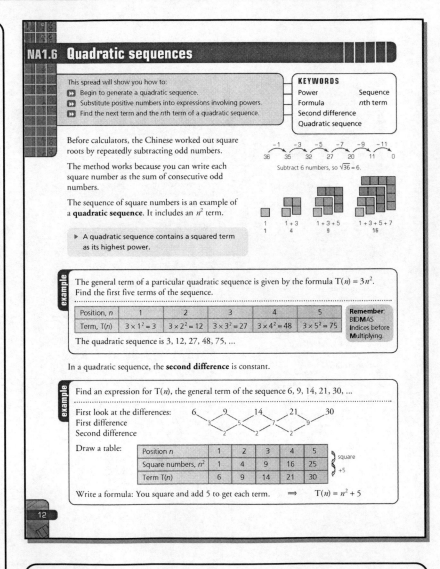

NA1.6 Quadratic sequences

This spread will show you how to:
- Begin to generate a quadratic sequence.
- Substitute positive numbers into expressions involving powers.
- Find the next term and the nth term of a quadratic sequence.

KEYWORDS
Power   Sequence
Formula nth term
Second difference
Quadratic sequence

Before calculators, the Chinese worked out square roots by repeatedly subtracting odd numbers.

The method works because you can write each square number as the sum of consecutive odd numbers.

The sequence of square numbers is an example of a **quadratic sequence**. It includes an n^2 term.

Subtract 6 numbers, so $\sqrt{36} = 6$.

▶ A quadratic sequence contains a squared term as its highest power.

example

The general term of a particular quadratic sequence is given by the formula $T(n) = 3n^2$. Find the first five terms of the sequence.

Position, n	1	2	3	4	5
Term, $T(n)$	$3 \times 1^2 = 3$	$3 \times 2^2 = 12$	$3 \times 3^2 = 27$	$3 \times 4^2 = 48$	$3 \times 5^2 = 75$

Remember: BIDMAS Indices before Multiplying.

The quadratic sequence is 3, 12, 27, 48, 75, ...

In a quadratic sequence, the **second difference** is constant.

example

Find an expression for $T(n)$, the general term of the sequence 6, 9, 14, 21, 30, ...

First look at the differences:
First difference
Second difference

Draw a table:

Position n	1	2	3	4	5
Square numbers, n^2	1	4	9	16	25
Term $T(n)$	6	9	14	21	30

square
+5

Write a formula: You square and add 5 to get each term. ⟹ $T(n) = n^2 + 5$

12

Plenary

Encourage students to find the general term of linear and quadratic sequences of fractions, for example:

$$\frac{5}{2}, \frac{5}{5}, \frac{5}{10}, \frac{5}{17}, \frac{5}{26}, ... \left(T(n) = \frac{5}{n^2 + 1} \right)$$

$$\frac{4}{2}, \frac{7}{8}, \frac{10}{18}, \frac{13}{32}, ... \left(T(n) = \frac{3n + 1}{2n^2} \right)$$

Further activities

Students can develop the ideas in this lesson by looking at 'shifted' sequences, where an operation takes place before squaring, for example $(n+1)^2$.

Differentiation

Extension questions:
▸ Questions 1 and 2 focus on generating quadratic sequences using differences.
▸ Questions 3 and 4 focus on using position-to-term rules to generate quadratic sequences.
▸ Questions 5–7 focus on finding second differences of sequences and using these to find the general term.

Core tier: focuses on the general term of a linear sequence.

Exercise NA1.6

1

> prime numbers 10, 40, 90, 160, 250, square numbers
> 3, 6, 9, 12, 15, 2, 4, 8, 16, 32, 64,
> triangular numbers
> 0, 3, 8, 15, 24, ... multiples of 11

Find all of the quadratic sequences in the box.

2 Generate the next five terms of each of these quadratic sequences.
  a 2, 5, 10, 17, 26, ...
  b ⁻1, 2, 7, 14, 23, ...
  c 4, 16, 36, 64, 100, ...
  d 2, 6, 12, 20, 30, ...
  e 10, 13, 18, 25, 34, ...

3 a For a particular sequence $T(n) = (n+2)(n+3)$.
   Generate the first five terms of this sequence.
  b Using differences, decide if the sequence is quadratic. Explain how you might have known this from the formula.

4 Copy and complete this crossword by finding the specified term for each clue.

Across
1 $T(n) = n^2 + 3$ ... 3rd term
3 $T(n) = n(n+1)$ ... 6th term
4 $T(n) = n^2$ ... 24th term
6 $T(n) = \frac{n^2}{2}$ ... 2nd term

Down
1 $T(n) = n^2 - 1$ ... 14th term
2 $T(n) = 2n^2$ ... 4th term
3 $T(n) = 3n^2 - 2$ ... 4th term
5 $T(n) = (n+3)(n+4)$ ... 5th term

5 a Copy and complete the table.

	Sequence	Value of second difference	Position-to-term formula
i	1, 4, 9, 16, 25, ...		
ii	2, 8, 18, 32, 50		
iii	3, 12, 27, 48, 75		
iv	4, 16, 36, 64, 100		
v	5, 20, 45, 80, 125		

  b Explain what you notice and use this to help you to find a general formula for
    i 10, 16, 26, 40, 58, ...
    ii 9, 39, 89, 159, 249, ...

6 For each sequence of diagrams, find a formula connecting the given quantities. Try to explain why this formula works.

  a
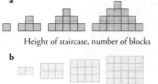
Height of staircase, number of blocks

  b

Height of rectangle, number of blocks

  c
Height of staircase, number of blocks
(Hint: Connect to **b**).
What is this sequence called?

7 Challenge
Find the mth term formula for:
  a 4, 9, 16, 25, 36, ...
  b 6, 12, 20, 30, 42, ...
  c $\frac{1}{9}, \frac{4}{16}, \frac{9}{25}, \frac{16}{36}, \frac{25}{49}, ...$

13

Exercise commentary

The questions assess the objectives on Framework Pages 151–153 and 155–157.

Problem solving
Questions 3, 5 and 6 provides the opportunity to assess the objectives on Framework Page 33.
Question 6 also assesses Page 7.

Group work
Students can work in pairs on the crossword in question 4, doing the substitutions mentally and discussing them.

Misconceptions
When substituting into formulae such as $3n^2$ students will often multiply by 3 before squaring. Emphasise the order of operations in BIDMAS.

Students may continue to confuse term-to-term and position-to-term formulae. This needs to be regularly reinforced.

Links
Quadratic equations link to work on graphs with curves.

Homework

NA1.6HW gives further practice in generating and generalising quadratic sequences, and provides mini-investigations resulting in square formulae.

Answers

1 10, 40, 90, 160, 250, ...; square numbers; triangular numbers; 0, 3, 8, 15, 24, ...
2 a 50, 81, 101, 145, 170 **b** 47, 78, 98, 142, 167
3 a 12, 20, 30, 42, 56 **b** Quadratic; expand brackets
4 Across: 12, 42, 576, 2; Down: 195, 32, 46, 72
5 a i 2, n^2 **ii** 4, $2n^2$ **iii** 6, $3n^2$ **iv** 8, $4n^2$ **v** 10, $5n^2$
  b i $2n^2 + 8$ **ii** $10n^2 - 1$
6 $B =$ No. blocks, $H =$ Height: **a** $B = H^2$ **b** $B = H(H+1)$ **c** $B = \frac{1}{2}(H^2 + H)$; triangular numbers
7 a $(n+1)^2$ **b** $n^2 + 3n + 2$ **c** $\frac{n^2}{(n+2)^2}$

Summary

The key objectives for this unit are:
- Make and justify estimates and approximations of calculations.
- Solve substantial problems by breaking them into simpler tasks.
- Generate terms of a sequence using term-to-term and position-to-term definitions of the sequence.
- Write an expression to describe the nth term of an arithmetic sequence.
- Give solutions to problems to an appropriate degree of accuracy.

Check out commentary

1. Some students will have difficulty deciding which values to choose for the calculation. This could be used as a useful discussion point – is 300 tubs 'near enough', or is it possible to calculate mentally with 250 tubs? Question 1b allows students to consider 'if 8.6 m is rounded to 10, is it necessary to round the 23?'

2. This problem requires students to realise that this task contains a number of smaller problems to solve (2×3, 1×4, ...).
   Questions then arise: are decimals allowed? Is a square a special kind of rectangle?

3. Reinforce that n represents the position of each term.
   In a(iv), BIDMAS may need to be highlighted to avoid students multiplying by 2 before squaring: I – indices before M – multiplication.
   In c encourage students not to use a calculator: if they find working in their head difficult, informal jottings can be used. Demonstrate that powers of two require a simple doubling strategy.

4. Part a mixes linear and quadratic sequences, so students must be encouraged to use differences, repeating until a constant difference is obtained. Part b examines the idea of a 'shifted' sequence which part (i) should lead into. The cube numbers in the second example may have been forgotten; students should be reminded of what they are in this case.

Plenary activity

Discuss methods for finding the position-to-term formula for linear and quadratic sequences. Ask students to write one example of each type of sequence and give it to a partner to find the position-to-term formula.

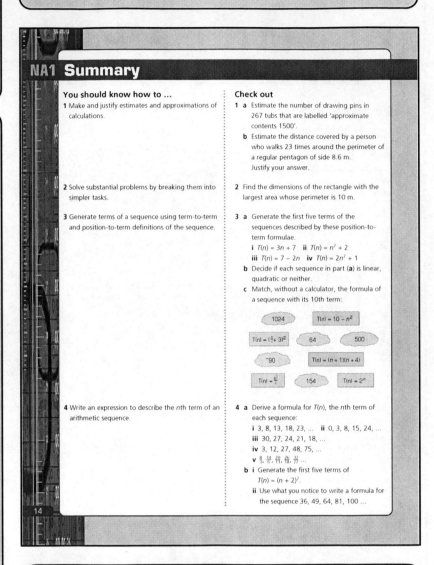

NA1 Summary

You should know how to ...

1 Make and justify estimates and approximations of calculations.

2 Solve substantial problems by breaking them into simpler tasks.

3 Generate terms of a sequence using term-to-term and position-to-term definitions of the sequence.

4 Write an expression to describe the nth term of an arithmetic sequence.

Check out

1 a Estimate the number of drawing pins in 267 tubs that are labelled 'approximate contents 1500'.
   b Estimate the distance covered by a person who walks 23 times around the perimeter of a regular pentagon of side 8.6 m. Justify your answer.

2 Find the dimensions of the rectangle with the largest area whose perimeter is 10 m.

3 a Generate the first five terms of the sequences described by these position-to-term formulae.
   i $T(n) = 3n + 7$ ii $T(n) = n^2 + 2$
   iii $T(n) = 7 - 2n$ iv $T(n) = 2n^2 + 1$
   b Decide if each sequence in part (a) is linear, quadratic or neither.
   c Match, without a calculator, the formula of a sequence with its 10th term:

   1024 $T(n) = 10 - n^2$
   $T(n) = (\frac{n}{2} + 3)^2$ 64   500
   ¯90 $T(n) = (n + 1)(n + 4)$
   $T(n) = \frac{n!}{2}$ 154 $T(n) = 2^n$

4 a Derive a formula for $T(n)$, the nth term of each sequence:
   i 3, 8, 13, 18, 23, ...   ii 0, 3, 8, 15, 24, ...
   iii 30, 27, 24, 21, 18, ...
   iv 3, 12, 27, 48, 75, ...
   v $\frac{8}{5}, \frac{14}{6}, \frac{20}{11}, \frac{26}{18}, \frac{32}{77}$ ...
   b i Generate the first five terms of $T(n) = (n + 2)^2$.
   ii Use what you notice to write a formula for the sequence 36, 49, 64, 81, 100 ...

14

Development

The number skills in this unit are used and developed throughout mathematics, as well as in further number units. The work on sequences is further developed in work on graphs.

Links

Algebraic skills are used throughout mathematics. Generalising sequences and justifying general terms will be useful in maths coursework later on.

Mental starters

Objectives covered in this unit:
▸ Visualise, describe and sketch 2-D shapes.
▸ Estimate and order acute, obtuse and reflex angles.
▸ Know or derive complements of 180.

Resources needed

* means class set needed

Essential:
S1.1OHP – Angle proofs
S1.2OHP – Angles in a triangle
S1.3OHP – Polygons
S1.4OHP – Parts of a circle
S1.5OHP – Constructing triangles
S1.6OHP – RHS triangles
Compasses
Protractor

Useful:
R1 digit cards
S1.6ICT – constructions

S1 Angles and bisectors

This unit will show you how to:

▸▸ Solve problems using properties of angles, of parallel and intersecting lines, and of triangles and other polygons.

▸▸ Explain how to find, calculate and use the interior and exterior angles of polygons.

▸▸ Know the definition of a circle and its parts.

▸▸ Explain why inscribed regular polygons can be constructed by equal divisions of a circle.

▸▸ Use straight edge and compasses to construct a triangle, given right angle, hypotenuse and side (RHS).

▸▸ Solve increasingly demanding problems and evaluate solutions.

▸▸ Present a concise, reasoned argument.

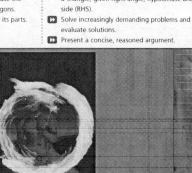

Circles are all around us

Before you start

You should know how to ...

1 Calculate angles on:
 a a straight line
 b in a triangle
 c at a point.

2 Know the names of polygons.

Check in

1 Calculate the unknown angles.

2 Name these shapes.

15

Unit commentary

Aim of the unit

This unit extends knowledge of angles in polygons and constructing triangles, and introduces circles.

Introduction

Review previous knowledge of angle properties. Focus on correct terminology, such as vertically opposite and supplementary.
Discuss the instruments used for constructing and measuring angles: protractor, compasses and ruler.
Emphasise the importance of bringing these to class.

Framework references

This unit focuses on:
▸ Teaching objectives pages: 183, 187, 195, 221 and 223.
▸ Problem-solving objectives pages: 17, 29, 31, and 33.

Differentiation

Core tier

Focuses on calculating angles in polygons and constructing bisectors.

Check in activity

Set some simple linear equations involving 180 and 360.
For example:
$5a = 180$
$2a + a + 42 = 180$
$3x + 5x + 112 + x = 360$

Mental starter

Review angle definitions and properties.
Draw two parallel lines on the board, and make different angles with a board ruler. Students should:
▸ Describe pairs of angles using terms such as *corresponding* and *alternate*.
▸ Estimate the angles.
  Discuss what is a reasonable estimate.

Useful resources

S1.1OHP shows geometrical proofs for the angles in a triangle and a quadrilateral.

Introductory activity

Refer to the mental starter.
Discuss the angles formed when a transverse line crosses a pair of parallel lines. In particular:
▸ Which angles are equal?
▸ Which angles add up to 180°?
Encourage students to use correct terminology: corresponding, alternate, supplementary and vertically opposite.

Discuss the angles in a triangle.
Recap that the angles in a triangle add up to 180°.

Demonstrate the geometrical proof in the example in the student book. The proof is illustrated in **S1.1OHP**.

Show how you can use the angles in a triangle to prove that the angles in a quadrilateral add up to 360°. This is also illustrated in **S1.1OHP**.

Show how you can use angle facts to solve problems. You could use this example:
($a = b = c = 89°$, $d = 64°$)

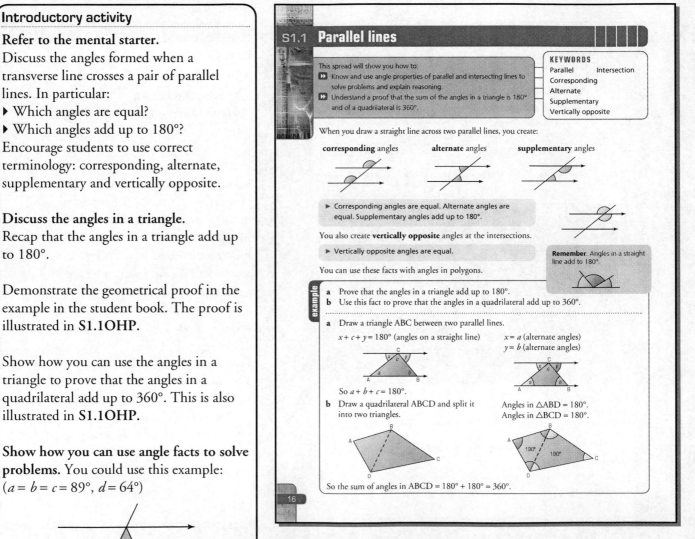

Encourage students to show geometrical reasoning at each stage, for example:
$b = a = 89°$ (vertically opposite angles).

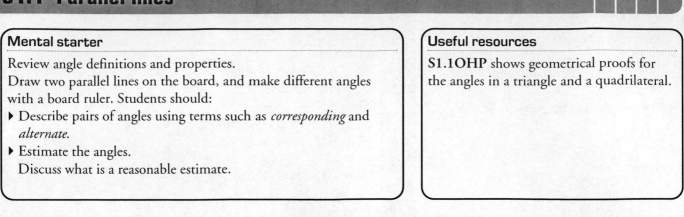

S1.1 Parallel lines

This spread will show you how to:
▸ Know and use angle properties of parallel and intersecting lines to solve problems and explain reasoning.
▸ Understand a proof that the sum of the angles in a triangle is 180° and of a quadrilateral is 360°.

KEYWORDS
Parallel   Intersection
Corresponding
Alternate
Supplementary
Vertically opposite

When you draw a straight line across two parallel lines, you create:

corresponding angles   **alternate** angles   **supplementary** angles

▸ Corresponding angles are equal. Alternate angles are equal. Supplementary angles add up to 180°.

You also create **vertically opposite** angles at the intersections.

▸ Vertically opposite angles are equal.

Remember. Angles in a straight line add to 180°.

You can use these facts with angles in polygons.

example

a Prove that the angles in a triangle add up to 180°.
b Use this fact to prove that the angles in a quadrilateral add up to 360°.

a Draw a triangle ABC between two parallel lines.
$x + c + y = 180°$ (angles on a straight line) $x = a$ (alternate angles)
                                                    $y = b$ (alternate angles)

So $a + b + c = 180°$.

b Draw a quadrilateral ABCD and split it into two triangles.

Angles in △ABD = 180°.
Angles in △BCD = 180°.

So the sum of angles in ABCD = 180° + 180° = 360°.

16

Plenary

Discuss questions 8, 9 and 10 from the exercise, which involve angles in a quadrilateral.

Further activities

Students can devise their own parallel line problems for a partner to solve.

Differentiation

Extension questions:
▸ Questions 1 and 2 involve a pair of parallel lines and a transverse.
▸ Questions 3–7 involve angles in more complex arrangements of lines.
▸ Questions 8–10 extend to quadrilaterals.

Core tier: focuses on parallel lines and triangles.

Exercise S1.1

Calculate the unknown angles, giving reasons for your answers.

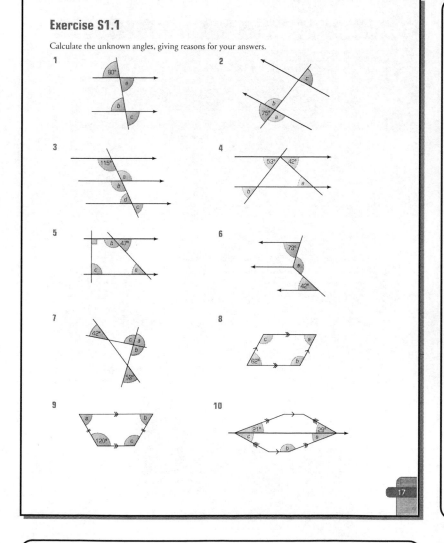

Exercise commentary

The questions assess the objectives on Framework Pages 183 and 187.

Problem solving

The questions in this exercise assess Framework Page 17. Also, they each require a mathematical justification, assessing Page 31.

Group work

The suggested further activity should be done in pairs.

Misconceptions

The most common mistake is to measure angles rather than calculate them. Emphasise the importance of reading the wording of the question carefully. Where three angles add up to 180° students may assume that they are equal. Encourage students to make no assumptions other than what is given.

Links

This topic links to synthesising information in algebraic form (Framework Page 27).

Homework

S1.1HW provides further practice at calculating angles and justifying methods.

Answers

1 All 80°   2 105°, 105°, 75°
3 115°, 115°, 65°, 65°   4 42°, 53°
5 47°, 133°, 90°   6 115°
7 85°, 95°, 95°   8 62°, 118°, 118°
9 60°, 60°, 120°   10 21°, 159°, 29°

Mental starter

Practise complements to 180.

Start with a single integer, and ask for the complement to 180. Progress to giving two integers.

Students can respond using digit cards.

Useful resources

R1 digit cards for the mental starter.

S1.2OHP shows the diagrams in the student book.

Introductory activity

Recap the angles of a triangle.
Look at various types of triangle, and ask questions like:
▸ What are the angles in an equilateral triangle?
▸ What are the equal angles in a right-angled isosceles triangle?

Ensure that students understand the terms *interior angle* and *exterior angle*.

Demonstrate a proof that the exterior angle of a triangle equals the sum of the two interior angles opposite.
S1.2OHP illustrates this proof using the diagram in the student book.

Show how you can use this fact to solve problems.
Discuss the examples in the student book, which are also illustrated on S1.2OHP.

Emphasise the importance of justifying each calculation, and use the examples to show how this can be done concisely.

Emphasise the importance of sketching a diagram.

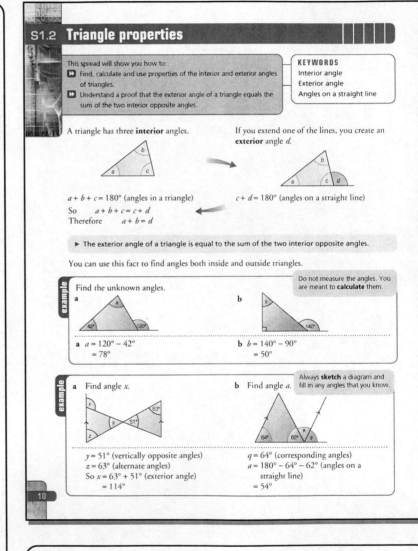

S1.2 Triangle properties

This spread will show you how to:
▸ Find, calculate and use properties of the interior and exterior angles of triangles.
▸ Understand a proof that the exterior angle of a triangle equals the sum of the two interior opposite angles.

KEYWORDS
Interior angle
Exterior angle
Angles on a straight line

A triangle has three **interior** angles.

If you extend one of the lines, you create an **exterior** angle d.

$a + b + c = 180°$ (angles in a triangle)
So $\quad a + b + c = c + d$
Therefore $\quad a + b = d$

$c + d = 180°$ (angles on a straight line)

▸ The exterior angle of a triangle is equal to the sum of the two interior opposite angles.

You can use this fact to find angles both inside and outside triangles.

example

Find the unknown angles.

Do not measure the angles. You are meant to **calculate** them.

a

b

a $a = 120° - 42°$
$\quad = 78°$

b $b = 140° - 90°$
$\quad = 50°$

example

a Find angle x.

b Find angle a.

Always **sketch** a diagram and fill in any angles that you know.

$y = 51°$ (vertically opposite angles)
$z = 63°$ (alternate angles)
So $x = 63° + 51°$ (exterior angle)
$\quad = 114°$

$q = 64°$ (corresponding angles)
$a = 180° - 64° - 62°$ (angles on a straight line)
$\quad = 54°$

18

Plenary

Recap the formula for the exterior angle of a triangle.
Discuss the relationship between the angles of this quadrilateral.

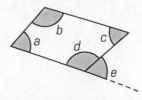

Further activities

Students can set angle problems for a partner to solve.

Emphasise that the problems should be solvable, and encourage students to give minimum but sufficient information in each diagram.

Differentiation

Extension questions:
▸ Questions 1 and 2 involve the relationship between the exterior and interior angles of a triangle.
▸ Questions 3–8 are more complicated problems involving angles.
▸ Questions 9 and 10 involve triangles and angles in parallel lines.

Core tier: focuses on angles in triangles and quadrilaterals.

Exercise S1.2

Find the unknown angles, giving reasons for your answers.

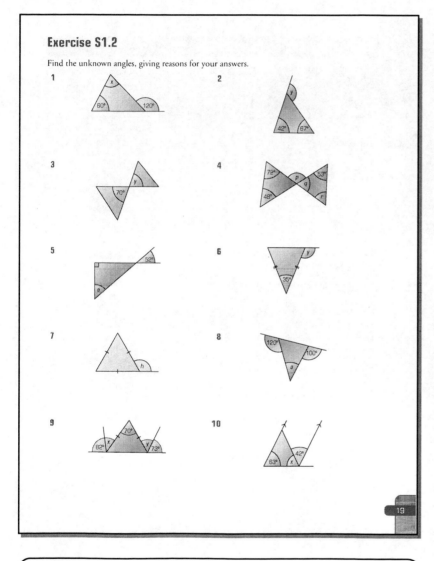

Exercise commentary

The questions assess the objectives on Framework Page 183.

Problem solving

The questions in this exercise assess the objectives on Framework Page 17. Justification of workings assesses Page 31.

Group work

Students can try the suggested further activity in pairs.

Misconceptions

Students may still try to measure the angles rather than calculate them. Emphasise that in questions like these, the diagrams are not drawn accurately. Students may mix up angles that add to 180° with those that add to 360°. Encourage students to draw a sketch to give an estimate of the answer.

Links

Geometrical reasoning has strong links with problem solving.

Homework

S1.2HW provides further practice in calculating angles in parallel lines and triangles.

Answers

1 60°   2 109°   3 70°
4 120°, 60°, 67°   5 38°
6 107.5°   7 120°   8 40°
9 43°, 52°   10 75°

Mental starter

Draw various polygons on the board, or use **S1.3OHP**.
Ask students to identify each of them.
Encourage the use of terms such as *irregular* and *regular* where appropriate.

Students can respond orally, or using mini-whiteboards.

Useful resources

Mini-whiteboards may be useful for the mental starter.
S1.3OHP shows a variety of regular and irregular polygons.

Introductory activity

Recap the sum of angles of a triangle.
Invite a student to show how to use this fact to determine the sum of the interior angles of a quadrilateral.

Show how this method can be extended to a polygon with a greater number of sides.
Divide each polygon into a number of triangles, as shown in the student book. You can use the shapes on **S1.3OHP**. Emphasise that the polygons can be irregular.

Encourage students to identify a pattern.
Ask questions like:
If a polygon has five sides, how many triangles can you split it into? What about 17 sides?
Then ask: *What will the angles add up to in each case?*
Generalise the result: $S = (n-2) \times 180°$.

Discuss the relationship between interior and exterior angles, and generalise:
Interior + exterior = 180°
For an *n*-sided polygon, sum of interior + exterior = $n \times 180°$
But sum of interior = $(n-2) \times 180°$
So sum of exterior = $2 \times 180° = 360°$

Extend to regular polygons. Show that the interior angle of a regular polygon is $\frac{S}{n}$, and illustrate with the example of the regular decagon in the student book.

S1.3 Calculating angles in polygons

This spread will show you how to:
▶ Find, calculate and use the interior and exterior angles of regular polygons.
▶ Find, calculate and use the sums of the interior and exterior angles of quadrilaterals, pentagons and hexagons.

KEYWORDS
Pentagon   Hexagon
Polygon
Interior angle
Exterior angle
Regular polygon

You can split a ...

... quadrilateral into two triangles

... pentagon into three triangles

... hexagon into four triangles.

The sum of the interior angles is $2 \times 180° = 360°$.

The sum of the interior angles is $3 \times 180° = 540°$.

The sum of the interior angles is $4 \times 180° = 720°$.

▶ A polygon with *n* sides can be split into $(n-2)$ triangles, each with an angle sum of 180°.
▶ The interior angle sum *S* of any polygon with *n* sides is given by the formula: $S = (n-2) \times 180°$.

Check the formula for a hexagon where $n = 6$:
$S = (6-2) \times 180° = 720°$.

▶ At each vertex of any polygon, the sum of the interior and exterior angles is 180°.

For a hexagon, the sum of six interior and six exterior angles is $6 \times 180°$.
But the sum of the interior angles is $4 \times 180°$, so the sum of the exterior angles is $2 \times 180° = 360°$.

▶ The sum of the exterior angles of any polygon is 360°.

In a regular polygon all the angles are equal.

▶ The interior angle of a regular polygon is $S \div n$ where S = the interior angle sum and n is the number of sides.

Remember:
Each of the interior angles in an equilateral triangle is $180° \div 3 = 60°$.

20

Plenary

Present a reverse problem:
You know that the interior angle of a regular polygon is 160°. How many sides does it have?
Encourage students to:
▶ find the exterior angle
▶ use the fact that the sum of exterior angles is 360°.

Further activities

In pairs or small groups, students can:
▸ Compile a table of results for the interior and exterior angles of different regular polygons.
▸ Explore any patterns or relationships.

Differentiation

Extension questions:
▸ Questions 1–3 require standard calculation of angles in a regular polygon.
▸ Questions 4–8 require the use of algebra and include irregular polygons.
▸ Questions 9–12 require students to calculate the number of sides of a regular polygon given the interior angle.

Core tier: focuses on the side, angle and symmetry properties of triangles.

Exercise S1.3

Calculate the unknown angles, giving reasons for your answers.

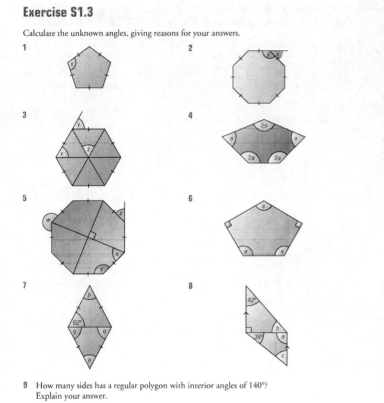

1
2
3
4
5
6
7
8

9 How many sides has a regular polygon with interior angles of 140°? Explain your answer.

10 How many sides has a regular polygon with interior angles of 135°? Explain your answer.

11 Can a regular polygon have an interior angle of 100°? Explain your answer.

12 **a** If the exterior angle of a regular polygon is 30°, how many sides does it have?
   b What are the sizes of the interior angles?

Exercise commentary

The questions assess the objectives on Framework Page 183.

Problem solving
The questions in this exercise assess the objectives on Framework Page 17. Using a mathematical justification assesses Page 31.

Group work
Questions 9–12 could be attempted in pairs.

Misconceptions
Students may assume that the interior angles of all polygons add up to the same amount. Encourage students to think about which sum to use.
Emphasise again that students should calculate, rather than measure, the angles.

Links
Geometrical reasoning has strong links with problem solving.

Homework

S1.3HW provides practice at calculating interior angles, and explores regular polygons further.

Answers

1 108° **2** 45°, 135° **3** All 60°
4 $a = 67.5°$, $2a = 135°$
5 225°, 67.5°, 135°, 45°
6 120° **7** 56°, 62° **8** 90°, 48°, 48°
9 9 sides **10** 8 sides
11 No, 360 ÷ 80 = 4.5
12 **a** 12 **b** 150°

Mental starter

Ask students to describe parts of a circle without actually using the name of the part.

This can be turned into a game (like 'Call my Bluff'), whereby other students can try guess the name of the part.

Useful resources

S1.4OHP illustrates the parts of a circle.

Compasses
Protractor

Introductory activity

Discuss circles.

Talk about where they appear in the real world, for example cogs and wheels, or the path traced out by a conker on the end of a string.

Ask the class to define a circle.

Encourage exact terminology and recap terms: *circumference*, *radius*, *centre*, and *diameter*.

Define other parts of a circle.

Invite volunteers to offer definitions for the terms: *arc*, *sector* and *segment*. Refer to the diagrams in the student book, which are also shown on S1.4OHP.

Briefly recap regular polygons.

Show, by sketching, how a regular polygon can be divided into congruent isosceles triangles that meet at the centre.

Demonstrate how you can construct a regular octagon within a circle. Ensure that each member of the class has access to compasses and a protractor.

Talk through the steps involved, as illustrated in the student book.

Challenge students to calculate the interior angle of a regular octagon, using the techniques learned in the previous lesson. Students can use the interior angle as a check for accuracy.

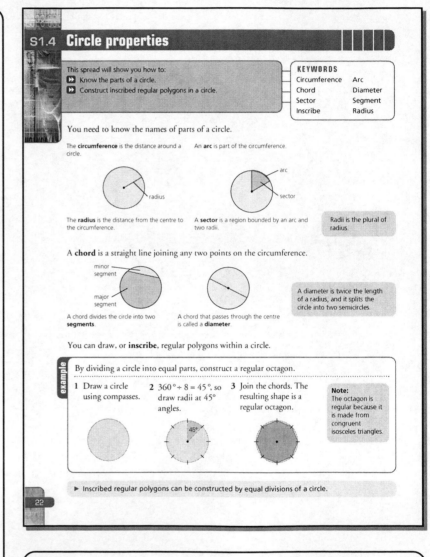

Plenary

Discuss question 2 of the exercise and generalise the result:
Angles in a semicircle are always 90°.

Further activities

Students can inscribe other regular polygons within a circle, for example a regular nonagon.

Differentiation

Extension questions:
▸ Question 1 consolidates knowledge of the parts of a circle.
▸ Question 2 is an investigation involving the angle in a semicircle.
▸ Question 3 requires students to inscribe a regular hexagon within a circle and question 4 involves unknown angles.

Core tier: focuses on the properties of quadrilaterals.

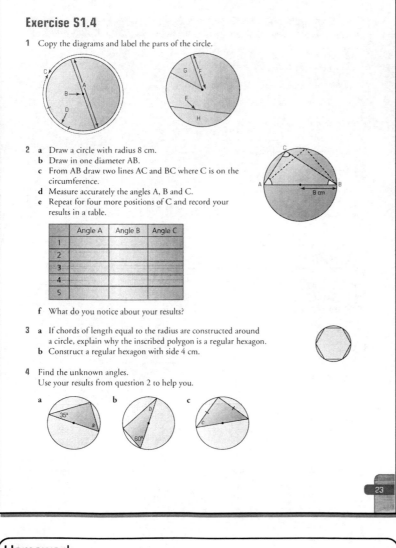

Exercise S1.4

1 Copy the diagrams and label the parts of the circle.

2 **a** Draw a circle with radius 8 cm.
   b Draw in one diameter AB.
   c From AB draw two lines AC and BC where C is on the circumference.
   d Measure accurately the angles A, B and C.
   e Repeat for four more positions of C and record your results in a table.

	Angle A	Angle B	Angle C
1			
2			
3			
4			
5			

   f What do you notice about your results?

3 **a** If chords of length equal to the radius are constructed around a circle, explain why the inscribed polygon is a regular hexagon.
   b Construct a regular hexagon with side 4 cm.

4 Find the unknown angles.
   Use your results from question 2 to help you.

   a 35°   **b** 60°   **c**

Exercise commentary

The questions assess the objectives on Framework Page 195.

Problem solving

Question 2 assesses the objectives on Framework Page 33. Question 3 is a word problem requiring reasoned argument, assessing Pages 17 and 31.

Group work

Question 2 and the related further activity could be attempted in pairs or small groups.

Misconceptions

Students may inscribe regular polygons inaccurately, for example by incorrectly dividing the centre of the circle. Encourage students to think how many sides the polygon has.
Emphasise the importance of knowing the names of the parts of a circle.

Links

This topic links to the circumference and area of a circle (Framework Pages 235 and 237).

Homework

S1.4HW provides further practice at inscribing regular polygons within a circle.

Answers

1 A diameter, B centre, C circumference, D arc, E chord, F radius, G sector, H segment
2 **g** Angle C is always 90°.
3 **a** Equilateral triangles are formed to make a hexagon.
4 **a** 55°   **b** 30°   **c** 45°

Mental starter

Discuss possible and impossible triangles. Ask:
Can you draw a triangle with sides of 1 cm, 2 cm and 10 cm?
Discuss why not, and encourage students to realise that the sum of
the two shorter sides must be greater than the longer side.
Now quicken the pace. Are these triangles possible:

3 cm, 4 cm, 5 cm	or	5 cm, 9 cm, 3 cm	or
20°, 30°, 70°	or	two obtuse angles ...?

Useful resources

S1.5OHP shows the construction
diagrams in the student book.

Compasses
Protractor

Introductory activity

Refer to the mental starter.
Emphasise that not all triangles are
possible, and list these two criteria:

▶ The angles must add to 180°.
▶ The sum of the two shorter sides must
exceed the longer side.

Recap the instruments of construction.
Ensure that the whole class has access to a
ruler, protractor and compasses.

**Discuss the information that you need to
construct a triangle.**
Students should appreciate that you need
three pieces of information. Emphasise
that if the triangle is to be **unique**, the
information needs to include at least
one side.

**Discuss the types of construction outlined
in the student book: ASA, SAS and SSS.**
Students should make these constructions
as you talk through them, step by step.
The diagrams for each construction are
shown on **S1.5OHP**.

Ensure that students use compasses for the
SSS example. Emphasise that they should
leave the arc lines showing on their
diagrams.

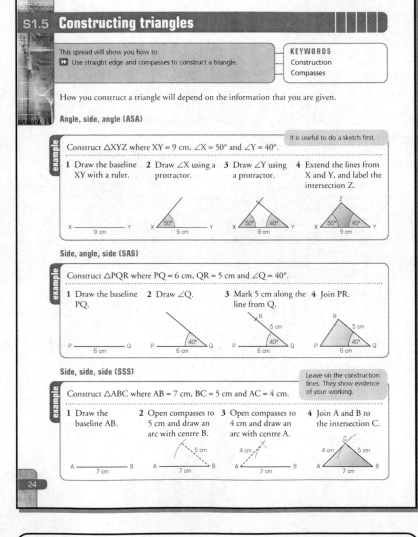

S1.5 Constructing triangles

This spread will show you how to:
▶▶ Use straight edge and compasses to construct a triangle.

KEYWORDS
Construction
Compasses

How you construct a triangle will depend on the information that you are given.

Angle, side, angle (ASA)

example

Construct △XYZ where XY = 9 cm, ∠X = 50° and ∠Y = 40°.
It is useful to do a sketch first.

1 Draw the baseline XY with a ruler.
2 Draw ∠X using a protractor.
3 Draw ∠Y using a protractor.
4 Extend the lines from X and Y, and label the intersection Z.

Side, angle, side (SAS)

example

Construct △PQR where PQ = 6 cm, QR = 5 cm and ∠Q = 40°.

1 Draw the baseline PQ.
2 Draw ∠Q.
3 Mark 5 cm along the line from Q.
4 Join PR.

Side, side, side (SSS)

Leave on the construction lines. They show evidence of your working.

example

Construct △ABC where AB = 7 cm, BC = 5 cm and AC = 4 cm.

1 Draw the baseline AB.
2 Open compasses to 5 cm and draw an arc with centre B.
3 Open compasses to 4 cm and draw an arc with centre A.
4 Join A and B to the intersection C.

24

Plenary

Discuss question 5, and outline the steps involved in constructing
a circumcircle.

Or

Discuss question 6, which relates to the mental starter. Identify the
impossible triangles in this question.

Further activities

Adapt question 5 to angle bisectors instead of perpendicular bisectors.
Ask students to describe the circle obtained, and define it as an **inscribed** circle (or incircle).

Differentiation

Extension questions:
▸ Questions 1–3 involve standard constructions.
▸ Questions 4 and 5 involve more difficult constructions, including a circumcircle.
▸ Question 6 explores possible and impossible triangles.
▸ Question 7 involves the construction of an inscribed circle.

Core tier: focuses on constructing bisectors.

Exercise S1.5

1 Construct △PQR with PQ = 6 cm, QR = 8 cm and RP = 7 cm.

2 a Construct △XYZ with XY = 5 cm, ∠X = 40°, ∠Y = 50°.
   b Measure angle Z.
   c What type of triangle is △XYZ?

3 Construct △ABC with AB = 6.2 cm, BC = 7.4 cm, ∠B = 43°.

4 Using only a pair of compasses construct:
   a △XYZ where XY = YZ = XZ = 5.4 cm
   b △PQR where PQ = QR = 4.2 cm and ∠Q = 60°.

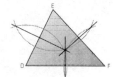

To construct an angle of 60°, draw a base line ①, draw an arc ② from X, draw a second arc ③ from Y.

   c What type of triangle is PQR?

5 a Construct △DEF where DE = 8 cm, EF = 9 cm, DF = 7 cm.
   b Bisect each side using compasses.

   c Label the point P where the bisectors meet and draw a circle, centre P, through DEF.

Note: This is known as a **circumcircle**.

6 Is it possible to construct these triangles?
   Give reasons for your answer.
   a △ABC: AB = 5 cm, BC = 3 cm, CA = 1 cm
   b △PQR: PQ = QR = RP = 4 cm
   c △XYZ: XZ = 5 cm, ∠X = 95°, ∠Y = 90°
   d △EFG: EF = 9 cm, ∠E = 90°, EG = 4 cm
   e △WXY: ∠X = 40°, ∠Y = 50°, ∠Z = 90°
   f △MNO where ∠M = 50°, ∠N = 50° and ∠O = 90°.

7 Draw △PQR, where PQ = QR = RP = 7 cm.
   Bisect each angle and use the point of intersection of the bisectors as the centre of an inscribed circle.

Exercise commentary

The questions assess the objectives on Framework Pages 221 and 223.

Problem solving
Question 5 is a substantial task that can be broken into smaller tasks, assessing the objectives on Framework Page 29. Question 6 requires a mathematical justification, assessing Page 31.

Group work
Question 5 and the further activity are suitable for working in pairs. Question 6 could be discussed in pairs.

Misconceptions
When performing constructions, students often measure angles and sides incorrectly. Encourage accuracy, particularly in lining up zeros and reading the correct scale. Students may label sides incorrectly, and a sketch would help considerably.

Links
This topic links to scale drawings (Framework Page 217), also congruence and similarity (Page 191).

Homework

S1.5HW is an investigation based around constructing triangles iteratively.

Answers

2 a 90°   b Right-angled
4 c Isosceles
6 a No   b Yes   c No   d Yes   e Yes
  f No

Mental starter

Have a selection of different triangles hidden behind a piece of paper. Reveal each one in turn slowly and ask students to guess which type they are.

Discuss the point at which triangles could be identified correctly.

Useful resources

S1.6OHP illustrates how to construct a right-angled triangle.

Compasses
Protractor

Introductory activity

Discuss right-angled triangles.
Discuss why they are useful and where they occur in the real world (for example architecture and interior design).

Define the term *hypotenuse.*
Emphasise that it is:
▸ the longest side
▸ opposite the right angle.

Recap how to construct a 90° angle.
Ensure that students have compasses to construct a right angle as you guide them through the steps.

Students should measure their angle with a protractor. Use correct terminology, such as *perpendicular.*
The first example in the student book shows how to construct the perpendicular from a point to a line.

Extend to constructing a right-angled triangle.
Discuss how to construct the triangle in the student book. The steps of the construction are illustrated on S1.6OHP.

Allow students time to construct this triangle themselves on plain paper.

Ask students to measure angle *Q* – this should be about 46°.

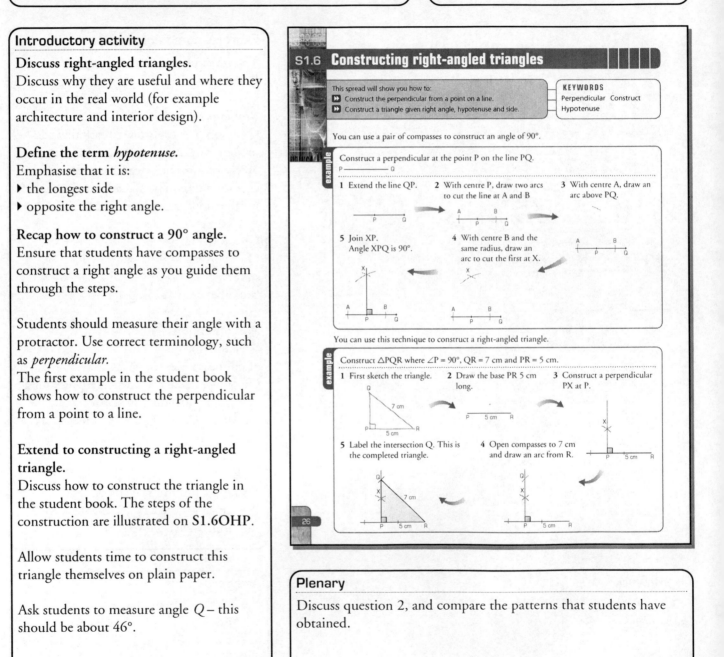

S1.6 Constructing right-angled triangles

This spread will show you how to:
▸ Construct the perpendicular from a point on a line.
▸ Construct a triangle given right angle, hypotenuse and side.

KEYWORDS
Perpendicular Construct
Hypotenuse

You can use a pair of compasses to construct an angle of 90°.

Construct a perpendicular at the point P on the line PQ.

P ——— Q

1 Extend the line QP. 2 With centre P, draw two arcs to cut the line at A and B 3 With centre A, draw an arc above PQ.

5 Join XP. Angle XPQ is 90°. 4 With centre B and the same radius, draw an arc to cut the first at X.

You can use this technique to construct a right-angled triangle.

Construct △PQR where ∠P = 90°, QR = 7 cm and PR = 5 cm.

1 First sketch the triangle. 2 Draw the base PR 5 cm long. 3 Construct a perpendicular PX at P.

5 Label the intersection Q. This is the completed triangle. 4 Open compasses to 7 cm and draw an arc from R.

26

Plenary

Discuss question 2, and compare the patterns that students have obtained.

Further activities

Students can attempt to construct a rectangle by constructing two right-angled triangles.

S1.6ICT allows students to generate constructions using an interactive geometry package.

Differentiation

Extension questions:

▸ Question 1 provides basic practice at constructing a series of related right-angled triangles.
▸ Question 2 is an iterative construction that leads to a pattern.
▸ Questions 3 and 4 are questions set in a practical context.

Core tier: focuses on constructing a perpendicular.

Exercise S1.6

1 **a** Draw a base line AB 10 cm long.
   b Construct a 90° angle at A.
   c From B find C_1 where $BC_1 = 11$ cm.
   d From B find C_2 where $BC_2 = 12$ cm.
   e From B find C_3 where $BC_3 = 13$ cm.
   f Measure **i** C_1C_2 **ii** C_2C_3, giving your answers to the nearest millimetre.

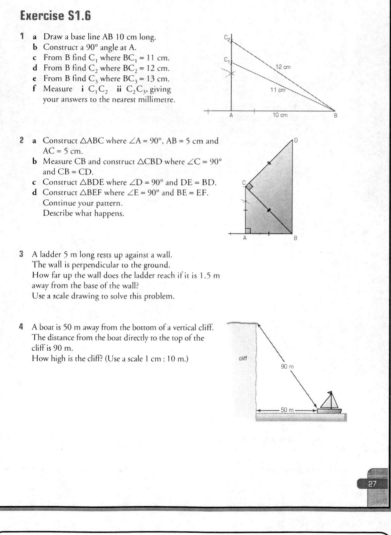

2 **a** Construct △ABC where ∠A = 90°, AB = 5 cm and AC = 5 cm.
   b Measure CB and construct △CBD where ∠C = 90° and CB = CD.
   c Construct △BDE where ∠D = 90° and DE = BD.
   d Construct △BEF where ∠E = 90° and BE = EF. Continue your pattern. Describe what happens.

3 A ladder 5 m long rests up against a wall. The wall is perpendicular to the ground. How far up the wall does the ladder reach if it is 1.5 m away from the base of the wall? Use a scale drawing to solve this problem.

4 A boat is 50 m away from the bottom of a vertical cliff. The distance from the boat directly to the top of the cliff is 90 m. How high is the cliff? (Use a scale 1 cm : 10 m.)

27

Exercise commentary

The questions assess the objectives on Framework Page 223.

Problem solving

Questions 3 and 4 use constructions to solve problems, assessing the objectives on Framework Page 17.

Group work

Students can attempt question 2 in pairs and discuss results.

Misconceptions

Students are often inaccurate in measuring angles and lengths. Discourage short cuts, such as using a protractor to make a 90° angle.

Students may confuse the techniques for construction, and mistakenly draw an arc in the wrong place. A sketch of the intended result would help.

Links

This topic links to scale drawings and maps (Framework Page 217).

Homework

S1.6HW provides practice at constructing right-angled triangles, including in the context of a scale diagram.

Answers

1 **f i** 2.1 cm   **ii** 1.7 cm
2 **b** CB = 7.1 cm
3 4.8 m
4 75 m

Summary

The key objectives for this unit are:
▸ Solve problems using properties of angles and polygons.
▸ Present a concise reasoned argument using diagrams and related explanatory text.

Plenary activity

Discuss what happens to an *n*-sided regular polygon as *n* increases.
▸ What happens to:
  ▸ the interior angle
  ▸ the exterior angle?
▸ What shape does the polygon approach?
Encourage students to visualise the shape without sketching it.

Check out commentary

1 and **2** Emphasise that the aim is to calculate, not measure, the angles.
Students may think that they have to find the angles in alphabetical order (**a** first).
Emphasise that the most important thing is to justify each stage of working.
Ensure that students are confident in the correct and appropriate use of terms such as 'opposite', 'corresponding' and 'alternate.'

3 Students may find it difficult to use compasses correctly. In particular, they may hold their compasses at the base near the point.
Encourage students not to make their circles too small, and emphasise that the end result will be more accurate with a larger diagram.

4 Encourage students to sketch their triangle first. This will give a good indication as to how much space to give to it, and whether the final result looks right.

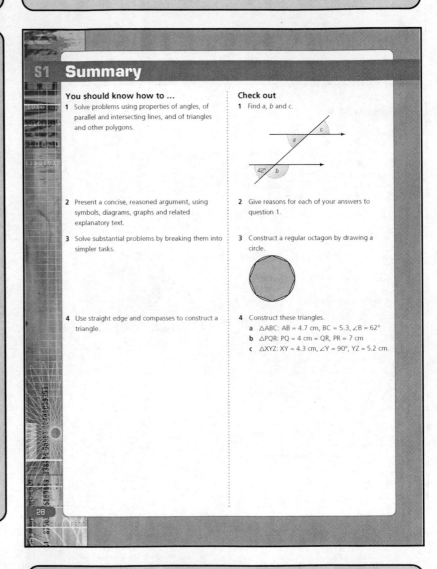

S1 Summary

You should know how to ...

1 Solve problems using properties of angles, of parallel and intersecting lines, and of triangles and other polygons.

2 Present a concise, reasoned argument, using symbols, diagrams, graphs and related explanatory text.

3 Solve substantial problems by breaking them into simpler tasks.

4 Use straight edge and compasses to construct a triangle.

Check out

1 Find a, b and c.

2 Give reasons for each of your answers to question 1.

3 Construct a regular octagon by drawing a circle.

4 Construct these triangles.
  a △ABC: AB = 4.7 cm, BC = 5.3, ∠B = 62°
  b △PQR: PQ = 4 cm, QR, PR = 7 cm
  c △XYZ: XY = 4.3 cm, ∠Y = 90°, YZ = 5.2 cm.

Development

Circles are further explored in S2, in which students will learn formulae for the area and circumference.
Construction techniques are applied to real-life problems in S4, which includes scale drawings, loci and bearings.

Links

Calculating angles links strongly to algebra and problem solving.
Geometrical construction has links to design and technology.

Mental starters

Objectives covered in this unit:
- Order, add, subtract, multiply and divide integers.
- Convert between fractions and decimals.
- Know or derive complements of 1.
- Calculate using knowledge of multiplication or division facts and place value.
- Apply mental skills to solve simple problems.

Resources needed

* means class set needed
Essential:
D1.2OHP – sample space and tree diagrams
D1.3OHP – sample space and tree diagrams
D1.6OHP – experimental probability

Useful:
Mini-whiteboards
R2 – decimal digit cards
R11 – tally chart
D1.6ICT – simulation

D1 Probability

This unit will show you how to:
- Use the vocabulary of probability in interpreting results involving uncertainty and prediction.
- Identify all the mutually exclusive outcomes of an experiment.
- Estimate probabilities from experimental data.
- Compare experimental and theoretical probabilities in a range of contexts.
- Find and record all possible outcomes.
- Appreciate the difference between mathematical explanation and experimental evidence.
- Solve increasingly demanding problems and evaluate solutions.
- Solve substantial problems by breaking them into simpler tasks.
- Present a concise reasoned argument.

Most games contain an element of chance.

Before you start

You should know how to ...
1 Find and justify simple probabilities based on equally likely outcomes.

2 Calculate with fractions and decimals.

3 Collate data into a tally chart.

Check in
1 A ten-sided dice numbered 1–10 is thrown. What is the probability of getting:
a 2 b 7 or 8?

2 Calculate:
a $1 - \frac{1}{3}$ b $1 - 0.24$ c $\frac{1}{6} \times \frac{1}{3}$

3 Draw a tally chart to show these frequencies of pets owned by a group of students.
Dog (D), Cat (C), Rabbit (R)
D C D R C D D D C D R C C D

Unit commentary

Aim of the unit
This unit focuses on the use of diagrams to represent outcomes associated with more than two events, and extends to experimental probability.
The unit includes practical activities for students to attempt, with the aim of illustrating how natural variation can affect probability, and highlights the importance of sample size.

Introduction
Discuss the meaning of probability. Outline different real-life situations, such as zero absenteeism on a particular day of the week, and ask students to estimate probabilities.

Framework references
This unit focuses on:
- Teaching objectives Pages: 277–285
- Problem-solving objectives Pages: 23, 29, 31, 33

Check in activity

Discuss different mental strategies when calculating with fractions and decimals, in particular adding and multiplying.

Differentiation

Core tier
Focuses on theoretical and experimental probability using simpler contexts.

Mental starter

Write out these numbers:

0.3 $\frac{1}{2}$ $^-0.2$ $^-\frac{1}{4}$ 1.5   0.7932   0.00001

Ask students to put the numbers into two groups: those that could represent a probability and those that cannot.

Use mini-whiteboards to show answers.

Discuss students' choices.

Useful resources

Mini-whiteboards may be useful for the mental starter.

Introductory activity

Recap basic probability ideas introduced in Year 7.

Begin to use formal language of probability and its correct notation, and encourage correct language for probability throughout. Referring to the starter activity, emphasise that probability is measured on a scale of 0 to 1.

Define sample space, trial, event and outcome. Discuss examples where outcomes are:

▸ equally likely, such as choosing a card at random from a set of digit cards.

▸ not equally likely, such as the eye colour of a student chosen at random.

Discuss the word random, and emphasise that for random events the outcome of a trial is uncertain.

Illustrate the formula for theoretical probability with the example in the student book.

Emphasise that the formula is only valid for events with equally likely outcomes.

Present a situation in which you know the probability of an event occurring, such as the probability of rain tomorrow being $\frac{2}{5}$. Ask for the probability of no rain, and generalise the result.

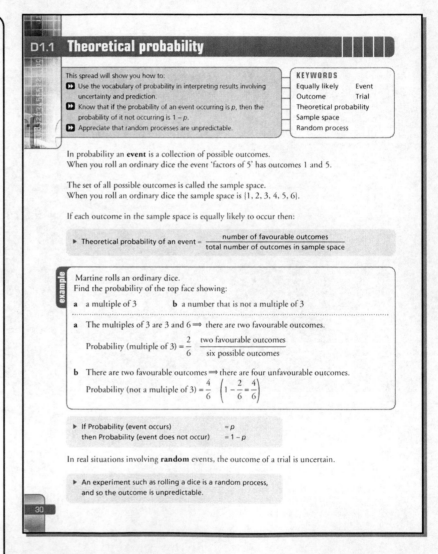

D1.1 Theoretical probability

This spread will show you how to:
▸ Use the vocabulary of probability in interpreting results involving uncertainty and prediction.
▸ Know that if the probability of an event occurring is p, then the probability of it not occurring is $1 - p$.
▸ Appreciate that random processes are unpredictable.

KEYWORDS
Equally likely   Event
Outcome   Trial
Theoretical probability
Sample space
Random process

In probability an **event** is a collection of possible outcomes.
When you roll an ordinary dice the event 'factors of 5' has outcomes 1 and 5.

The set of all possible outcomes is called the sample space.
When you roll an ordinary dice the sample space is {1, 2, 3, 4, 5, 6}.

If each outcome in the sample space is equally likely to occur then:

▸ Theoretical probability of an event = $\dfrac{\text{number of favourable outcomes}}{\text{total number of outcomes in sample space}}$

example

Martine rolls an ordinary dice.
Find the probability of the top face showing:
a a multiple of 3   **b** a number that is not a multiple of 3

a The multiples of 3 are 3 and 6 $\Rightarrow$ there are two favourable outcomes.

Probability (multiple of 3) = $\dfrac{2}{6}$ $\dfrac{\text{two favourable outcomes}}{\text{six possible outcomes}}$

b There are two favourable outcomes $\Rightarrow$ there are four unfavourable outcomes.

Probability (not a multiple of 3) = $\dfrac{4}{6}$ $\left(1 - \dfrac{2}{6} = \dfrac{4}{6}\right)$

▸ If Probability (event occurs) $= p$
then Probability (event does not occur) $= 1 - p$

In real situations involving **random** events, the outcome of a trial is uncertain.

▸ An experiment such as rolling a dice is a random process, and so the outcome is unpredictable.

30

Plenary

Recap correct terminology of probability.

Discuss events A to E in question 5: what real situations could these refer to?

Further activities

Extend question 6 to find out how many of each colour there would be if there were 40 sweets, 100 sweets, or 1000 sweets.

Differentiation

Extension questions:

▶ Question 1 requires students to list outcomes.
▶ Questions 2–4 focus on listing outcomes and calculating theoretical probabilities.
▶ Questions 5 and 6 concern the probability of an event not occurring.

Core tier: focuses on the theoretical probability of a single event.

Exercise D1.1

1 For each of these trials, list all the outcomes in the sample space.
   a The school year of a student chosen at random from an 11–16 secondary school.
   b The way a drawing pin points when dropped.
   c The score from a single dart thrown at a dartboard. (Ignore doubles, trebles and bulls'-eyes)
   d The colour of traffic lights at a busy junction.
   e The result of a football match.
   f The suit chosen when picking a card from a pack of ordinary playing cards.

2 For each of the trials outlined in question 1:
   ▶ State whether or not all the outcomes are equally likely.
   ▶ If they are equally likely, write down the probability of each outcome occurring.
     If they are not equally likely give a reason why not.

3 For each of these trials explain why you cannot easily list all the outcomes in the sample space.
   a The colour of the next car to arrive in the school car park.
   b The first name of a person chosen at random in the high street.

4 For each of these events:
   i List the favourable outcomes.
   ii Use your answer to a to write down the theoretical probability that the event occurs.
     a Obtaining an even number when rolling an ordinary dice.

   b Obtaining a prime number when rolling an ordinary dice.
   c Choosing an ace from a pack of ordinary playing cards.
   d Choosing a picture card from an ordinary pack of playing cards.
   e Choosing a diamond from an ordinary pack of playing cards.
   f Choosing the correct answer in a multiple-choice test question that has five possible answers.

5 Copy and complete the table, working out the probabilities that the events do not occur.

Event	A	B	C	D	E
Probability (event occurs)	$\frac{1}{4}$	0.82	$\frac{2}{3}$	$\frac{5}{12}$	0.36
Probability (event does not occur)					

6 Marsha has a bag of coloured sweets. They are red, blue, green or yellow. She has partly completed a table showing the probabilities of choosing a particular colour.
   a Copy and complete the table.

Outcome	Red	Blue	Green	Yellow
Probability (event occurs)	0.45			0.1
Probability (event does not occur)		0.8		

   b There are 20 sweets in total. How many of each colour are there?

Exercise commentary

The questions assess the objectives on Framework Pages 277 and 279.

Problem solving

Question 3 provides an opportunity to assess Framework Page 23.

Group work

Questions 1 to 3 can be attempted in small groups.

Misconceptions

Students will often use the formula for probability inappropriately. Emphasise that it only works if the outcomes are equally likely, and ensure that students list **all** possible outcomes.
A common mistake is to write a probability as a fraction upside-down. Emphasise that probability cannot be greater than 1.

Links

As most of the work at this level is context-based, the topic links strongly to problem-solving (Framework Page 23).

Homework

D1.1HW looks at the probability of winning prizes at different stalls at a school fete.

Answers

1 a 7 8 9 10 11; b point up/down; c 1–20, doubles, trebles, 25, 50;
   d red, red and amber, amber, green; e win lose draw;
   f hearts clubs diamonds spades
2 f $\frac{1}{4}$
4 i $\frac{1}{2}$ (2, 4, 6); ii $\frac{1}{2}$ (2, 3, 5); iii $\frac{1}{13}$; iv $\frac{3}{13}$; v $\frac{1}{4}$; vi $\frac{1}{5}$
5 $\frac{3}{4}$, 0.18, $\frac{1}{3}$, $\frac{7}{12}$, 0.64
6 a 0.55, 0.2, 0.25, 0.75, 0.9 b 9, 4, 5, 2

31

D1.2 Recording outcomes

Mental starter

Give a selection of fractions and decimals:

For example: $\frac{1}{2}$, $\frac{3}{4}$, $\frac{2}{3}$, $\frac{5}{8}$, 0.4, 0.65, 0.72, 0.375

For each number, ask what fraction or decimal will be needed to make a total of 1.

Students can respond with mini-whiteboards.

Useful resources

Mini-whiteboards may be useful for the mental starter.

D1.2OHP shows a sample space diagram and a tree diagram.

Introductory activity

Discuss situations where outcomes are mutually exclusive, such as 'has blue eyes' and 'does not have blue eyes', or 'has blond hair' and 'does not have blond hair'.

Choose two criteria appropriate for the class, and do a class exercise.

▸ Count the number of students possessing each of the features.

▸ Convert into a probability.

Emphasise that this represents the probability of a student chosen **at random** possessing this feature.

Use notation P(E) for the probability of an event E.

Discuss methods of recording outcomes:

▸ Sample space diagram

▸ Tree diagram

D1.2OHP illustrates the example in the student book. Emphasise that to calculate theoretical probabilities you need to record **all** of the outcomes.

Draw up a sample space diagram and tree diagram for the data from the class exercise.

Emphasise that you can use either diagram when there are only two events, but for three or more events you will need a tree diagram.

D1.2 Recording outcomes

This spread will show you how to:
- ▸▸ Identify all the mutually exclusive outcomes of an experiment.
- ▸▸ Find and record all possible outcomes in a systematic way.
- ▸▸ Know that the sum of probabilities of all mutually exclusive outcomes is 1.

KEYWORDS
Mutually exclusive
Outcome   Systematic
Sample space diagram
Tree diagram   Event

▸ Two events are **mutually exclusive** if they cannot occur at the same time.

In a bag of marbles some are yellow and the rest are blue.

The events 'choose a yellow marble' and 'choose a blue marble' are mutually exclusive.

If p(yellow) = 0.7, then p(blue) = 0.3

▸ The sum of probabilities of all the mutually exclusive outcomes is 1.

A second bag contains marbles that are either green or white or red. One marble is chosen from each of the two bags. Here are two ways to record all possible outcomes.

1. A **sample space diagram** is a table that shows all the possible outcomes.

	Green	White	Red
Yellow	Y and G	Y and W	Y and R
Blue	B and G	B and W	B and R

2. A **tree diagram** shows each event on a set of branches.

The six possible outcomes are found by travelling along the branches.

One possible outcome is (Y, R), shown by the orange line.

32

Plenary

Ask students to draw both types of diagram to show outcomes when throwing two dice.

Recap on the reasons for choosing a sample space diagram and a tree diagram.

Further activities

Extend question 4 for more than three coins.

Look for patterns in the number of combinations.

Differentiation

Extension questions:
▸ Question 1 requires students to draw a sample space diagram within a simple context.
▸ Questions 2 and 3 focus on sample space diagrams for two events.
▸ Questions 4 and 5 extend to three events and focus on tree diagrams.

Core tier: focuses on sample space diagrams.

Exercise D1.2

1 Hari buys an ice-cream. He can choose to have one flake, two flakes or none.
He can choose nut topping, chocolate sauce, strawberry sauce or no topping.
Draw a sample space diagram to show all the combinations of ice-cream that Hari can choose.

2 Two tetrahedral (four-sided) dice are rolled. One is numbered 1, 2, 3, 4. The other is numbered 2, 4, 6, 8.
Draw a sample space diagram to show all possible outcomes when:
a the base numbers are added together
b the base numbers are multiplied together.

3 On a restaurant set menu you can choose a starter and a main course.

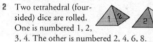

The starters are melon, prawn cocktail or soup.
The main courses are lemon chicken, baked trout, lasagne or vegetarian flan.
a Draw a sample space diagram to show the different combinations that someone could choose.
b How many different combinations are there?
c Write down how you could work out the total number of combinations without drawing the sample space diagram.

d How many combinations would there be if there were:
  i 4 starters and 5 main courses
  ii 7 starters and 6 main courses
  iii 4 starters, 6 main courses and 3 puddings?

4 Jerome has three coins: a 1p, 2p and 10p.
He throws all the coins in the air and notes whether they show heads or tails.
a Copy and complete the tree diagram to show all the different ways the coins could land.

b How many combinations are there?
c How many of these combinations will give you two heads and a tail?

5 Jenny has three bags of coins.
In one bag she has 1p and 2p coins.
In a second bag she has 5p and 10p coins.
In a third bag she has 5p and 20p coins.

Jenny chooses one coin from each of the three bags.
Draw a tree diagram to show all the possible combinations of coins that Jenny can choose.

33

Exercise commentary

The questions assess the objectives on Framework Pages 279 and 281.

Problem solving

Questions 1, 3 and 5 assess the objectives on Framework Page 23.

Group work

The suggested further activity, extending question 4, can be attempted in a small group.

Misconceptions

Students may identify outcomes only partially. For example, in question 2 (a) it is important to distinguish between the outcomes 2 + 4 and 4 + 2 as different ways to achieve a sum of 6.
In question 4 students do not always draw a complete set of branches from each of the branches in the preceding event.

Links

This topic links directly to problem-solving (Framework Page 23).

Homework

D1.2HW requires students to construct sample space diagrams, and use them to find probabilities.

Answers

1 12 combinations
2 Both diagrams with 16 outcomes
3 b 12   c 3 × 4   d 20, 42, 72
4 b 8;   c 3

Mental starter

Recap multiplication of two decimal fractions.
Give products where the numbers are no more than 1 decimal place, such as 0.7×0.4.
Students can respond using mini-whiteboards or **R2** decimal digit cards. Discuss strategies to find the products.

Useful resources

Mini-whiteboards or
R2 decimal digit cards may be useful for the mental starter.
D1.3OHP shows the diagrams in the student book.

Introductory activity

Recap the uses of tree diagrams.
Emphasise that they can be used to list outcomes for more than two events. However, they can be untidy if there are a large number of outcomes.

Discuss the first example.
Individual probabilities for each outcome are the same, so a sample space diagram is appropriate. The diagram is shown on **D1.3OHP**.

Discuss the second example.
Each colour has a different probability, so the probability of each of the six possible combinations is different.

Emphasise that if individual probabilities for each outcome are not the same, then a tree diagram is more sensible.

D1.3OHP shows the tree diagram from the second example.

Discuss how to allocate probabilities to a tree diagram, and in particular:
▸ the sum of probabilities on each set of branches is 1
▸ when you have one event **and** a successive event you multiply (along the branches).

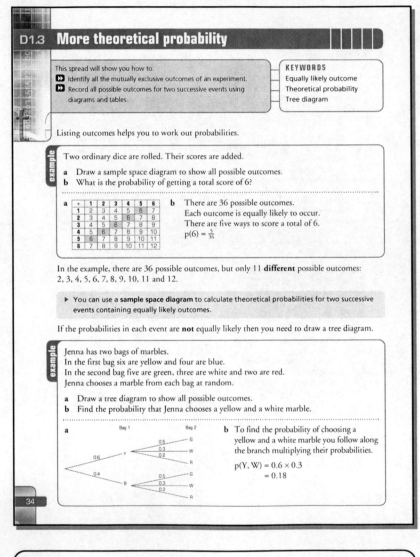

D1.3 **More theoretical probability**

This spread will show you how to:
▸▸ Identify all the mutually exclusive outcomes of an experiment.
▸▸ Record all possible outcomes for two successive events using diagrams and tables.

KEYWORDS
Equally likely outcome
Theoretical probability
Tree diagram

Listing outcomes helps you to work out probabilities.

example
Two ordinary dice are rolled. Their scores are added.
a Draw a sample space diagram to show all possible outcomes.
b What is the probability of getting a total score of 6?

a
+	1	2	3	4	5	6
1	2	3	4	5	6	7
2	3	4	5	6	7	8
3	4	5	6	7	8	9
4	5	6	7	8	9	10
5	6	7	8	9	10	11
6	7	8	9	10	11	12

b There are 36 possible outcomes.
Each outcome is equally likely to occur.
There are five ways to score a total of 6.
$p(6) = \frac{5}{36}$

In the example, there are 36 possible outcomes, but only 11 **different** possible outcomes: 2, 3, 4, 5, 6, 7, 8, 9, 10, 11 and 12.

▸ You can use a **sample space diagram** to calculate theoretical probabilities for two successive events containing equally likely outcomes.

If the probabilities in each event are **not** equally likely then you need to draw a tree diagram.

example
Jenna has two bags of marbles.
In the first bag six are yellow and four are blue.
In the second bag five are green, three are white and two are red.
Jenna chooses a marble from each bag at random.

a Draw a tree diagram to show all possible outcomes.
b Find the probability that Jenna chooses a yellow and a white marble.

b To find the probability of choosing a yellow and a white marble you follow along the branch multiplying their probabilities.
$p(Y, W) = 0.6 \times 0.3$
$= 0.18$

34

Plenary

Reinforce the main features of a tree diagram.
When outcomes are shown in a tree diagram, you need to multiply **along** the branches to find the probability of successive events.

Further activities

Carry out an experiment based on question 6.
▸ Do drawing pins tend to land point up or point down?
▸ Are the combinations in the expected proportions?
Link this activity to Year 7 work on experimental probability.

Differentiation

Extension questions:
▸ Question 1 requires students to simply list outcomes in a sample space diagram.
▸ Questions 2–5 focus on using diagrams to calculate probabilities.
▸ Question 6 extends to more than two events, and involves investigating arrangements.

Core tier: focuses on calculating probabilities from a sample space diagram.

Exercise D1.3

1. Tom is having pizza for his tea. He is allowed two extra toppings on his pizza. He chooses from ham, mushroom, pineapple, sweetcorn and extra cheese.
   a. Draw a sample space diagram to show all the possible combinations of toppings Tom could choose. (He could choose two lots of the same topping.)
   b. How many outcomes are there?

2. Use the sample space diagram on the opposite page to find the following probabilities of these events when two ordinary dice are rolled and their scores are added.
   a. The score is odd.
   b. The score is less than 5.
   c. The score is a factor of 12.

3. Ursula rolls a dice and tosses a coin.

   a. Draw a sample space diagram to show all possible outcomes.
   If she gets a factor of 6 she will wash her hair. If she gets a tail she will also use conditioner.
   b. What is the probability that Ursula will wash her hair and use conditioner?

4. Use the tree diagram on the opposite page to find the probabilities of these events when one marble is chosen from each of Jenna's two bags of marbles.
   a. A yellow and a green marble.
   b. A blue and a red marble.

5. Thierry collects coins in two jars.

   In one jar there are ten 1p coins and forty 2p coins.
   In the second jar there are seventy 1p coins and thirty 2p coins.
   Thierry chooses, at random, one coin from each of the two jars.
   a. Draw a tree diagram showing the different combinations of coins that Thierry could choose. Write on your tree diagram the probability of choosing each coin.
   b. Work out the probability that Thierry chooses coins with a total value of:
      i 2p   ii 3p   iii 4p

6. Maria empties a box containing three drawing pins on the floor.
   a. Draw a tree diagram to show if they land point up (U) or point down (D).
   b. One possible outcome is UUU. How many ways are there that the drawing pins could land?
   c. How many ways would there be if Maria had
      i four drawing pins
      ii five drawing pins
      iii six drawing pins?
   d. Write down a quick method that Maria could use to calculate the number of ways the drawing pins could land for any number of pins.

35

Exercise commentary

The questions assess the objectives on Framework Pages 279 and 281.

Problem solving

The questions in this exercise assess the objective on Framework Page 23.
Question 6 assesses Page 33, as students are asked to make a generalisation.

Group work

Work in pairs or small groups on the extension to question 6 outlined in further activities.

Misconceptions

Students often add probabilities when they should multiply.
Encourage students to check their method, particularly if their answer is greater than 1.

Links

This topic links to problem-solving (Framework Page 23).

Homework

D1.3HW provides practice at constructing and using a tree diagram.

Answers

1 b 25 combinations
2 a $\frac{1}{2}$ b $\frac{1}{6}$ c $\frac{1}{3}$
3 $\frac{1}{3}$
4 a 0.3   b 0.08
5 b i 0.14 ii 0.56 iii 0.24
6 b 8   c 16, 32, 64   d 2^n where n is number of drawing pins.

Mental starter

Brainstorm ten typical names for pet dogs and write them down. Ask students to estimate the probability of choosing a name with the letter r in it.

Write a further ten names of pets on the board (use the students' pets' names) and repeat the probability estimate for the letter r.

Useful resources

Mini-whiteboards may be useful for the mental starter.

R11 – tally chart

Introductory activity

Recap experimental probability from Year 7, and link to the mental starter.

Discuss the task outlined in the student book. A blank tally chart can be found on **R11**.

Describe how an experiment may sometimes be the only way to estimate the probability of an outcome.

Emphasise that you can only find an **estimate** of probability, since all limericks will be different.

Discuss what might happen to the probability of choosing a five-letter word if a different limerick were used.

Discuss how you could obtain a reliable estimate of the probability.

How many limericks would you need? This idea will be discussed in more depth in the next lesson.

Discuss the formula for estimated probability.

In particular, discuss the terminology: 'successes' and 'trials'.

Highlight the similarities with the formula for theoretical probability.

D1.4 Experimental probability

This spread will show you how to:
- Estimate probabilities based on experimental data.

KEYWORDS
Tally chart
Experimental probability

Outcomes are not often equally likely.
You can estimate the probability of an outcome using an **experiment**.

Laura and Lucy are doing a project on the length of words in limericks.
They want to estimate the probability of choosing a five-letter word in a limerick.
Laura wrote a limerick.

Laura's limerick

There was a young lady from Dover
Who once found a fifteen-leaf clover
Her friend said 'Oh dear –
But your eyes have gone queer!
You counted each leaf five times over.'

Lucy compiled a frequency table for the number of letters in each word.

Number of letters	1	2	3	4	5	6	7
Tally	II	I	JHT	JHT JHT IIII	JHT I	II	II
Frequency	2	1	5	14	6	2	2

There are 32 words in the limerick.
Six words have five letters.

$$\text{p(5 letter word)} = \frac{6}{32}$$
$$= 0.1875$$
$$\simeq 0.2$$

This is only an estimate because it is based on the data that the girls collected from Laura's limerick.

Laura and Lucy **estimate** that the probability of choosing a five-letter word at random from any limerick is 0.2.

In an experiment:
▸ Experimental probability = $\dfrac{\text{number of successes}}{\text{total number of trials}}$

36

Plenary

Discuss the length of words in newspapers and magazines.
Relate to reading age, target readership and other factors.

Make up a limerick of your own, and compile a tally chart of the number of letters in each word.
Compare the distribution of the numbers of letters both with question 1 and the example.

Extension questions:
▸ Question 1 is similar to the example in the student book.
▸ Questions 2 and 3 focus on estimating experimental probability from data in context.
▸ Question 4 extends to calculating probability with a two-way table.

Core tier: focuses on estimating probability from experimental data.

Exercise D1.4

1 Peter wrote a limerick.

> There once was a woman from Ealing
> Who walked upside down on a ceiling.
> She fell on her neck
> And said 'Oh my heck,
> What a peculiar feeling'.

a Draw a tally chart to show the length of each word.
b What is the probability that a word chosen at random will have
  i   3 letters
  ii 4 letters
  iii 6 or more letters?

2 Suzy counted the number of drawing pins in each of 40 boxes.

Here are her results:

35 37 42 36 36 35 40 41 41 39
41 42 34 37 38 39 39 93 42 36
41 42 40 38 37 38 37 34 36 35
42 35 35 39 36 42 42 35 41 39

a Suzy studied the results and decided that the result 93 had been incorrectly recorded. Give a reason for Suzy to reach this decision.
b Rewrite 93 as 39 and collate all the results in a tally chart.
c Use your chart to determine the probability of choosing a box of drawing pins containing:
  i   38 pins
  ii 42 pins
  iii 37 or fewer pins.

3 The table shows the number of goals scored by Warren Juniors Football Club in their last 20 games.

Goals scored	Number of games
0	3
1	6
2	4
3	2
4	3
5	1
6	0
7	0
8	1

a Estimate the probability that in their next game Warren JFC will score:
  i   1 goal
  ii 3 goals or fewer
  iii 5 or more goals.
b Explain why you think your answer to iii may be unrealistic.

4 A class of 32 students noted how many brothers and sisters they had. Their results are summarised in the table.

| | | Brothers | | |
		0	1	2	3
S	0	4	5	1	0
i	1	7	6	3	0
s	2	0	4	0	0
t	3	1	0	0	1

One student from the class is chosen at random.
What is the probability that this student will:
a be an only child
b have one brother and no sisters
c have one sister and two brothers
d have no brothers
e be in a family of seven children?

37

The questions assess the objectives on Framework Page 283.

Problem solving
The questions in this exercise assess Framework Page 23.

Group work
Students can work together on the limericks suggested in the further activity.

Misconceptions
In frequency tables with a quantitative variable, students sometimes use the wrong column to calculate probabilities. Encourage students to realise the link between probability and frequency.
In the formula for estimated probability, students may misinterpret the term 'successes'. Emphasise that this just means 'favourable outcomes'.

Links
This work links to recording and tabulating data using tally charts and frequency tables (Framework Page 253).

Homework

D1.4HW is based on an experiment to determine whether a dice is fair or biased, and to estimate probabilities.

Answers

1 $\frac{5}{28}$, $\frac{8}{28}$, $\frac{6}{28}$
2 a too different   c $\frac{3}{40}$, $\frac{7}{40}$, $\frac{17}{40}$
3 $\frac{6}{20}$, $\frac{15}{20}$, $\frac{2}{20}$; no 6 or 7 goals scored
4 a $\frac{4}{32}$ b $\frac{5}{32}$ c $\frac{3}{32}$ d $\frac{12}{32}$ e $\frac{1}{32}$

Mental starter

Instruct each student to:
▶ throw a coin ten times
▶ note the number of tails
▶ estimate the probability of a tail
Alternatively students could use dice and the number 6.
Discuss why the results are different.

Useful resources

Mini-whiteboards may be useful for the mental starter.
R11 tally chart may also be useful for the mental starter.

Introductory activity

Refer to the mental starter.
Emphasise how experiments like this can only give an **estimate** of probability.

Discuss the task outlined in the students' book.
Focus on the objective: to determine the probability that a **randomly** chosen word will contain the letter s.

Discuss why the results are all different, and ask which data should be chosen to provide a reliable estimate of probability.

Encourage the response that a better estimate would come from combining all the results.

Discuss how an even more reliable estimate of probability may be obtained.
In particular, suggest carrying out a large number of trials.

Discuss the practical limitations of carrying out a large number of trials (time, effort and money), and why a compromise is sometimes necessary.

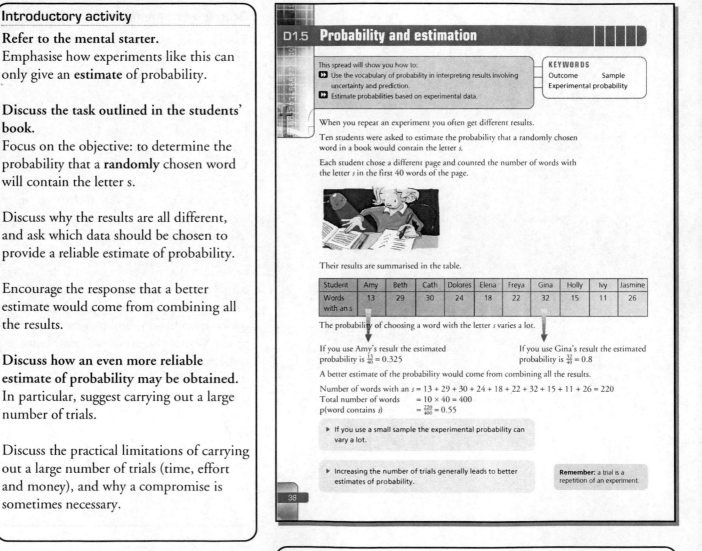

D1.5 Probability and estimation

This spread will show you how to:
▶ Use the vocabulary of probability in interpreting results involving uncertainty and prediction.
▶ Estimate probabilities based on experimental data.

KEYWORDS
Outcome Sample
Experimental probability

When you repeat an experiment you often get different results.

Ten students were asked to estimate the probability that a randomly chosen word in a book would contain the letter s.

Each student chose a different page and counted the number of words with the letter s in the first 40 words of the page.

Their results are summarised in the table.

Student	Amy	Beth	Cath	Dolores	Elena	Freya	Gina	Holly	Ivy	Jasmine
Words with an s	13	29	30	24	18	22	32	15	11	26

The probability of choosing a word with the letter s varies a lot.

If you use Amy's result the estimated probability is $\frac{13}{40} = 0.325$

If you use Gina's result the estimated probability is $\frac{32}{40} = 0.8$

A better estimate of the probability would come from combining all the results.

Number of words with an s = 13 + 29 + 30 + 24 + 18 + 22 + 32 + 15 + 11 + 26 = 220
Total number of words = $10 \times 40 = 400$
p(word contains s) = $\frac{220}{400} = 0.55$

▶ If you use a small sample the experimental probability can vary a lot.

▶ Increasing the number of trials generally leads to better estimates of probability.

Remember: a trial is a repetition of an experiment.

38

Plenary

Use question 2 as a basis for discussion on how you might determine which letter of the alphabet is the most common. Include comments on which letters you could leave out and why.

Further activities

Draw graphs for questions 1 and 2 to show how the probabilities change with each trial.

Differentiation

Extension questions:
▸ Question 1 requires students to perform a simple experiment.
▸ Question 2 describes a rather more complex experiment with an increased number of variable factors.
▸ Question 3 is a game for two players based on probability.

Core tier: focuses on natural variation within experimental data.

Exercise D1.5

1 **a** Toss a coin ten times and write down how many times you get a tail.
Use this to estimate the probability of getting a tail.
b Toss the coin a further ten times and use all 20 throws to estimate the probability of throwing a tail.
c Toss the coin a further ten times and use all 30 throws to estimate the probability of throwing a tail.
d Repeat the experiment until you have thrown the coin 100 times.
Write all ten estimated probabilities in a table.
Comment on how the estimated probabilities change as the number of trials increases.

2 Choose a book, preferably a novel, and select a page at random.
a i Select a passage of 100 words from the page and count the number of words that contain the letter *t*.
ii Use your data to estimate the probability that a word in the book will contain the letter *t*.
b Repeat this experiment a further four times, each time selecting a different page at random.
Comment on the different estimates of probability you found each time.
c Use your data to write down a better estimate of the probability that a word chosen at random will contain the letter *t*.
d How could you improve your estimate further?

3 This is a dice game for two players.
Rules
Player 1 rolls a dice.
▸ If the dice shows a six, player 1 scores nothing and it is player 2's turn.
▸ If the dice shows a number other than a six, player 1 scores 1 point.
Player 1 can roll the dice as many times as they like, and will score 1 point each time, as long as they don't get a 6 on the dice.
At any time, player 1 can say 'bank' instead of rolling the dice.
When this happens:
▸ The score for this go is added to player 1's banked total.
▸ It is then player 2's go to roll the dice.
The first player to bank 10 points is the winner.
Remember – only banked scores count towards the winning total!

When you think you have worked out some good strategies, you could try altering the rules.

39

Exercise commentary

The questions assess the objectives on Framework Pages 277 and 283.

Problem solving
Questions 1 and 2 are substantial problems that can be broken down into smaller tasks, assessing Framework Page 29.

Group work
Questions 1 and 2 may be completed in small groups or as a class exercise.
Question 3 is a game that requires two students to work together.

Misconceptions
Students often assume that if their results in a random experiment are different from what they expected, then they are wrong. Emphasise that variation is normal, and will be most obvious with a small sample.

Links
Variation within experimental bounds has links to practical work in science.

Homework

D1.5HW requires students to design a dice game.

Answers

1 As number of trials increases estimated probabilities should get close to $\frac{1}{2}$.
2 **d** Increase the number of trials.

Mental starter

Write a selection of fractions on the board for students to convert to decimals.

Encourage students to decide whether mental division or use of a calculator is more appropriate in each case.

Students can respond using mini-white boards.

Useful resources

Mini-whiteboards may be useful for the mental starter.

D1.6OHP contains the table and graph from the student book.

Introductory activity

Recap the meanings of theoretical and experimental probability.

Emphasise that theoretical probability is based on assumptions, whereas experimental probability is based on the results of a number of trials.

Discuss the task outlined in the students' book.

D1.6OHP contains the table in the student book, with an extra blank row for probability.

Complete this table as a class exercise, ensuring that students are confident in calculating experimental probability.

Show with a line graph how the probability changes after each trial.

Encourage students to describe how, with an increasing number of trials, the probability approaches a particular value. Emphasise that this value is a good **estimate** of the probability of choosing a blue marble.

D1.6OHP contains the line graph in the student book.

Discuss how you may use experimental probability to determine bias.

Discuss bias in simple cases, such as a loaded dice.

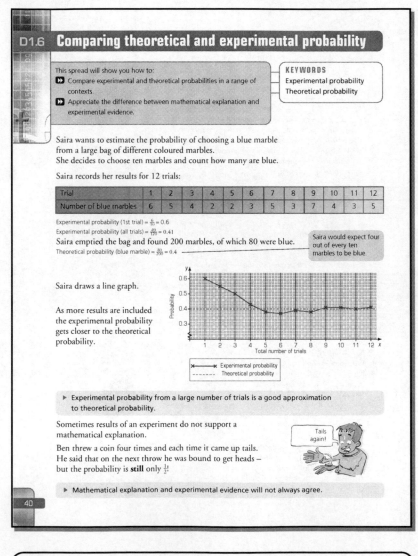

D1.6   Comparing theoretical and experimental probability

This spread will show you how to:
- ▶ Compare experimental and theoretical probabilities in a range of contexts.
- ▶ Appreciate the difference between mathematical explanation and experimental evidence.

KEYWORDS
Experimental probability
Theoretical probability

Saira wants to estimate the probability of choosing a blue marble from a large bag of different coloured marbles.
She decides to choose ten marbles and count how many are blue.

Saira records her results for 12 trials:

Trial	1	2	3	4	5	6	7	8	9	10	11	12
Number of blue marbles	6	5	4	2	2	3	5	3	7	4	3	5

Experimental probability (1st trial) = $\frac{6}{10}$ = 0.6
Experimental probability (all trials) = $\frac{49}{120}$ = 0.41
Saira emptied the bag and found 200 marbles, of which 80 were blue.
Theoretical probability (blue marble) = $\frac{80}{200}$ = 0.4

Saira would expect four out of every ten marbles to be blue.

Saira draws a line graph.

As more results are included the experimental probability gets closer to the theoretical probability.

✕——✕ Experimental probability
---- Theoretical probability

▶ Experimental probability from a large number of trials is a good approximation to theoretical probability.

Sometimes results of an experiment do not support a mathematical explanation.

Ben threw a coin four times and each time it came up tails.
He said that on the next throw he was bound to get heads – but the probability is **still** only $\frac{1}{2}$!

Tails again!

▶ Mathematical explanation and experimental evidence will not always agree.

40

Plenary

Use question 3 to discuss the idea of independent events, with reference to replacement and non-replacement.

Further activities

Students can try question 3 as an experiment in a small group, and determine how close their theoretical and experimental probabilities are.

D1.6ICT explores simulations, requiring students to develop their own statistical models.

Differentiation

Extension questions:
- Questions 1 and 2 compare experimental and theoretical probability to determine bias in the context of dice.
- Question 3 introduces the idea of independent events in the context of playing cards.
- Question 4 focuses on independent events in a lottery game.

Core tier: focuses on the use of experimental trials to determine bias.

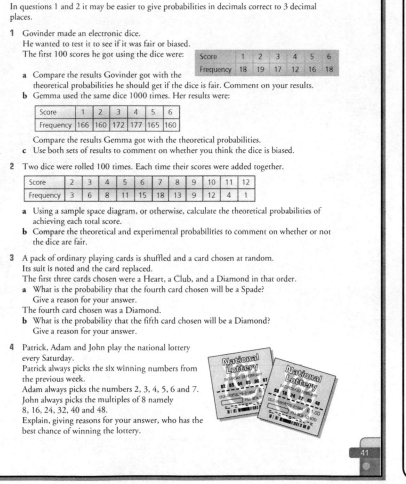

Exercise D1.6

In questions 1 and 2 it may be easier to give probabilities in decimals correct to 3 decimal places.

1   Govinder made an electronic dice.
    He wanted to test it to see if it was fair or biased.
    The first 100 scores he got using the dice were:

Score	1	2	3	4	5	6
Frequency	18	19	17	12	16	18

    a   Compare the results Govinder got with the theoretical probabilities he should get if the dice is fair. Comment on your results.
    b   Gemma used the same dice 1000 times. Her results were:

Score	1	2	3	4	5	6
Frequency	166	160	172	177	165	160

    Compare the results Gemma got with the theoretical probabilities.
    c   Use both sets of results to comment on whether you think the dice is biased.

2   Two dice were rolled 100 times. Each time their scores were added together.

Score	2	3	4	5	6	7	8	9	10	11	12
Frequency	3	6	8	11	15	18	13	9	12	4	1

    a   Using a sample space diagram, or otherwise, calculate the theoretical probabilities of achieving each total score.
    b   Compare the theoretical and experimental probabilities to comment on whether or not the dice are fair.

3   A pack of ordinary playing cards is shuffled and a card chosen at random.
    Its suit is noted and the card replaced.
    The first three cards chosen were a Heart, a Club, and a Diamond in that order.
    a   What is the probability that the fourth card chosen will be a Spade?
        Give a reason for your answer.
    The fourth card chosen was a Diamond.
    b   What is the probability that the fifth card chosen will be a Diamond?
        Give a reason for your answer.

4   Patrick, Adam and John play the national lottery every Saturday.
    Patrick always picks the six winning numbers from the previous week.
    Adam always picks the numbers 2, 3, 4, 5, 6 and 7.
    John always picks the multiples of 8 namely 8, 16, 24, 32, 40 and 48.
    Explain, giving reasons for your answer, who has the best chance of winning the lottery.

Exercise commentary

The questions assess the objectives on Framework Pages 283 and 285.

Problem solving
The questions in this exercise are word problems involving probability, assessing Framework Page 23. They each involve a degree of interpretation, assessing Page 31.

Group work
Students could work together on question 3 and extend it as described in the further activity.

Misconceptions
In question 3 students may not grasp that, in this case, the outcome of a trial is independent of preceding trials. You may use this opportunity to discuss whether the situation would be different if the card were not replaced each time.

Links
This topic requires the use of mathematical argument to support experimental data, which is an essential skill for coursework.
The theme of simulation is further explored in D4.

Homework

D1.6HW is based on a simulation involving a tetrahedral dice.

Answers

1 a 4 seems on low side   c Unbiased
2 Fair
3 a $\frac{1}{4}$ b $\frac{1}{4}$
4 All equally likely

Summary

The key objective for this unit is:
▸ Use the fact that the sum of all mutually exclusive outcomes is 1 when solving problems.

Plenary activity

Draw a circle divided into four equal sectors, with one sector coloured blue. State that p(blue) $= \frac{1}{3}$.

Discuss whether the diagram represents the statement. If it does not, ask what needs to be changed.

Check out commentary

1 Students may mistakenly give the probability of a yellow sector as $\frac{1}{3}$, since there are three colours to choose from. Reinforce the meaning of the formula, in particular the number of favourable outcomes and the total number of outcomes.

In part **a**, emphasise that p (not yellow) $= 1 - $ p (yellow).

In part **b**, encourage students to ask, 'what is a quarter of 8?'

Part **c** relies on an understanding of mutually exclusive outcomes, which is a key objective at this level.

2 Students will be fairly new to the terminology, so emphasise that a sample space diagram is really just a table. Reiterate the use of the formula for theoretical probability in part **a**, and the formula for experimental probability in part **b**.

Part **c** requires a reasoned comparison with a conclusion, which is an evolving skill at this level. Emphasise that theoretical and experimental probabilities are very unlikely to be identical.

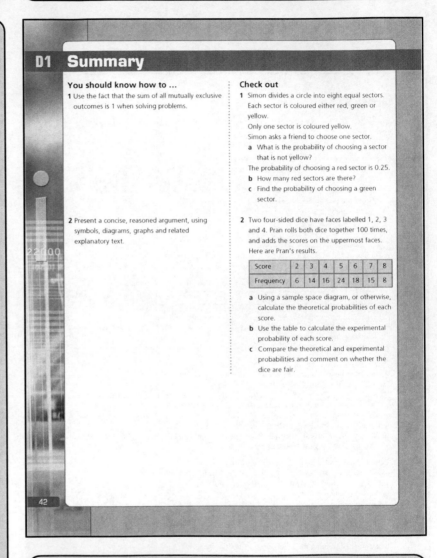

D1 Summary

You should know how to ...

1 Use the fact that the sum of all mutually exclusive outcomes is 1 when solving problems.

2 Present a concise, reasoned argument, using symbols, diagrams, graphs and related explanatory text.

Check out

1 Simon divides a circle into eight equal sectors. Each sector is coloured either red, green or yellow.
Only one sector is coloured yellow.
Simon asks a friend to choose one sector.
a What is the probability of choosing a sector that is not yellow?
The probability of choosing a red sector is 0.25.
b How many red sectors are there?
c Find the probability of choosing a green sector.

2 Two four-sided dice have faces labelled 1, 2, 3 and 4. Pran rolls both dice together 100 times, and adds the scores on the uppermost faces. Here are Pran's results.

Score	2	3	4	5	6	7	8
Frequency	6	14	16	24	18	15	8

a Using a sample space diagram, or otherwise, calculate the theoretical probabilities of each score.
b Use the table to calculate the experimental probability of each score.
c Compare the theoretical and experimental probabilities and comment on whether the dice are fair.

42

Development

The themes of this unit are developed further in D4.

Links

This work uses the equivalence of fractions and decimals. Experimental probability links to data projects, and also to work in other subjects such as science.

Mental starters

Objectives covered in this unit:
▶ Convert between fractions, decimals and percentages.
▶ Find fractions and percentages of quantities.
▶ Calculate using knowledge of multiplication and division facts.
▶ Use factors to multiply and divide mentally.
▶ Multiply and divide a two-digit number by a one-digit number.

Resources needed

Essential:
N2.1OHP – proportions of shapes

Useful:
R1 – digit cards
R2 – decimal digit cards
R3 – fraction cards
R6 – number lines
Mini-whiteboards
N2.6F – comparing proportions
N2.5ICT – percentages

N2 Fractions, decimals and percentages

This unit will show you how to:

▶▶ Know that a recurring decimal is an exact fraction.
▶▶ Use efficient methods to add, subtract, multiply and divide fractions.
▶▶ Cancel common factors before multiplying or dividing.
▶▶ Recognise when fractions or percentages are needed to compare proportions.
▶▶ Solve problems involving percentage changes.
▶▶ Use proportional reasoning to solve a problem.

▶▶ Extend mental methods of calculation, working with decimals, fractions and percentages.
▶▶ Make and justify estimates and approximations.
▶▶ Multiply and divide by decimals.
▶▶ Enter numbers on a calculator and interpret the display in context.
▶▶ Solve increasingly demanding problems and evaluate solutions.
▶▶ Solve substantial problems by breaking them into simpler tasks.
▶▶ Give solutions to problems to an appropriate degree of accuracy.

Nutritional information is often expressed using decimals and percentages.

Before you start

You should know how to ...
1 Find the HCF and LCM by prime factor decomposition.

2 Convert between fractions, decimals and percentages.

3 Find simple fractions and percentages of amounts.

Check in
1 a Write 36 and 48 as the product of their prime factors.
  b Use your answers to a to find:
    i the HCF of 36 and 48
    ii the LCM of 36 and 48

2 Express each of these numbers in the other two forms (percentage or decimal or fraction)
  a $\frac{9}{20}$ b 0.145   c 26%

3 Work out:
  a i $\frac{1}{8}$ of 248 mg   ii $\frac{4}{15}$ of £300
  b i 23% of 40   ii $17\frac{1}{2}$% of £240

43

Unit commentary

Aim of the unit
This unit aims to extend understanding of fractions, decimals and percentages, and to apply these techniques to problems involving proportion.
Adding and subtracting fractions draws on previous knowledge involving HCF and LCM.

Introduction
Discuss the use of percentages, decimals and fractions as different mathematical 'languages'.
Focus on when it is appropriate to use each form and why it is useful to convert between them.

Framework references
This unit focuses on:
▶ Teaching objectives pages: 61–69, 75, 77
▶ Problem-solving objectives pages: 3, 7, 29, 31, 33, 35

Check in activity

Display a variety of percentages, decimals and fractions on the board, some of which are equivalent.
Ask students to match the equivalent quantities.

Differentiation

Core tier
Focuses on simpler cases of fractions, decimals and percentages, and does not include inverse percentage problems.

Mental starter

Fraction countdown

Display six digits: 3 7 8 2 9 4

Give students a target decimal, such as 0.4

Students must write a fraction, using two of the digits, that is equivalent or as close as possible to the target decimal.

Start with simple decimals that will have an exact equivalent fraction, and progress to decimals that will give an approximation.

Students can respond with **R3** fraction cards, or mini-whiteboards.

Useful resources

R3 fraction cards or **mini-whiteboards** may be useful for the mental starter.

N2.1OHP illustrates proportions of shapes.

Introductory activity

Discuss equivalent fractions.

Use shaded shapes to highlight the equivalence of various fractions.

N2.1OHP shows the pair of shapes in the student book, and one other pair.

Discuss how you could use equivalent fractions in context.

The first example in the student book illustrates how to express one number as a fraction of another number. Show that you can either simplify $\frac{60}{360}$:

▶ in a single step: $\frac{60}{360} = \frac{1}{6}$

▶ in a sequence of steps:

$\frac{60}{360} = \frac{30}{180} = \frac{15}{90} = \frac{3}{18} = \frac{1}{6}$

Emphasise that the first method is quicker, but requires finding the **HCF** of 60 and 360 – briefly recap this concept from unit NA1.

Refer to the mental starter.

Ensure by using examples that students know how to convert:

▶ from a fraction to a decimal (using a mental, written or calculator method)

▶ from a decimal to a fraction (using a mental or written method).

Discuss how you can use these techniques to order fractions, either by converting to decimals or using equivalent fractions.

Use the last example in the student book to introduce **recurring decimals**.

Write various recurring decimals on the board and ask students to abbreviate them using the dot notation.

Show with simple examples that recurring decimals are equivalent to fractions.

N2.1 Fractions and decimals

This spread will show you how to:

▶▶ Use fraction notation to describe a proportion of a shape.

▶▶ Express one number as a fraction of another.

▶▶ Order fractions.

▶▶ Know that a recurring decimal is an exact fraction.

KEYWORDS

Proportion Decimal
Recurring

You use fractions to describe a proportion of a whole.

To work out the fraction shaded purple ...   ... divide the shape into equal parts

$\frac{6}{16}$ of the shape is shaded, or $\frac{3}{8}$

You can express a number as a fraction of another number.

example

Class 8Y were asked to name their favourite subject.
This pie chart shows the results.
What fraction of students prefers Art?

Add up the angles:	$60° + 90° + 40° + 110° = 300°$
Find the missing angle:	$360° - 300° = 60°$
Express as a fraction:	$\frac{60}{360} = \frac{1}{6}$ ⟹ $\frac{1}{6}$ of students in class 8Y prefer Art.

Converting to decimals is a good way to order fractions.

example

Which is the larger, $\frac{4}{7}$ or $\frac{2}{3}$?

Convert to decimals:
$\frac{4}{7} = 4 ÷ 7 = 0.5714 ...$
$\frac{2}{3} = 2 ÷ 3 = 0.6666 ...$
Compare the first digit:
$\frac{2}{3}$ is larger than $\frac{4}{7}$.

You could use equivalent fractions instead:
$\frac{4}{7} = \frac{12}{21}$ $\frac{2}{3} = \frac{14}{21}$
Compare numerators: $\frac{2}{3}$ is larger than $\frac{4}{7}$.

The last example contained two **recurring** decimals.
$\frac{2}{3} = 0.66666 ... = 0.\dot{6}$ $\frac{4}{7} = 0.5714285714 ... = 0.\dot{5}7142\dot{8}$

In a recurring decimal the sequence of digits will repeat forever.

▶ You can write a recurring decimal as an exact fraction.

44

Plenary

Repeat the fraction countdown game from the mental starter, but with these numbers:

4   11   9   13   12   1

Students must make a target recurring decimal, such as $0.0\dot{9}$.

Encourage students to use a calculator. They can respond using mini-whiteboards.

Further activities

Students can work in pairs to investigate recurring decimals.

If you know that $\frac{1}{3} = 0.3333333\ldots$ how many other recurring decimals can you find?

For example,

$\frac{2}{3} = 2 \times 0.33333\ldots = 0.66666\ldots$

$\frac{1}{9} = 0.333333 \div 3 = 0.11111\ldots$

Differentiation

Extension questions:
▸ Question 1 requires students to order pairs of fractions.
▸ Questions 2–4 focus on expressing a number as a fraction of another number in context.
▸ Question 5 explores recurring decimals.

Core tier: focuses on expressing proportions as fractions.

Exercise N2.1

1 Identify the larger fraction in each pair.
  a $\frac{3}{8}$ $\frac{5}{14}$
  c $\frac{4}{9}$ $\frac{4}{7}$
  e $\frac{3}{11}$ $\frac{5}{16}$
  b $\frac{3}{5}$ $\frac{7}{12}$
  d $\frac{2}{3}$ $\frac{7}{11}$
  f $\frac{1}{9}$ $\frac{2}{13}$

2 A square has side 5 cm.
  A, B, C and D are the midpoints of each side of the square.

  a i Calculate the area of the whole square.
    ii Calculate the area of the shaded part.
    iii What fraction of the whole square is unshaded?
  b The diagram shows the same square shaded differently.

  By dividing the square up into equal parts express the shaded part as a fraction of the whole in its simplest terms.

3 The gradient of the line segment AB is $\frac{2}{3}$.

On graph paper draw axes from 0 to 10.
  a On the same axes, draw line segments with these gradients:
    i $\frac{1}{2}$ ii $\frac{3}{5}$ iii $\frac{1}{6}$
  b Put the fractions $\frac{1}{2}$, $\frac{3}{5}$ and $\frac{1}{6}$ in order of size from smallest to largest.
  c What can you say about the gradients that you drew in part **a** as the fractions get bigger?

4 The pie chart represents the votes cast for six candidates in a school council election.

What fraction of the votes was cast for each candidate? Express these answers as:
  a a fraction
  b a decimal

5 Puzzle
  Find a fraction that is equivalent to each of these recurring decimals.
  a $0.\dot{4}2857\dot{1}$
  b $0.\dot{3}8461\dot{5}$

Exercise commentary

The questions assess the objectives on Framework Pages 61 and 65.

Problem solving
Questions 2, 3 and 4 explore connections across a range of contexts, assessing the objectives on Framework Page 7.

Group work
Question 5 can be attempted by students working in pairs.

Misconceptions
When writing a proportion as a fraction, students may write the denominator incorrectly.
For example, if 4 people vote yes and 7 people vote no, a common mistake is to write the fraction of people voting yes as $\frac{4}{7}$. Remind students that the denominator is the total.

Links
Ordering fractions by graphing them links to gradients (Framework Page 167).

Homework

N2.1HW is an investigation based on proportions expressed as fractions.

Answers

1 a $\frac{3}{8}$ b $\frac{3}{5}$ c $\frac{4}{9}$ d $\frac{2}{3}$ e $\frac{5}{16}$ f $\frac{2}{13}$
2 a i 25 cm^2 ii 15.625 cm^2 iii $\frac{3}{8}$ b $\frac{5}{64}$
3 b $\frac{1}{6}, \frac{1}{2}, \frac{3}{5}$ c They get steeper.
4 a $\frac{1}{20}, \frac{1}{5}, \frac{67}{360}, \frac{49}{360}, \frac{8}{45}, \frac{1}{4}$ 0.05, 0.2, 0.2, 0.1, 0.2, 0.25
5 a $\frac{3}{7}$ b $\frac{5}{13}$

Mental starter

Using a 1–9 dice or the random number function on a calculator, select four digits at random.
Make two fractions from these four digits, and write them with a + or − sign between them.
Students work out the answer and could respond using mini-whiteboards.

Useful resources

Mini-whiteboards are useful for the mental starter.

Introductory activity

Refer to the mental starter.
Discuss strategies for adding two fractions with different denominators.
Illustrate how to find a common denominator using one of the sums in the starter activity as an example.

Write down an example that cannot easily be solved mentally.
You could use $\frac{7}{18} - \frac{4}{21}$.
Invite discussion on how to find a common denominator.
Emphasise that this is equivalent to finding a common multiple, and preferably the **lowest common multiple (LCM)**.

Briefly recap strategies for finding the LCM. Demonstrate:
▸ The factor tree method
▸ Repeated division.
Emphasise that the LCM contains the highest power of each prime factor.

Summarise the rest of the procedure.
▸ Rewrite as equivalent fractions.
▸ Combine the fractions.
▸ Simplify, converting to a mixed number where appropriate.

Extend to combining three fractions.

Briefly discuss adding and subtracting fractions by converting them to decimals.
Discuss the problem of then reconverting them back at the end.

N2.2 Adding and subtracting fractions

This spread will show you how to:
▶ Add and subtract fractions.

KEYWORDS
Denominator LCM
Equivalent

You can add or subtract fractions easily when they have the same denominator.

When the denominators are different, you need to change them to equivalent fractions.

$\frac{1}{7} + \frac{3}{7} = \frac{4}{7}$

example
Calculate $\frac{7}{12} + \frac{9}{20} - \frac{2}{3}$.
▶ First find the LCM of the denominators, 12, 20 and 3:
$12 = 2 \times 2 \times 3$
$20 = 2 \times 2 \times 5$
$3 = 3$
$LCM = 2^2 \times 3 \times 5 = 60$
▶ Rewrite as equivalent fractions: $\frac{7}{12} = \frac{35}{60}$ $\frac{9}{20} = \frac{27}{60}$ $\frac{2}{3} = \frac{40}{60}$
▶ Combine the fractions: $\frac{35}{60} + \frac{27}{60} - \frac{40}{60} = \frac{22}{60}$
▶ Simplify: $\frac{22}{60} = \frac{11}{30}$

Remember:
You find the LCM by multiplying the highest power of each prime factor.

When the denominators are larger you may need to use a written method to find the LCM.

example
Calculate $\frac{17}{28} + \frac{29}{70}$, leaving your answer as a mixed number.
▶ Find the LCM of 28 and 70 by prime factor decomposition:

2) 28	2) 70
2) 14	5) 35
7) 7	7) 7
1	1

$28 = 2^2 \times 7$
$70 = 2 \times 5 \times 7$
$LCM = 2^2 \times 5 \times 7 = 140$

▶ Convert to equivalent fractions: $\frac{17}{28} = \frac{85}{140}$ $\frac{29}{70} = \frac{58}{140}$
▶ Add the fractions: $\frac{85}{140} + \frac{58}{140} = \frac{143}{140}$
▶ Convert to a mixed number: $\frac{143}{140} = 1\frac{3}{140}$

Alternatively, you could convert to decimals:
$\frac{17}{28} = 0.6071$ (to 4 d.p.)
$\frac{29}{70} = 0.4143$ (to 4 d.p.)
$\frac{17}{28} + \frac{29}{70} = 0.6071 + 0.4143$
$= 1.0214$
This method is not as accurate in this case.

46

Plenary

Extend the mental starter activity, but now allowing the denominator to have two digits.
This will entail finding six random digits for each sum.

Further activities

Extend the Fraction Pyramids in question 10.

What happens to the top number when you add on an extra term to the bottom layer ($\frac{1}{64}$, $\frac{1}{128}$ ...)?

Students can investigate in pairs.

Differentiation

Extension questions:

▸ Question 1 practises fraction to decimal conversion.

▸ Questions 2–9 focus on adding and subtracting fractions, including word problems in context

▸ Question 10 involves successive addition of fractions with increasingly large denominators

Core tier: focuses on adding and subtracting fractions with smaller denominators.

Exercise N2.2

1 Convert each of these fractions into decimals, giving your answer to 3 decimal places where appropriate.

a $\frac{3}{8}$ **b** $\frac{6}{25}$ **c** $\frac{9}{20}$ **d** $\frac{5}{9}$ **e** $\frac{2}{11}$

2 Calculate:

a $\frac{3}{4} + \frac{2}{5}$ **b** $\frac{4}{5} + \frac{1}{3}$ **c** $\frac{3}{4} + \frac{3}{7}$

d $\frac{5}{9} - \frac{2}{5}$ **e** $2\frac{7}{10} + 1\frac{1}{4}$ **f** $1\frac{5}{6} - \frac{7}{9}$

3 **a** List the prime factors of these numbers:

 i 36 **ii** 54 **iii** 120 **iv** 435

 b Use your answers to part **a** to answer these questions.

 i What is the LCM of 36 and 54?

 ii What is the LCM of 54, 120 and 435?

 iii What is the lowest common multiple of all four numbers?

4 Calculate these, expressing each answer in its simplest form.

a $\frac{4}{9} + \frac{5}{6}$ **b** $\frac{5}{18} + \frac{17}{20}$

c $\frac{24}{45} - \frac{13}{24}$ **d** $\frac{31}{41} + \frac{17}{14} - \frac{2}{15}$

e $5\frac{2}{13} - 3\frac{4}{9}$ **f** $6\frac{23}{25} + \frac{3}{4} - 10\frac{5}{11}$

5 A plank of wood is $3\frac{2}{15}$ m long. A piece of length $1\frac{3}{8}$ m is sawn off. What length of wood is left?

6 Use the fractions in the box to complete these number sentences:

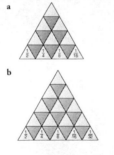

☐ + ☐ = ☐

☐ − ☐ = ☐

☐ + ☐ = ☐

☐ − ☐ = ☐

7 Emilio says that if you repeatedly subtract $\frac{2}{31}$ from $\frac{5}{8}$ you will eventually reach ⁻1. Is he correct? Show working to justify your answer.

8 Joe worked out that he spent $\frac{3}{8}$ of his spare time watching TV, $\frac{1}{3}$ with friends and $\frac{2}{5}$ in other activities. Show that this is impossible and explain your answer.

9

a For each question **i**, **ii** and **iii** choose two fractions from the cloud that have the difference indicated.

 i $\frac{49}{162}$ **ii** $\frac{43}{155}$ **iii** $\frac{14}{891}$

b What is the sum of:

 i the two largest fractions

 ii the two smallest fractions?

10 In a Fraction Pyramid, the number in each triangle is the sum of the two triangles directly below.
Copy and complete these diagrams.

a

b

Exercise commentary

The questions assess the objectives on Framework Page 67.

Problem solving

Questions 5–10 are of a problem-solving nature and assess the objectives on Framework Page 7.

Group work

Question 6 is a puzzle that can be attempted in pairs.

Misconceptions

Students may get the strategies for finding the LCM and the HCF mixed up. Emphasise that the common denominator is a multiple, so it will not be smaller than the original denominators.

Remind students to give answers as mixed numbers where appropriate.

Links

This topic links to multiples, factors and primes (Framework Page 55).

Homework

N2.2HW provides practice in adding and subtracting fractions.

Answers

1 **a** 0.375 **b** 0.24 **c** 0.45 **d** 0.556 **e** 0.182

2 **a** $1\frac{3}{20}$ **b** $1\frac{2}{15}$ **c** $1\frac{1}{28}$ **d** $\frac{7}{45}$ **e** $3\frac{19}{20}$ **f** $1\frac{1}{18}$

3 **a i** $2^2 \times 3^2$ **ii** 2×3^3 **iii** $2^3 \times 3 \times 5$ **iv** $3 \times 5 \times 29$

 b i 108 **ii** 31 320 **iii** 31 320

4 **a** $1\frac{5}{18}$ **b** $1\frac{23}{180}$ **c** $\frac{-1}{120}$ **d** $1\frac{151}{1230}$ **e** $1\frac{83}{117}$ **f** $-2\frac{313}{1100}$ **5** $1\frac{61}{120}$ m

6 $\frac{1}{3} + \frac{3}{8} = \frac{17}{24}$, $\frac{5}{6} - \frac{2}{5} = \frac{13}{30}$, $\frac{1}{4} + \frac{2}{3} = \frac{11}{12}$, $\frac{4}{5} - \frac{1}{2} = \frac{3}{10}$ **7** No **8** $\frac{3}{8} + \frac{1}{3} + \frac{2}{5} > 1$

9 **a i** $\frac{1}{2} - \frac{16}{81}$ **ii** $\frac{6}{10} - \frac{10}{31}$ **iii** $\frac{16}{81} - \frac{2}{11}$ **b i** $1\frac{16}{35}$ **ii** $\frac{338}{891}$

10 **a** $\frac{27}{16}, \frac{9}{8}, \frac{9}{16}, \frac{3}{4}, \frac{3}{8}, \frac{3}{16}$ **b** $\frac{81}{32}, \frac{27}{16}, \frac{27}{32}, \frac{9}{8}, \frac{9}{16}, \frac{9}{32}, \frac{3}{4}, \frac{3}{8}, \frac{3}{16}, \frac{3}{32}$

Mental starter

Ask questions relating to proportions of quantities, such as:

▶ What is $\frac{2}{7}$ of 140?

▶ How many $\frac{1}{16}$ths are there in $\frac{1}{2}$?

▶ What is 16% of 25?

▶ What is 0.35×250?

▶ What is 5% of 7 kg?

Students can respond using mini-whiteboards or decimal digit cards.

Useful resources

Mini-whiteboards may be useful for the mental starter.

R2 – decimal digit cards.

Introductory activity

Refer to the mental starter.

Briefly discuss strategies for calculating proportions of a quantity, when the proportion is expressed as a decimal, fraction or percentage.

Use examples with units of measure.

Discuss how to multiply a fraction by an integer.

Emphasise that any fraction can be split into a unit fraction multiplied by an integer: $\frac{5}{8} = \frac{1}{8} \times 5$

Students may need reminding that $\times \frac{1}{k}$ is equivalent to $\div k$.

Discuss the order of operations.

You could divide first or multiply first, or you could use an equivalent decimal.

Discuss how to multiply a fraction by a fraction.

Justify the method by splitting into unit fractions:

$$\frac{3}{5} \times \frac{2}{3} = \frac{1}{5} \times 3 \times \frac{1}{3} \times 2$$
$$= 3 \times 2 \div 5 \div 3$$
$$= \frac{3 \times 2}{5 \times 3}$$

Generalise the method: you multiply the numerators together, and the denominators together.

Discuss how to divide by a fraction.

Remind students that $\div \frac{1}{k}$ is equivalent to $\times k$, and then discuss progressively complex examples:

$6 \div \frac{1}{5}$, $6 \div \frac{2}{5}$, $\frac{5}{7} \div \frac{3}{4}$

These examples are in the student book.

Generalise the method: dividing by $\frac{a}{b}$ is the same as multiplying by $\frac{b}{a}$.

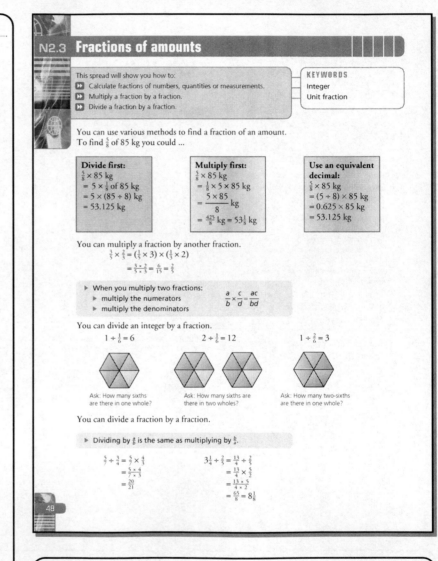

Plenary

Link the three general results to algebra:

▶ Calculating a fraction of a quantity:
$\frac{a}{b} \times Q = \frac{aQ}{b}$

▶ Multiplying a fraction by a fraction:
$\frac{a}{b} \times \frac{c}{d} = \frac{ac}{bd}$

▶ Dividing a fraction by a fraction:
$\frac{a}{b} \div \frac{c}{d} = \frac{a}{b} \times \frac{d}{c} = \frac{ad}{bc}$

Further activities

Students can invent fraction sequences like the ones in question 8 for a partner to solve.
The term-to-term rule should involve multiplying or dividing by a fraction.

Differentiation

Extension questions:
▶ Question 1 provides practice at multiplying a fraction by an integer.
▶ Questions 2–10 focus on multiplying and dividing by a fraction.
▶ Questions 11 is a word problem that involves multiplying fractions.

Core tier: focuses on multiplying and dividing by fractions, but only includes dividing an integer by a fraction.

Exercise N2.3

1 Calculate:
a $\frac{4}{5}$ of £220   b $\frac{3}{8} \times 4$
c $1\frac{1}{4} \times 22$ m   d $\frac{11}{24}$ of 3 days
e $\frac{3}{7} \times 63$ f $27 \times \frac{2}{9}$

2 Puzzle
There are 550 sweets in a jar.
$\frac{2}{5}$ of these are for Hani, and $\frac{1}{2}$ are for Jane.
The rest remain in the jar.
a Who gets most: Hani or Jane? Explain your answer.
b What fraction of the sweets remains in the jar?
c How many sweets remain in the jar?
Samantha gets $\frac{3}{11}$ of those that remain in the jar.
d How many sweets does Samantha get?

3 Work out each of these amounts:
a $\frac{3}{5}$ of £33   b $\frac{5}{13}$ of 450 m
c 0.4×25 cm   d $\frac{1}{17}$ of £42
e 23% of 540 g   f 0.8×145 kg
g 52% of £563   h 0.875×560 mm
i $\frac{4}{11} \times £45$

4 Calculate these products, leaving your answer as a mixed number where appropriate.
a $\frac{5}{9} \times 27$ b $\frac{3}{8} \times 44$
c $\frac{7}{24} \times 15$ d $28 \times \frac{17}{21}$
e $2\frac{5}{8} \times 6$ f $2\frac{7}{10} \times 5$

5 In a competition, Julie won $\frac{2}{13}$ of £304 and Jasmine won $\frac{3}{17}$ of £230.
Who won the most money? Show all your working.

6 Boris says: 'If you multiply a number by another number, you will always increase the amount you started with.'
Mikhail disagrees. Suggest a reason why Mikhail might disagree.

7 Work out:
a $\frac{2}{5}$ of $\frac{3}{4}$ b $\frac{4}{3}$ of $\frac{6}{7}$ c $0.4 \times \frac{8}{9}$
d $\frac{3}{7} \div \frac{1}{4}$ e 30% $\div \frac{5}{6}$ f $0.3 \div 52$
g $\frac{6}{11}$ of $\frac{6}{5}$ h $0.75 \div \frac{6}{7}$ i $\frac{3}{10} \times \frac{5}{7}$

8 Puzzle
Here is a fraction sequence:
$\frac{3}{4}, \frac{3}{8}, \frac{3}{16}, \frac{3}{32}, \_\_, \_\_$
You can find each term by multiplying the previous term by $\frac{1}{2}$:
$\frac{3}{4} \times \frac{1}{2} = \frac{3}{8}$, $\frac{3}{8} \times \frac{1}{2} = \frac{3}{16}$, ...
The next two terms will be:
$\frac{3}{32} \times \frac{1}{2} = \frac{3}{64}$
$\frac{3}{64} \times \frac{1}{2} = \frac{3}{128}$
Find the next two terms of these fraction sequences:
a $\frac{1}{2}, \frac{3}{5}, \frac{3}{10}, \frac{9}{50}, \_\_, \_\_$
b $3, \frac{2}{3}, 2, \frac{4}{3}, \frac{8}{3}, \_\_, \_\_$
c $\frac{3}{4}, \frac{4}{5}, \frac{3}{5}, \frac{12}{25}, \_\_, \_\_$

9 A rectangular window is $2\frac{1}{4}$ m high and $1\frac{1}{3}$ m wide.
Calculate its area.

10 The area of a rectangle is 340 cm^2 and its length is $15\frac{2}{3}$ cm.
What is the width of the rectangle?
Give your answer as a mixed number.

11 Investigation
$\frac{2}{5} \times 11 = \frac{22}{5} = 4\frac{2}{5}$
$\frac{3}{7} \times 15 = \frac{45}{7} = 6\frac{3}{7}$
a What is the smallest integer that can be multiplied by $\frac{2}{9}$ so that the result is a mixed number whose fractional part is also $\frac{2}{9}$?
b Investigate for any starting fraction. Explain your answer.

49

Exercise commentary

The questions assess the objectives on Framework Pages 67 and 69.

Problem solving
Questions 2, 5 and 8 assess the objectives on Framework Page 7.
Question 6 assesses Page 35.
Question 11 is an investigation, assessing Page 33.

Group work
Question 11 can be attempted in pairs.

Misconceptions
Students may have difficulty in understanding that dividing by a fraction is the same as multiplying by its reciprocal. Emphasise the multiplicative nature of fractions as operators.

Links
This topic links to multiplying and dividing algebraic fractions (Framework Page 119).

Homework

N2.3HW provides practice at multiplying and dividing with mixed numbers in the context of the area of a rectangle.

Answers

1 a £176 b $1\frac{1}{2}$ c $27\frac{1}{2}$ m d $1\frac{3}{8}$ days e 27 f 6
2 a Jane b $\frac{1}{10}$ c 55 d 15
3 a £19.80 b $173\frac{1}{13}$ m c 10 cm d £2.47 e 124.2 g f 116 kg g £292.76 h 490 mm i £16.36
4 a 15 b $16\frac{1}{2}$ c $4\frac{3}{8}$ d $22\frac{2}{3}$ e $15\frac{3}{4}$ f $13\frac{1}{2}$
5 Julie (£46.77 compared with £40.59)
6 Not if you multiply by a number between 0 and 1.
7 a $\frac{3}{10}$ b $1\frac{1}{7}$ c $\frac{16}{45}$ d $1\frac{5}{7}$ e $\frac{12}{25}$ f $\frac{3}{520}$ g $\frac{36}{55}$ h $\frac{7}{8}$ i $\frac{3}{14}$
8 a $\frac{27}{500}, \frac{234}{25\,000}$ b $3\frac{5}{9}, 9\frac{13}{27}$ c $\frac{36}{125}, \frac{432}{3125}$ 9 $2\frac{7}{10}$ cm^2 10 $22\frac{6}{77}$ cm
11 a 10 b Integer is 1 more than a multiple of the fraction's denominator.

49

Mental starter

Display the spider diagram.

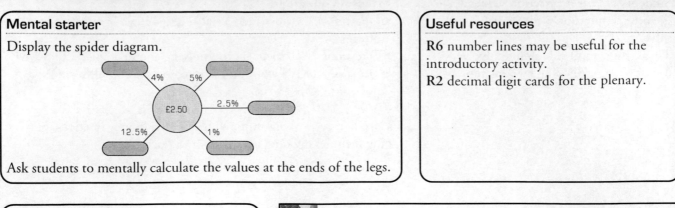

Ask students to mentally calculate the values at the ends of the legs.

Useful resources

R6 number lines may be useful for the introductory activity.
R2 decimal digit cards for the plenary.

Introductory activity

Discuss the equivalence of fractions, decimals and percentages.

Emphasise that students should memorise the important equivalences (such as $0.1 = \frac{1}{10} = 10\%$), and know techniques for calculating others.

Use **R6** number lines, marked from 0–1, to illustrate the equivalences.

Discuss methods for calculating a percentage of an amount.

Use 23% of 60 kg as an example, and work it out:

▸ By multiplying first $(23 \times 60) \div 100$
▸ By dividing first $(23 \div 100) \times 60$
▸ By converting to a decimal (23×0.6).

Emphasise the importance of:

▸ Approximating: *'23% is roughly a quarter, which makes 15 kg.'*
▸ Including any units in the answer (13.8 **kg**, not just 13.8).

Discuss how to work out a percentage increase or decrease.

Refer to the example in the student book and show that you can:

▸ Add the increase/subtract the decrease at the end
▸ Calculate the new amount in a single calculation.

Reinforce the single calculation method by using a number line.

You can use **R6** number lines, and mark them from 0 to 200%. Then ask questions like:

If you decrease an amount by 18%, what is this equivalent to?
$(\times 82\% = 0.82)$

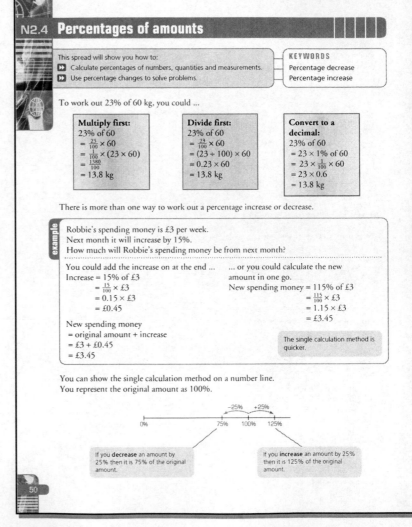

Plenary

Consolidate the single calculation method by asking questions such as:

▸ What decimal would you multiply by to increase an amount by 30%?
▸ What decimal would you multiply by to decrease an amount by 16%?

Students can respond using decimal digit cards.

Further activities

If students have access to ICT equipment, they could consider how a spreadsheet or graphical calculator may be used to perform compound calculations efficiently.

Differentiation

Extension questions:

▸ Question 1 provides straight-forward practice at calculating a percentage of a quantity.
▸ Questions 2 and 3 focus on percentage increase and decrease in context.
▸ Questions 4–6 extend to more complex percentage problems, including compound percentages.

Core tier: focuses on using mental, written and calculator techniques to find a percentage of a quantity.

Exercise N2.4

1 Find these percentages of amounts. In each case, choose a mental, written or calculator method.
   a 13% of £650   **b** 43% of 80 cl
   c 17.5% of 25 m   **d** 16.4% of €400
   e 7.5% of £2.40   **f** 0.03% of 1600 kg
   g 0.8% of 3 million **h** 1.3% of 8 g

2 For each pair of items, work out which is cheaper. Show your working.
   a A television costing £546, or the same television originally priced at £674 with 18% discount.
   b A DVD costing £124 including VAT (value added tax) or the same DVD for £99 plus VAT at 17.5%.

3 Show your workings in each of these problems. You may use a calculator.
   a Henry's father offered him two choices: a £1.20 per week increase in his spending money or a 12% increase. Henry currently receives £9 per week.
      i Which should Henry choose to receive the greatest increase?
      ii What amount would he need to receive now for the 12% increase to equal the £1.20 increase?
   b A railway engineer calculates that the points on a railway track will increase in width by 0.03% on a very hot summer day. If the width of the points is normally 90 mm calculate:
      i the increase in width that would occur
      ii the new width.
   c At a local youth club, the secretary states that the membership subscriptions in the previous 12 months decreased from £3610 by 14%. Calculate the amount the youth club needs to raise to bring the income to the level of the previous year.

4 Three years ago, Adam and Jayne received a gift of £500 from their aunt. They put it into a bank account and saved it.
   After one year, the value of their savings grew by 5%.
   a How much did Adam and Jayne have in their account after one year?
   After the second year, the new value of their savings increased again by 5%.
   b How much did they have in their account after two years?
   c How much money do Adam and Jayne have in their bank account now?

5 Investigation
   Bus fares rose by 8% in January.
   They rose by a further 12% in February.
   Azad says 'The new bus fare is 20% more than the old fare'.
   Sabina disagrees.
   Explain why Sabina disagrees.

6 Investigation
   Morrissey wants to hire a tractor to plough his land.
   Here are the rates from three different hire companies.

TRACTOR FOR HIRE
£80 + £13 PER DAY

A-TRACTOR
(£90 + Number of Days 1% of £900)

TRACTORSAURUS
£50 + (Number of days x 5% of £350)

   The maximum hire period for a tractor is 20 days.
   a If Morrissey takes 10 days to plough his land, from which company should he hire a tractor?
   b Investigate which company Morrissey should use for different lengths of time.

Exercise commentary

The questions assess the objectives on Framework Page 77.

Problem solving

Questions 4, 5 and 6 are substantial problems and assess the objectives on Framework Page 29.
Question 5 assesses Page 35.

Group work

Questions 6 and 7 could be attempted in pairs.

Misconceptions

Students may find difficulty in calculating a percentage change by using a single decimal multiplier. Emphasise that the original amount is 100% (= 1.00). An increase will involve a decimal greater than 1, and a decrease will involve a decimal smaller than 1.

Links

This work links to proportionality (Framework Page 79).

Homework

N2.4HW reinforces the techniques of calculating a percentage increase or decrease.

Answers

1 **a** £84.50 **b** 34.4 cl **c** 4.375 m **d** €65.60 **e** £0.18 **f** 0.48 kg
  g 24 000 **h** 0.104 g
2 **a** £546   **b** £99 + VAT
3 **a i** £1.20 **ii** £10   **b i** 0.027 mm   **ii** 90.027 mm   **c** £505.40
4 **a** £525   **b** £551.25   **c** £578.81   5 The rise is 20.96%.
6 **a** Tractor for hire
  b 1–6 days A-Tractor, 7–16 days Tractor for hire, 17–20 days Tractorsaurus

Mental starter

What was I?

Ask questions such as:

▶ I am 20, which is $\frac{1}{4}$ of my original value. What was I?

▶ I am 60, which is $\frac{2}{3}$ of my original number. What was I?

Students can respond using digit cards.

Useful resources

R1 digit cards may be useful for the mental starter.

Introductory activity

Refer to the mental starter.

Discuss strategies for finding the original number.

Move swiftly on to percentages, and ask questions such as:

75% of an amount is £120.

What is the original amount?

Students may be able to do this mentally (75% converts to $\frac{3}{4}$, then:

£120 ÷ 3 = £40, then × 4 = £160).

Emphasise that the original amount is 100%.

Discuss the unitary method for finding an original amount.

Illustrate with an example how you:

▶ Divide to find 1%

▶ Multiply by 100.

Discuss the method of using inverse operations.

You will need to revise the single calculation method from N2.4.

Show with an example that if:

New amount = original amount × decimal multiplier, then ...

Original amount = new amount ÷ decimal multiplier.

Formalise the inverse operations method using algebra.

Let the original price be *N*, and show that you can display the working in a more condensed form using algebra.

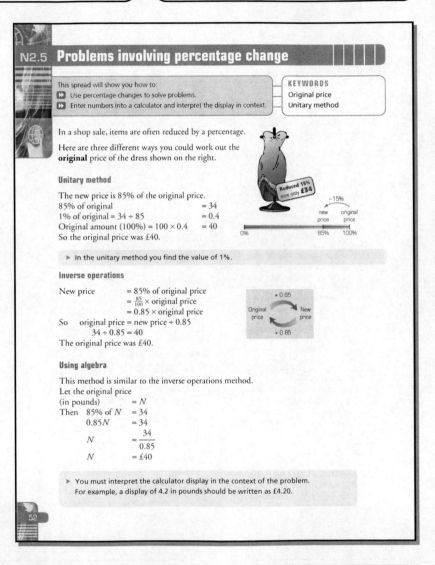

N2.5 Problems involving percentage change

This spread will show you how to:
- Use percentage changes to solve problems.
- Enter numbers into a calculator and interpret the display in context.

KEYWORDS
Original price
Unitary method

In a shop sale, items are often reduced by a percentage.

Here are three different ways you could work out the **original** price of the dress shown on the right.

Unitary method

The new price is 85% of the original price.

85% of original	= 34
1% of original = 34 ÷ 85	= 0.4
Original amount (100%) = 100 × 0.4	= 40

So the original price was £40.

▶ In the unitary method you find the value of 1%.

Inverse operations

$$\begin{aligned} \text{New price} &= 85\% \text{ of original price} \\ &= \tfrac{85}{100} \times \text{original price} \\ &= 0.85 \times \text{original price} \end{aligned}$$

So original price = new price ÷ 0.85

34 ÷ 0.85 = 40

The original price was £40.

Using algebra

This method is similar to the inverse operations method.

Let the original price (in pounds) = *N*

$$\begin{aligned} \text{Then } 85\% \text{ of } N &= 34 \\ 0.85N &= 34 \\ N &= \frac{34}{0.85} \\ N &= £40 \end{aligned}$$

▶ You must interpret the calculator display in the context of the problem. For example, a display of 4.2 in pounds should be written as £4.20.

52

Plenary

Adapt the starter activity, but use percentages rather than fractions. For example, you could ask:

'I am 180, which is 95% of my original value. What was I?'

Students working in pairs can investigate interest rates. Looking at different rates (4%, 6%, 8%, 10%), calculate how long it will take for an initial amount to double. Students could show their results in a graph.

N2.5ICT requires students to use a spreadsheet to calculate percentages of amounts and percentage change.

Differentiation

Extension questions:
▸ Question 1 provides practice at increasing and decreasing amounts by a percentage
▸ Questions 2–6 are percentage problems involving money and weight, including inverse percentages
▸ Questions 7 and 8 involve compound percentages.

Core tier: focuses on finding a percentage increase or decrease, given the original amount.

Exercise N2.5

1 Use a mental, written or calculator method to work out the new amount.
  a Increase £46 by 3%
  b Decrease 130 °C by 20%
  c Decrease 3.5 m by 1%
  d Increase 1 h 10 min by 5%
  e Decrease 12.3 s by 15%
  f Increase 70 cl by 4%
  g Increase £3000 by 1.6%
  h Increase £9.30 by 26%

2 Mr. Smith increased all the prices in his sweet shop by 10%.
  Copy and complete this price list.

Item	Original price	New price
Bag of mints	40p	44p
Small assorted sweets	£1.20	
Large assorted sweets	£1.65	
Box of chocolates	£2.82	

3 15 300 people attended a pop concert. However this was only 73% of the attendance the previous night.
  Work out the previous night's attendance (to the nearest person).

4 Wendy is given an 8% pay rise each year. Her starting salary was £9500 per year. How many years will Wendy have to work before her salary is at least:
  a £11 000 b £13 000 c £15 000?

5 The weight of a three-month old baby is 6.08 kg.
  This is 190% of her weight at birth.
  Calculate the baby's birth weight.

6 Charlotte is saving her spending money to buy a new microwave oven for her Great Aunt Lucy.
  Charlotte has saved £87, but this is only 60% of the amount she needs.
  a How much does the microwave oven cost?
  The store manager agrees to reduce the price of the microwave by 15%.
  b How much more money does Charlotte need to save?

7 Puzzle
  Earl went to the bank and was pleasantly surprised to find that he had £66.55 in his savings account.
  He had opened his account three years ago and had left it untouched ever since.
  The bank added on 10% to the value at the end of each year.
  How much money did Earl put into the bank originally?
  Show your workings clearly.

8 a During the first few weeks of its life, a squid increases its body weight by 6% each day. A squid is born with a body weight of 180 g.
    i How much will it weigh after one day?
    ii How much will it weigh after three days?
    iii How many days will it take the squid to double its body weight?
  b Steve invests £1000 in SuperSquid Inc. Each year his money increases by 8.4%. After eight years what is the value of his investment?

Exercise commentary

The questions assess the objectives on Framework Page 77.

Problem solving

Questions 2–8 are problems involving percentages and money, and assess the objectives on Framework Page 3.

Group work

The further activity should be tackled in pairs. Encourage students to evolve strategies for working out compound percentages quickly and efficiently.

Misconceptions

Students often do not readily identify inverse percentage problems.
Encourage students to seek clues in the wording of the problem. If it asks for the original amount, it is an inverse percentage problem.

Links

This topic links to proportionality (Framework Page 79).

Homework

N2.5HW requires students to find the original price of items in a sale.

Answers

1 a £47.38 b 104°C c 3.465 m
  d 1 hr 13 min 30 sec e 10.455 sec f 72.8 cl
  g £2952 h £11.72
2 £1.32, £1.82, £3.10 3 20 959
4 a 2 years b 5 years c 6 years 5 3.2 kg
6 a £145 b £36.25 7 £50
8 a i 191 g ii 214 g iii 12 days b £1906.49

Mental starter

Find the fraction

Ask students to find a fraction equivalent to a variety of decimals, including simple cases of recurring decimals.

Decimals could include 0.8, 0.375, 0.6, and also numbers greater than 1.
Students could respond using mini-whiteboards.

Useful resources

Mini-whiteboards are useful for the mental starter.

Introductory activity

Refer briefly to the mental starter.
Discuss the equivalence of fractions, percentages and decimals, and recap with examples how to convert between one form and the other.

Discuss the meaning of the word 'proportion'. Encourage the response that it is a part related to a whole, and illustrate with real examples:
17 out of 20 in a test, four days per week, five people in every thousand ...

Highlight the similarity between proportions and fractions, and hence also decimals and percentages.

Illustrate proportions in context.
You can draw a bar chart, and ask for the proportion in a particular range.
Encourage a variety of responses: decimals, fractions and percentages.

Discuss the use of proportions to make comparisons. You can use the example on baked beans in the student book.

Extend to percentage change.
Illustrate with examples how percentage change compares the size of a change to the original amount.
Give an example of a percentage increase and also a decrease.

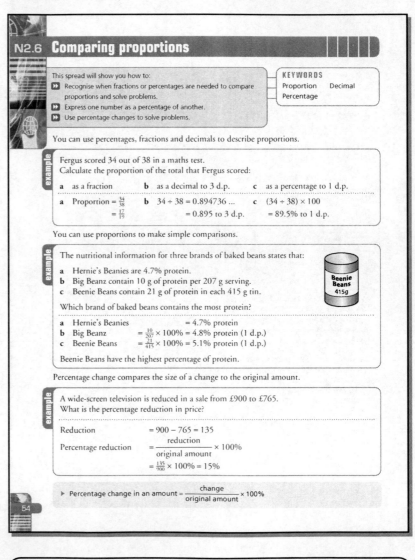

N2.6 Comparing proportions

This spread will show you how to:
- Recognise when fractions or percentages are needed to compare proportions and solve problems.
- Express one number as a percentage of another.
- Use percentage changes to solve problems.

KEYWORDS
Proportion Decimal
Percentage

You can use percentages, fractions and decimals to describe proportions.

example
Fergus scored 34 out of 38 in a maths test.
Calculate the proportion of the total that Fergus scored:
a as a fraction **b** as a decimal to 3 d.p. **c** as a percentage to 1 d.p.

a Proportion $= \frac{34}{38}$ **b** $34 \div 38 = 0.894736 ...$ **c** $(34 \div 38) \times 100$
$= \frac{17}{19}$ $= 0.895$ to 3 d.p. $= 89.5\%$ to 1 d.p.

You can use proportions to make simple comparisons.

example
The nutritional information for three brands of baked beans states that:
a Hernie's Beanies are 4.7% protein.
b Big Beanz contain 10 g of protein per 207 g serving.
c Beenie Beans contain 21 g of protein in each 415 g tin.

Which brand of baked beans contains the most protein?

a Hernie's Beanies $= 4.7\%$ protein
b Big Beanz $= \frac{10}{207} \times 100\% = 4.8\%$ protein (1 d.p.)
c Beenie Beans $= \frac{21}{415} \times 100\% = 5.1\%$ protein (1 d.p.)

Beenie Beans have the highest percentage of protein.

Percentage change compares the size of a change to the original amount.

example
A wide-screen television is reduced in a sale from £900 to £765.
What is the percentage reduction in price?

Reduction $= 900 - 765 = 135$
Percentage reduction $= \frac{reduction}{original\ amount} \times 100\%$
$= \frac{135}{900} \times 100\% = 15\%$

▶ Percentage change in an amount $= \frac{change}{original\ amount} \times 100\%$

54

Plenary

Refer back to the mental starter.
Discuss the equivalences that students should know – use a number line to help you.

Further activities

N2.6F contains extra questions on percentage change and proportion in context.

Differentiation

Extension questions:
▸ Question 1 provides basic practice in expressing a proportion as a fraction.
▸ Questions 2–5 focus on proportions expressed as a percentage, including percentage change.
▸ Question 6 continues the theme of proportion expressed as a percentage, but with larger numbers.

Core tier: focuses on using proportion to make comparisons.

Exercise N2.6

1 What fraction of:
   a 300 pupils is 46 pupils
   b £2.40 is 68p
   c 80 cm is 20 mm
   d €9 is €0.80
   e 140 CDs is 72 CDs
   f 10 000 apples is 3464 apples
   g 1 day is 1 minute
   h 6 days and 8 hours is 6 hours?
   Express your answers as fractions in their simplest form.

2 What proportion of each of these shapes is shaded? Give your answer as
   i a fraction   ii a percentage.

   a   b

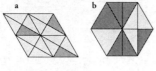

   c   d

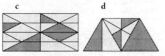

3 Barry, Jenny and Faye all work on Saturdays in the local supermarket. They have all been given a pay increase. The amounts are given as hourly rates:

BARRY	JENNY
£3.10 to £4.60	£3.30 to £4.80

FAYE
£3.00 to £4.50

   a Work out the percentage wage increase for each of these workers.
   b Is the pay rise fair? Comment on your answer.

4 Greengage School ran a campaign to encourage students to eat more vegetables in its canteen. The table shows the weight of each vegetable sold on one day both before and after the campaign.

	Before	After
Cabbage	12 kg	13 kg
Broccoli	5 kg	6 kg
Carrots	7.5 kg	8 kg
Cucumber	8 kg	9.4 kg
Onions	8.3 kg	9.2 kg
Lettuce	5.4 kg	7 kg

   a Work out the percentage increase in sales for each type of vegetable.
   b Is it fair to say that the vegetable with the greatest percentage increase is the more popular? Comment on your answer.

5 a To pass his maths exam, Joachim needs to score 45 marks out of 80. What percentage of the marks does he need to pass?
   b David earns £28 345 a year. He pays £7936 in tax. What percentage of his earnings does he pay in tax?

6 Copy and complete this table showing the yearly earnings of all five employees at Cheapo Clothes shop.

Employee	Earnings (£)	Percentage of total wage bill
Mark		18.75%
John	£8 000	12.5%
Richard	£7 400	
Roy	£12 800	20%
Carl		
Total	£64 000	100%

55

Exercise commentary

The questions assess the objectives on Framework Pages 75 and 77.

Problem solving
Questions 3 and 4 require students to provide a concise, reasoned argument, assessing the objectives on Framework Page 31.

Group work
Questions 3 and 4 can be discussed in pairs, as they require a degree of interpretation.

Misconceptions
A common mistake is to ignore mixed units when expressing a proportion, such as in question 1b and 1c.
Ensure that students read each question thoroughly.
Suggest converting to the smaller unit to avoid unnecessary decimals.

Links
This topic links to proportionality and proportional reasoning (Framework Page 79).

Homework

N2.6HW is based on proportion problems.

Answers

1 a $\frac{23}{150}$ b $\frac{17}{60}$ c $\frac{1}{40}$ d $\frac{4}{45}$ e $\frac{18}{35}$ f $\frac{433}{1250}$ g $\frac{1}{1440}$ h $\frac{3}{76}$

2 a i $\frac{3}{16}$ ii 18.75% b i $\frac{7}{12}$ ii 58.3%
  c i $\frac{5}{16}$ ii 31.25% d i $\frac{11}{18}$ ii 61.1%

3 a 48.4%, 45.5%, 50% b No

4 a 8.3%, 20%, 6.7%, 17.5%, 10.8%, 29.6% b No

5 a 56.25% b 28%

6 Mark £12 000, Richard 11.6%, Carl £23 800

Summary

The key objectives for this unit are:
▸ Use efficient methods to add, subtract, multiply and divide fractions
▸ Use proportional reasoning to solve a problem, choosing the correct numbers to take as 100%

Plenary activity

Set this problem:
Which is better:
£1000 invested at 20% per year, or £2000 invested at 10%?
Encourage students to justify their responses with calculation (the first option overtakes the second option after 8 years).

Check out commentary

1　Students should be encouraged to discuss the different strategies they have learned using the four operations +, −, × and ÷ with fractions.

2　Students may incorrectly assume that they are required to find 63% of 9135 people.
Encourage students to write down what information they have been given and what they are being asked to find. Similarly with the 'speed, distance, time' problem: ask students to clarify or write down the information given and the problem to be solved.

3　This could be used a problem where students work in pairs or small teams to a time limit – this would involve students discussing what tasks were required and interpreting the problem in their own words.

4　Students should be encouraged to justify the figures they choose for the estimate bearing in mind the purpose for which the calculation might be used.

N2 Summary

You should know how to ...

1 Use efficient methods to add, subtract, multiply and divide fractions.

2 Use proportional reasoning to solve a problem, choosing the correct numbers to take as 100%.

3 Solve substantial problems by breaking them into simpler tasks.

4 Give solutions to problems to an appropriate degree of accuracy.

Check out

1 Work out:
a $\frac{5}{11} - \frac{4}{5}$　　b $\frac{7}{8} + \frac{13}{17}$
c $3\frac{5}{6} - 1\frac{7}{17}$　　d $6\frac{2}{3} + 14\frac{6}{11}$
e $\frac{7}{9} \times \frac{4}{9}$　　f $\frac{3}{11} + \frac{3}{4}$
g $2\frac{1}{5} \times 3\frac{5}{6}$　　h $5\frac{1}{4} \div 3\frac{2}{7}$

2 a In a local election 63% of the population of a village who were eligible to vote did so. This was 9135 people.
　i How many people were actually eligible to vote?
　ii How many of the eligible population did not vote?
b If a car travels 50 km in 40 minutes, how far would it travel in one hour?

3 A father wrote a will that shared out his £32 000 savings to each of his five children equally.
Each child spent 23% of their share and saved $\frac{1}{5}$. They each gave the remainder to charity.
a What percentage of the £32 000 was
　i saved ii given to charity?
b If the saved amounts were invested at an interest rate of 5% how much would each child have saved at the end of one year?

4 The average amount of water drunk per day by each student in Year 8 is 3.68 litres.
Estimate the number of litres you would need to supply for a year if there are 143 students in Year 8.

56

Development

The theme of proportion is taken up in N3, which focuses on multiplicative relationships.
A5 addresses proportion problems through the use of algebra.
D2 includes the use of percentages in statistics.

Links

Proportional reasoning is a major theme in the maths curriculum. Proportion is used in various subjects, such as science, food technology and design.

Mental starters

Objectives covered in this unit:
- Order, add, subtract, multiply and divide integers.
- Know and use squares, cubes, roots and index notation.

Resources needed

* means class set needed
Essential:
A2.1OHP – rules of indices
A2.2OHP – 'power cable'
A2.4OHP – spider diagram
A2.5OHP – Venn diagram

Useful:
Scrap paper
Card
OHP calculator
Mini-whiteboards*
A2.5F – factorising
A2.4ICT – formulae

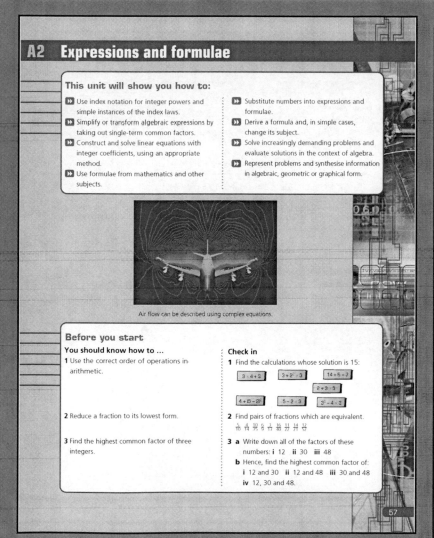

A2 Expressions and formulae

This unit will show you how to:

- Use index notation for integer powers and simple instances of the index laws.
- Simplify or transform algebraic expressions by taking out single-term common factors.
- Construct and solve linear equations with integer coefficients, using an appropriate method.
- Use formulae from mathematics and other subjects.
- Substitute numbers into expressions and formulae.
- Derive a formula and, in simple cases, change its subject.
- Solve increasingly demanding problems and evaluate solutions in the context of algebra.
- Represent problems and synthesise information in algebraic, geometric or graphical form.

Air flow can be described using complex equations.

Before you start

You should know how to ...
1 Use the correct order of operations in arithmetic.

2 Reduce a fraction to its lowest form.

3 Find the highest common factor of three integers.

Check in
1 Find the calculations whose solution is 15:

$3 \cdot 4 + 3$ $3 + 2^2 \cdot 3$ $14 + 5 \div 3$
$2 + 9 \cdot 9$
$4 + 13 - 2^3$ $5 - 2 \cdot 3$ $3^3 - 4 \cdot 3$

2 Find pairs of fractions which are equivalent.
$\frac{5}{10} \quad \frac{6}{14} \quad \frac{10}{16} \quad \frac{9}{15} \quad \frac{8}{16} \quad \frac{11}{24} \quad \frac{12}{42}$

3 a Write down all of the factors of these numbers: **i** 12 **ii** 30 **iii** 48
  b Hence, find the highest common factor of:
  i 12 and 30 **ii** 12 and 48 **iii** 30 and 48
  iv 12, 30 and 48.

57

Unit commentary

Aim of the unit

This unit revisits and extends a range of algebraic skills, starting with the order of operations and the rules of indices. Expansion of brackets leads into the reverse process, factorisation. Finally all the skills practised are used in re-arranging expressions, and constructing and solving equations.

Introduction

Discuss these key words, used in algebraic contexts:

solve, expand, substitute, simplify, evaluate.

Describe how students will be developing their skills under these headings, and also learning to factorise. This will allow students to become more sophisticated in their use of algebra, which is an important maths language.

Framework references

This unit focuses on:
- Teaching objectives pages: 57, 59, 114–115, 116–117, 123–125
- Problem-solving objectives pages: 7, 9, 19

Differentiation

Core tier

Focuses on simplifying expressions and using simple formulae.

Check in activity

Write $\frac{16}{40}$ on the board. Ask quick-fire questions round the class, for example:

Express the numerator using indices. Is there more than one answer?

What is the HCF of the numerator and denominator?

True or false ... the numerator is equal to $3 + 1 \times 4$?

Students can then pose questions of their own for the class.

Mental starter

Write on the board:

$$2^5 = 32$$

Encourage students to explain this equation. What does 2^5 mean? Practice mental evaluation of indices with quick-fire questions around the room. Extend to 1^{30}, 0^{17}, $(^-2)^4$, $\frac{1}{2}^2$.

Give students one minute to write index questions whose answer is between 1 and 20 (for example 3^2, 2^4). Who wrote the most?

Useful resources

Mini-whiteboards may be useful for the mental starter and the plenary activity.
A2.1 OHP – rules of indices

Introductory activity

Encourage students to work in pairs to evaluate 2^{10} and report back on their thinking.

Discuss the strategies used. What other powers of 2 did they evaluate during their calculation? Mark these on a 'power cable':

2^1	2^2	2^3	2^4	2^5	2^6	2^7	2^8	2^9	2^{10}
2	4	8	16	32	64	128	256	512	1024

Encourage students to use the values from the power cable to evaluate $2^2 \times 2^4$. What do they notice? Can they suggest another way of getting the answer (look at 2^6). Give them other calculations to evaluate, to reinforce the point before looking at the example in the student book.

Introduce division in the same way, for example, $2^7 \div 2^2$ and then examples such as $(2^2)^3$ to develop the other rules.

Write on the board:

$$3^5 = 243$$
$$3^6 = 729$$
$$3^7 = 2187$$

Encourage students to use this information to find $3^4 \times 3^2$, $\dfrac{3^7}{3^2}$, one-third of 3^5.

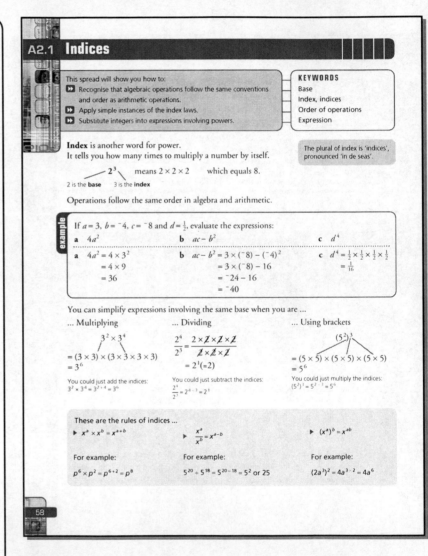

Plenary

Write these incorrect statements on the board:

$$3^x \times 3^y = 9^{x+y} \qquad (3x^2)^3 = 3x^6$$
$$6^x \times 2^y = 12^{x+y} \qquad 4^p \div 4^{-q} = 4^{p-q}$$

Encourage students to discuss in pairs why these are incorrect. Extend to a whole class discussion, reinforcing the index rules, and that these only apply if the same base is used.

Further activities

Students could use a calculator to investigate powers of $^-1$, for example $(^-1)^{11}$, $(^-1)^{150}$, so that they can predict the solution based on an odd or even power, possible generalising this as a formula.

Differentiation

Extension questions:
▸ Questions 1–3 focus on evaluating indices.
▸ Questions 4–6 focus on applying the index rules.
▸ Questions 7–11 focus on developing the index rules, extending to solving equations and graph work.

Core tier: focuses on the order of algebraic operations.

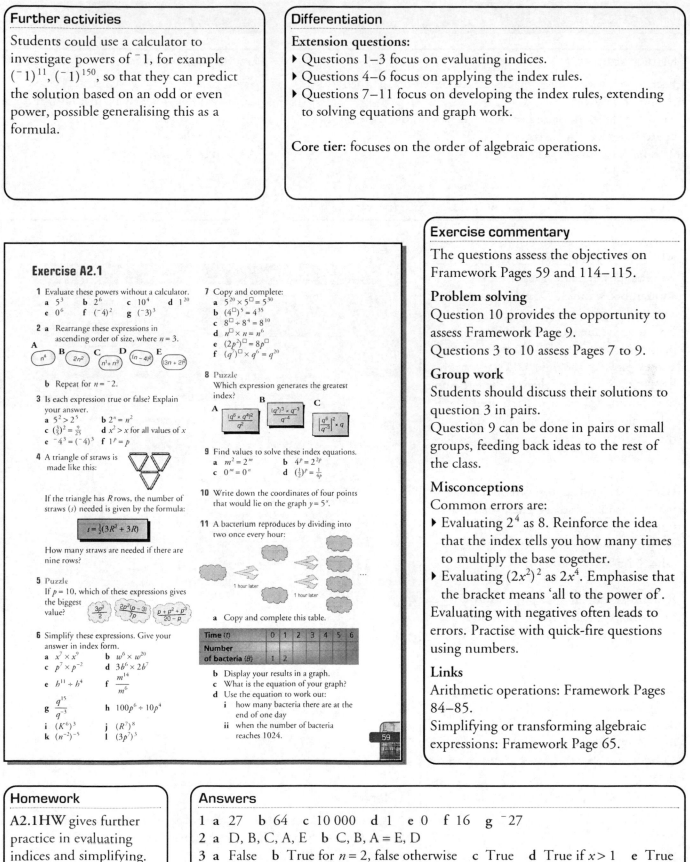

Exercise A2.1

1 Evaluate these powers without a calculator.
 a 5^3 b 2^6 c 10^4 d 1^{20}
 e 0^6 f $(^-4)^2$ g $(^-3)^3$

2 a Rearrange these expressions in ascending order of size, where $n = 3$.
 A n^4 B $2n^2$ C $n^4 + n^3$ D $(n-4)^2$ E $(3n+2)^2$

 b Repeat for $n = ^-2$.

3 Is each expression true or false? Explain your answer.
 a $5^2 > 2^5$ b $2^n = n^2$
 c $(\frac{3}{5})^2 = \frac{9}{25}$ d $x^2 > x$ for all values of x
 e $^-4^3 = (^-4)^3$ f $1^p = p$

4 A triangle of straws is made like this:

If the triangle has R rows, the number of straws (s) needed is given by the formula:

$$s = \tfrac{1}{2}(3R^2 + 3R)$$

How many straws are needed if there are nine rows?

5 Puzzle
If $p = 10$, which of these expressions gives the biggest value? $\frac{3p^3}{2}$ $\frac{2p^2(p-3)}{7p}$ $\frac{p+p^2+p^3}{20-p}$

6 Simplify these expressions. Give your answer in index form.
 a $x^7 \times x^9$ b $w^6 \times w^{20}$
 c $p^7 \times p^{-2}$ d $3b^6 \times 2b^7$
 e $h^{11} \div h^4$ f $\frac{m^{14}}{m^6}$
 g $\frac{q^{15}}{q^{-3}}$ h $100p^6 \div 10p^4$
 i $(K^6)^3$ j $(R^7)^8$
 k $(n^{-2})^{-5}$ l $(3p^7)^3$

7 Copy and complete:
 a $5^{20} \times 5^{\square} = 5^{30}$
 b $(4^{\square})^5 = 4^{35}$
 c $8^{\square} \div 8^4 = 8^{10}$
 d $n^{\square} \times n = n^6$
 e $(2p^2)^{\square} = 8p^{\square}$
 f $(q^7)^{\square} \times q^6 = q^{20}$

8 Puzzle
Which expression generates the greatest index?
 A $\frac{(q^6 \times q^4)^2}{q^2}$ B $\frac{(q^7)^3 \times q^{-3}}{q^{-4}}$ C $\left[\frac{q^6}{q^{-3}}\right]^2 \times q$

9 Find values to solve these index equations.
 a $m^2 = 2^m$ b $4^p = 2^{2p}$
 c $0^m = 0^n$ d $(\frac{1}{2})^p = \frac{1}{4^p}$

10 Write down the coordinates of four points that would lie on the graph $y = 5^x$.

11 A bacterium reproduces by dividing into two once every hour:

 1 hour later   1 hour later   ...

 a Copy and complete this table.

Time (t)	0	1	2	3	4	5	6
Number of bacteria (B)	1	2					

 b Display your results in a graph.
 c What is the equation of your graph?
 d Use the equation to work out:
  i how many bacteria there are at the end of one day
  ii when the number of bacteria reaches 1024.

59

Exercise commentary

The questions assess the objectives on Framework Pages 59 and 114–115.

Problem solving
Question 10 provides the opportunity to assess Framework Page 9.
Questions 3 to 10 assess Pages 7 to 9.

Group work
Students should discuss their solutions to question 3 in pairs.
Question 9 can be done in pairs or small groups, feeding back ideas to the rest of the class.

Misconceptions
Common errors are:
▸ Evaluating 2^4 as 8. Reinforce the idea that the index tells you how many times to multiply the base together.
▸ Evaluating $(2x^2)^2$ as $2x^4$. Emphasise that the bracket means 'all to the power of'.
Evaluating with negatives often leads to errors. Practise with quick-fire questions using numbers.

Links
Arithmetic operations: Framework Pages 84–85.
Simplifying or transforming algebraic expressions: Framework Page 65.

Homework

A2.1HW gives further practice in evaluating indices and simplifying.

Answers

1 a 27 b 64 c 10 000 d 1 e 0 f 16 g $^-27$
2 a D, B, C, A, E b C, B, A = E, D
3 a False b True for $n = 2$, false otherwise c True d True if $x > 1$ e True
 f True for $p = 1$, false otherwise
4 135 5 $\frac{3p^3}{2}$
6 a x^{16} b w^{26} c p^5 d $6b^{13}$ e h^7 f m^8 g q^{18} h $10p^2$ i K^{18}
 j R^{56} k n^{10} l $27p^{21}$
7 a 10 b 7 c 14 d 5 e 3, 6 f 2 8 B 9 a 2 b Any c Any d 4
11 a 4, 8, 16, 32, 64 c $B = 2^t$ d i 16 777 216 ii After 10 hours

Mental starter

Each student needs two small pieces of paper, one labelled 'true', the other 'false'.

To recap the work on indices from the last lesson, write on the board statements such as:

2^5 is bigger than 5^2

$3^3 \times 3^2 = 3^6$

$(\frac{3}{4})^2 = \frac{6}{8}$

Students should hold up 'true' or 'false' for each statement.

Useful resources

Scrap paper

Mini-whiteboards may be useful for the plenary activity.

OHP calculator

A2.2 OHP – 'power cable'

Introductory activity

Use the 'power cable' idea from the student book and A2.2 OHP to introduce zero and negative indices.

Encourage students to explain what happens to the values on the power cable as they move to the right ($\times 2$), and to the left ($\div 2$). Fill in the values to the left along the bottom of the cable. Then fill in the values along the top of the cable, following the pattern of decreasing the index by 1 each time.

Use an OHP calculator to demonstrate how to evaluate indices using the $\boxed{x^y}$ button. Encourage the students to use their calculators to find the values of other bases to the power zero.

Encourage students to spot patterns on the power cable for 2, for example 2^{-3} is the reciprocal of 2^3, and describe them in their own words.

Students investigate fractional indices $\frac{1}{2}$ and $\frac{1}{3}$ in question 10. It may be helpful to demonstrate how to use the fractional index key $\boxed{a^{b/c}}$ on the calculator.

To reinforce their discoveries, demonstrate that

$$4^{\frac{1}{2}} \times 4^{\frac{1}{2}} = 4^1$$

So $4^{\frac{1}{2}}$ must be 2, since $2 \times 2 = 4$.

Hence $4^{\frac{1}{2}} = \sqrt{4}$.

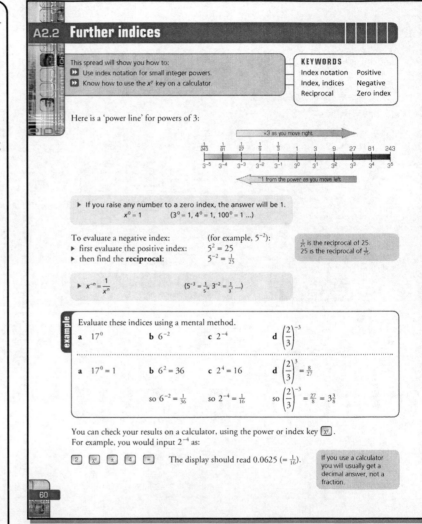

A2.2 Further indices

This spread will show you how to:
- Use index notation for small integer powers.
- Know how to use the x^y key on a calculator.

KEYWORDS
Index notation   Positive
Index, indices   Negative
Reciprocal   Zero index

Here is a 'power line' for powers of 3:

$\times 3$ as you move right

| $\frac{1}{243}$ | $\frac{1}{81}$ | $\frac{1}{27}$ | $\frac{1}{9}$ | $\frac{1}{3}$ | 1 | 3 | 9 | 27 | 81 | 243 |

| 3^{-5} | 3^{-4} | 3^{-3} | 3^{-2} | 3^{-1} | 3^0 | 3^1 | 3^2 | 3^3 | 3^4 | 3^5 |

-1 from the power as you move left

▶ If you raise any number to a zero index, the answer will be 1.
$x^0 = 1$ ($3^0 = 1$, $4^0 = 1$, $100^0 = 1$ ...)

To evaluate a negative index: (for example, 5^{-2}):
▶ first evaluate the positive index: $5^2 = 25$
▶ then find the **reciprocal**: $5^{-2} = \frac{1}{25}$

$\frac{1}{25}$ is the reciprocal of 25.
25 is the reciprocal of $\frac{1}{25}$.

▶ $x^{-n} = \frac{1}{x^n}$ ($5^{-3} = \frac{1}{5^3}$, $3^{-2} = \frac{1}{3^2}$ ...)

example

Evaluate these indices using a mental method.
a 17^0 **b** 6^{-2} **c** 2^{-4} **d** $\left(\frac{2}{3}\right)^{-3}$

a $17^0 = 1$ **b** $6^2 = 36$ **c** $2^4 = 16$ **d** $\left(\frac{2}{3}\right)^3 = \frac{8}{27}$

so $6^{-2} = \frac{1}{36}$ so $2^{-4} = \frac{1}{16}$ so $\left(\frac{2}{3}\right)^{-3} = \frac{27}{8} = 3\frac{3}{8}$

You can check your results on a calculator, using the power or index key $\boxed{x^y}$.
For example, you would input 2^{-4} as:

$\boxed{2}$ $\boxed{x^y}$ $\boxed{\pm}$ $\boxed{4}$ $\boxed{=}$ The display should read 0.0625 ($= \frac{1}{16}$).

If you use a calculator you will usually get a decimal answer, not a fraction.

60

Plenary

Assess the students' understanding of zero, negative and fractional indices through quick-fire questions such as:

18^0 7^{-2} $64^{\frac{1}{2}}$

Try some questions 'in reverse', such as:

What is the value of x in:

$5^x = \frac{1}{25}$ $16^x = 4$ $17^x = 1$?

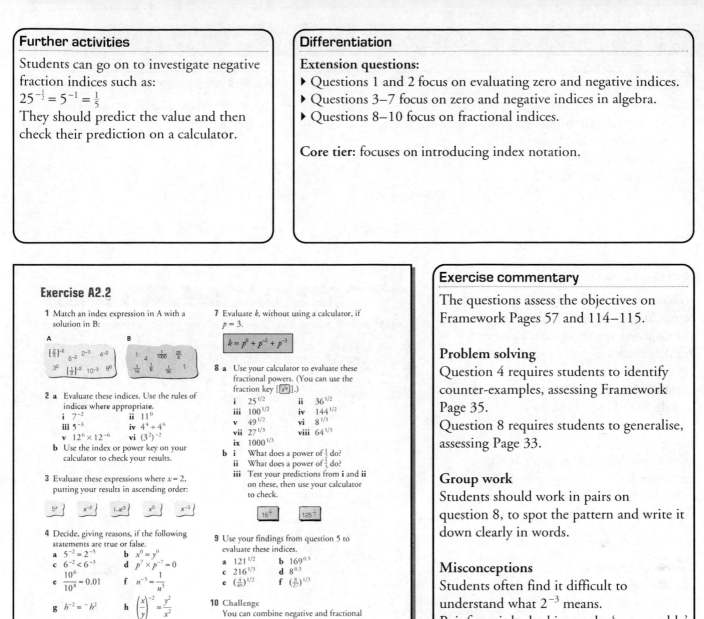

Exercise A2.2

1 Match an index expression in A with a solution in B:

A

$\left(\frac{2}{5}\right)^{-2}$ 6^{-2} 2^{-3} 4^{-2}
3^0 $\left(\frac{1}{2}\right)^{-2}$ 10^{-3} 8^0

B

1 4 $\frac{1}{1000}$ $\frac{25}{4}$
$\frac{1}{16}$ $\frac{1}{8}$ $\frac{1}{36}$ 1

2 a Evaluate these indices. Use the rules of indices where appropriate.
  i 7^{-2} ii 11^0
  iii 5^{-3} iv $4^4 \div 4^6$
  v $12^6 \times 12^{-6}$ vi $(3^2)^{-2}$
  b Use the index or power key on your calculator to check your results.

3 Evaluate these expressions where $x = 2$, putting your results in ascending order:

5^x x^{-2} $(-x^3)$ x^0 x^{-3}

4 Decide, giving reasons, if the following statements are true or false.
  a $5^{-2} = 2^{-5}$ b $x^0 = y^0$
  c $6^{-2} < 6^{-3}$ d $p^7 \times p^{-7} = 0$
  e $\frac{10^6}{10^8} = 0.01$ f $n^{-3} = \frac{1}{n^3}$
  g $h^{-2} = {}^-h^2$ h $\left(\frac{x}{y}\right)^{-2} = \frac{y^2}{x^2}$

5 Find the value of p in each case:
  a $5^p = \frac{1}{125}$ b $6^p = 0$
  c $\left(\frac{1}{2}\right)^p = 8$ d $7^{-p} = 49$

6 a Write as many expressions involving indices with the answer $\frac{1}{64}$ as you can.
  b Repreat for a number of your choice.

7 Evaluate k, without using a calculator, if $p = 3$.

$$k = p^0 + p^{-2} + p^{-3}$$

8 a Use your calculator to evaluate these fractional powers. (You can use the fraction key $[\boxed{a^x}]$.)
  i $25^{1/2}$ ii $36^{1/2}$
  iii $100^{1/2}$ iv $144^{1/2}$
  v $49^{1/2}$ vi $8^{1/3}$
  vii $27^{1/3}$ viii $64^{1/3}$
  ix $1000^{1/3}$
  b i What does a power of $\frac{1}{2}$ do?
    ii What does a power of $\frac{1}{3}$ do?
    iii Test your predictions from i and ii on these, then use your calculator to check.

$16^{\frac{1}{7}}$ $125^{\frac{1}{3}}$

9 Use your findings from question 5 to evaluate these indices.
  a $121^{1/2}$ b $169^{0.5}$
  c $216^{1/3}$ d $8^{0.3}$
  e $\left(\frac{4}{49}\right)^{1/2}$ f $\left(\frac{8}{27}\right)^{1/3}$

10 Challenge
You can combine negative and fractional powers.
Look at these examples:
$25^{-1/2} \rightarrow$ reciprocal of $25^{1/2} \rightarrow \frac{1}{5}$
$64^{-1/3} \rightarrow$ reciprocal of $64^{1/3} \rightarrow \frac{1}{4}$
  a Evaluate these:
    i $49^{-1/2}$ ii $64^{-1/2}$
    iii $1000^{-1/3}$ iv $125^{-1/3}$
  b Check on your calculator.

61

Answers

1 $\left(\frac{2}{5}\right)^{-2} = \frac{25}{4}$, $6^{-2} = \frac{1}{36}$, $2^{-3} = \frac{1}{8}$, $4^{-2} = \frac{1}{16}$, $3^0 = 1$, $\left(\frac{1}{2}\right)^{-2} = 4$, $10^{-3} = \frac{1}{1000}$, $8^0 = 1$

2 a i $\frac{1}{49}$ ii 1 iii $\frac{1}{125}$ iv $\frac{1}{16}$ v 1 vi $\frac{1}{81}$ 3 $^-8$, $\frac{1}{8}$, $\frac{1}{4}$, 1, 25

4 a False   b True   c False   d False   e True   f True   g False   h True

5 a $^-3$ b 1 c $^-3$ d $^-2$ 7 $1\frac{4}{27}$

8 a i 5 ii 6 iii 10 iv 12 v 7 vi 2 vii 3 viii 4 ix 10
  b i Square root   ii Cube root   iii $4, 5$

9 a 11 b 13 c 6 d 2 e $\frac{2}{7}$ f $\frac{2}{3}$ 10 a i $\frac{1}{7}$ ii $\frac{1}{8}$ iii $\frac{1}{10}$ iv $\frac{1}{5}$

Mental starter

Starting with an expression such as $2x^2$ or $3x + 4y$, encourage students to suggest other ways of writing these, 'reversing' the simplification process.
For example:

$$2x^2 = 2 \times x \times x = 2 \times x^2 = x^2 + x^2$$
$$3x + 4y = 3 \times x + 4 \times y = x + x + x + y + y + y + y$$

Useful resources

Mini-whiteboards may be useful for the starter activity.

Introductory activity

Recap methods of simplifying expressions from Year 7: collecting like terms, and multiplying terms together to remove multiplication signs.

Emphasise that students need to work carefully and think about what they are doing, to avoid mistakes.

Encourage students to identify first whether the expression involves addition/subtraction or multiplication/division.

For example $3x + 4y$ $3x \times 4y$

The multiplication expression is easier, because you just need to carry on multiplying. The addition expression cannot be simplified because there are no like terms to collect.

Discuss the examples in the student book, giving further examples where necessary to reinforce students' learning.
In particular, emphasise that cancelling algebraic fractions is the same as cancelling numerical fractions. Use this to highlight the fact the letters represent numbers.
In the examples, suggest that the numbers are dealt with first, and then the letters.

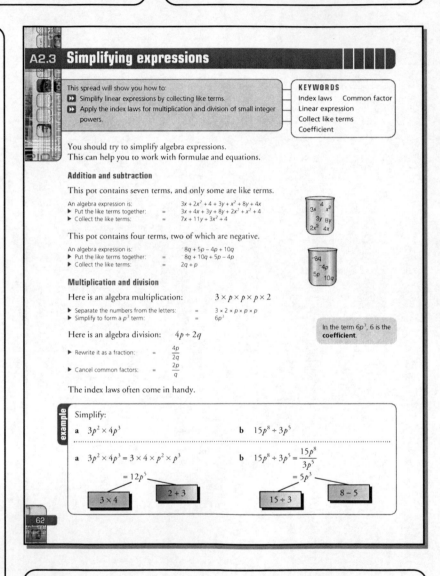

Plenary

Write a series of expressions such as $5x + 3x^2$, $6x + 5x$, on the board or on flashcards.

Encourage students to decide whether each expression can be simplified, and if so, demonstrate how they would do this.

Further activities

Students can draw geometrical shapes (beginning with a square), writing their own expressions for the dimensions. They can then write expressions for the perimeter and area, simplifying these to derive formulae.

Differentiation

Extension questions:

▸ Questions 1 and 2 offer practice at simplification of expressions.
▸ Questions 3–5 focus on using simplification to solve problems.
▸ Question 6 focuses on simplifying expressions for perimeter and area.

Core tier: focuses on collecting like terms in problem-solving contexts such as magic squares.

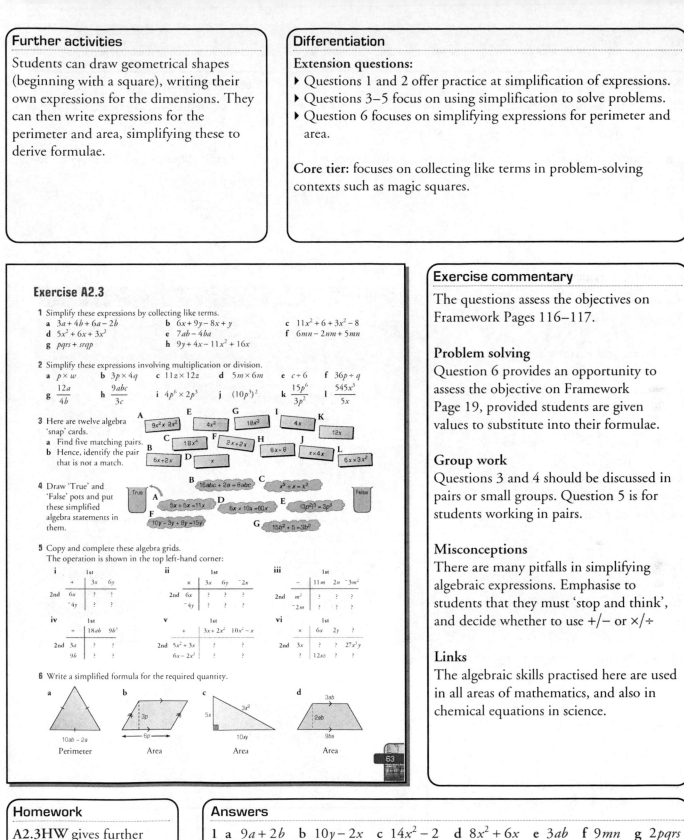

Exercise A2.3

1 Simplify these expressions by collecting like terms.
  a $3a + 4b + 6a - 2b$ **b** $6x + 9y - 8x + y$ **c** $11x^2 + 6 + 3x^2 - 8$
  d $5x^2 + 6x + 3x^2$ **e** $7ab - 4ba$ **f** $6mn - 2nm + 5mn$
  g $pqrs + srqp$ **h** $9y + 4x - 11x^2 + 16x$

2 Simplify these expressions involving multiplication or division.
  a $p \times w$ **b** $3p \times 4q$ **c** $11z \times 12z$ **d** $5m \times 6m$ **e** $c \div 6$ **f** $36p \div q$
  g $\frac{12a}{4b}$ **h** $\frac{9abc}{3c}$ **i** $4p^6 \times 2p^3$ **j** $(10p^3)^2$ **k** $\frac{15p^6}{3p^2}$ **l** $\frac{545x^3}{5x}$

3 Here are twelve algebra 'snap' cards.
  a Find five matching pairs.
  b Hence, identify the pair that is not a match.

4 Draw 'True' and 'False' pots and put these simplified algebra statements in them.

5 Copy and complete these algebra grids.
The operation is shown in the top left-hand corner:

6 Write a simplified formula for the required quantity.

a Perimeter
b Area
c Area
d Area

Exercise commentary

The questions assess the objectives on Framework Pages 116–117.

Problem solving
Question 6 provides an opportunity to assess the objective on Framework Page 19, provided students are given values to substitute into their formulae.

Group work
Questions 3 and 4 should be discussed in pairs or small groups. Question 5 is for students working in pairs.

Misconceptions
There are many pitfalls in simplifying algebraic expressions. Emphasise to students that they must 'stop and think', and decide whether to use +/− or ×/÷

Links
The algebraic skills practised here are used in all areas of mathematics, and also in chemical equations in science.

Homework

A2.3HW gives further practice in simplifying algebraic expressions.

Answers

1 a $9a + 2b$ **b** $10y - 2x$ **c** $14x^2 - 2$ **d** $8x^2 + 6x$ **e** $3ab$ **f** $9mn$ **g** $2pqrs$
  h $20x + 9y - 11x^2$
2 a pw **b** $12pq$ **c** $132z^2$ **d** $30m^2$ **e** $\frac{c}{6}$ **f** $\frac{36p}{q}$ **g** $\frac{3a}{b}$ **h** $3ab$ **i** $8p^9$
  j $100p^6$ **k** $5p^4$ **l** $109x^2$
3 a $9x^2 \times 2x^2 = 18x^4$, $4x^2 = x \times 4x$, $18x^3 = 6x \times 3x^2$, $4x = 2x + 2x$, $8x \div 8 = x$
  b $12x$, $6x + 2x$
4 True: $5x + 6x = 11x$, $10y - 3y + 8y = 15y$, $15b^2 \div 5 = 3b^2$
  False: $16abc \div 2a = 8abc$, $x^2 + x = x^3$, $6x \times 10x = 60x$, $(3p^2)^3 = 3p^6$
6 a $P = 30ab - 6a$ **b** $A = 18p^2$ **c** $A = 25x^2y$ **d** $A = 12a^2b^2$

A2.4 Expanding brackets

Mental starter

Write on the board expressions such as:

$$3(x+2) \qquad x(x+y) \qquad p^2(p^3 - q)$$

Encourage students to demonstrate how they expand the brackets.

Useful resources

Mini-whiteboards may be useful for the mental starter.

A2.4OHP shows the spider diagram in the student book.

Introductory activity

Recap expanding simple brackets from Year 7.

Use numerical examples to reinforce the techniques used in expanding brackets.
For example
$$8(3 + 4) = 8 \times 7 = 56$$
or $\qquad = 8 \times 3 + 8 \times 4 = 56$
So $8(x + 4) = 8 \times x + 8 \times 4 = 8x + 32$

Use similar numerical examples to demonstrate multiplying a bracket by a negative number,
For example
$$^-3(2 + 6) \quad = ^-3 \times 8 = ^-24$$
or $\qquad = ^-3 \times 2 + ^-3 \times 6$
$\qquad = ^-6 + ^-18 = ^-24$
So $^-8(x - 3) = ^-8 \times x + ^-8 \times ^-3$
$\qquad = ^-8x + 24$

Discuss factorisation as the inverse of expansion

For example $10x + 20 = 10(x + 2)$

Keep to simple examples.
A2.4OHP shows the spider diagram in the student book.

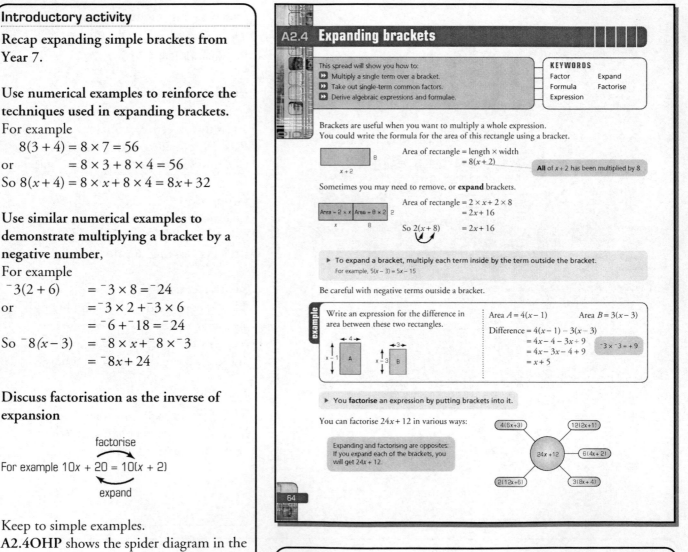

Plenary

Discuss any difficulties students encountered in the exercise, and how they solved them.

Discuss and try out students' ideas for expanding the combination of brackets in question 8. Reinforce the method by comparing the problem with a numerical example, such as $(10 + 2)(10 + 2)$.

Further activities

Students can develop question 8 by investigating the expansion of combinations of brackets, using a variety of examples.

A2.4ICT allows students to construct formulae using an Excel spreadsheet.

Differentiation

Extension questions:
▸ Question 1 focuses on expanding brackets where further simplification is not needed.
▸ Questions 2–5 focus on expansions involving negatives, where further simplification is needed.
▸ Questions 6–8 focus on factorisation as the 'reverse' of expansion.

Core tier: focuses on expanding a bracket in simple expressions.

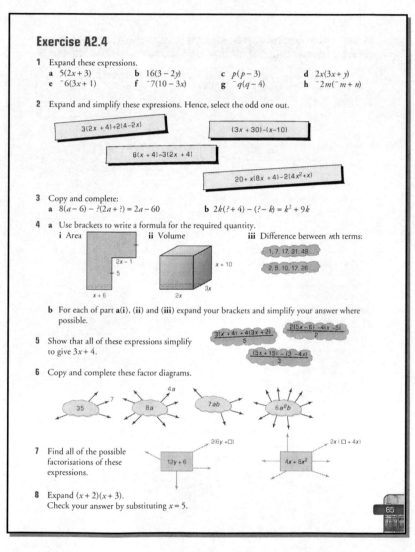

Exercise A2.4

1 Expand these expressions.
  a $5(2x+3)$ b $16(3-2y)$ c $p(p-3)$ d $2x(3x+y)$
  e $^-6(3x+1)$ f $^-7(10-3x)$ g $^-q(q-4)$ h $^-2m(^-m+n)$

2 Expand and simplify these expressions. Hence, select the odd one out.

$3(2x+4)+2(4-2x)$

$(3x+30)-(x-10)$

$8(x+4)-3(2x+4)$

$20+x(8x+4)-2(4x^2+x)$

3 Copy and complete:
  a $8(a-6)-?(2a+?)=2a-60$ b $2k(?+4)-(?-k)=k^2+9k$

4 a Use brackets to write a formula for the required quantity.
   i Area   ii Volume   iii Difference between mth terms:
                                       1, 7, 17, 31, 49
                                       2, 5, 10, 17, 26

  b For each of part a(i), (ii) and (iii) expand your brackets and simplify your answer where possible.

5 Show that all of these expressions simplify to give $3x+4$.

$\dfrac{3(x+4)+4(3x+2)}{5}$ $\dfrac{2(5x-6)-4(x-5)}{2}$ $\dfrac{(5x+15)-(3-4x)}{3}$

6 Copy and complete these factor diagrams.

   35   8a (with 7, 4a)   7ab   6a²b

7 Find all of the possible factorisations of these expressions.

   12y+6   → 2(6y+□)
   4x+8x² → 2x(□+4x)

8 Expand $(x+2)(x+3)$.
  Check your answer by substituting $x=5$.

Exercise commentary

The questions assess the objectives on Framework Pages 116–117.

Problem solving

Question 4 requires expressions to be constructed in algebraic form, assessing the objective on Framework Page 27.

Group work

Students should work in pairs for question 2, to discuss which is the odd one out, and question 3, to discuss whether there is more than one solution. Questions 6 and 7 should also be attempted in pairs, to ensure that all possibilities are considered.

Misconceptions

Students frequently make errors in algebraic simplification. Remind them that they must think carefully about the operations they are using.
The negative sign outside a bracket is another common pitfall. Emphasise that it reverses the sign of all the terms in the bracket.

Links

The algebraic skills practised here are used in all areas of mathematics.

Homework

A2.4HW gives further practice in expanding brackets, and also in the reverse process, factorising.

Answers

1 a $10x+15$ b $48-32y$ c p^2-3p d $6x^2+2xy$ e $^-18x-6$
  f $^-70+21x$ g $^-q^2+4q$ h $2m^2-2mn$
2 All $2x+20$ except $(3x+30)-(x-10)$ 3 a 3, 4   b k, k^2
4 b i $20x+55$ ii $6x^3+60x^2$ iii n^2-2 8 x^2+5x+6

Mental starter

Write some numbers on the board:

For example,   12   16   36   48

Ask students to write down all the factors of each number. Which is the highest common factor?

Repeat for other sets of numbers.

Useful resources

Mini-whiteboards may be useful for the mental starter.

A2.5OHP is a Venn diagram.

Introductory activity

Recap finding the HCF from Year 7.

Emphasise that to factorise fully you have to divide by the HCF of all the terms in the expression.

Demonstrate how to draw a Venn diagram of all the factors, and how to use it to find the HCF.

A2.5OHP shows the Venn diagram in the student book. You can use this to illustrate how to factorise $3x^2 + 9xy$.

Go over the bulleted examples in the student book.

Use the examples to illustrate that the common factor could be a letter or a number, or a combination of the two.

Emphasise that you can check your factorisation by expanding your answer, and seeing if you get back to what you started with.

Discuss whether you can use substitution to check a factorisation, and ask which values of the letter it would be sensible to substitute.

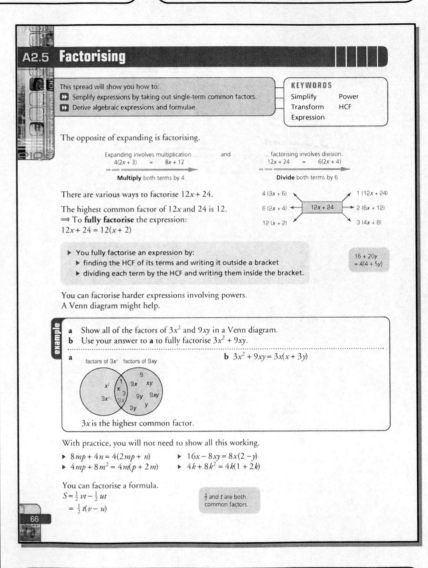

A2.5 Factorising

This spread will show you how to:
- Simplify expressions by taking out single-term common factors.
- Derive algebraic expressions and formulae

KEYWORDS
Simplify   Power
Transform   HCF
Expression

The opposite of expanding is factorising.

Expanding involves multiplication ...   and   ... factorising involves division.
$4(2x + 3)$ = $8x + 12$ $12x + 24$ = $6(2x + 4)$
Multiply both terms by 4   **Divide** both terms by 6

There are various ways to factorise $12x + 24$.

The highest common factor of $12x$ and 24 is 12.
⇒ To **fully factorise** the expression:
$12x + 24 = 12(x + 2)$

$4(3x + 6)$ $1(12x + 24)$
$6(2x + 4)$ ← $12x + 24$ → $2(6x + 12)$
$12(x + 2)$ $3(4x + 8)$

▸ You fully factorise an expression by:
  ▸ finding the HCF of its terms and writing it outside a bracket
  ▸ dividing each term by the HCF and writing them inside the bracket.

$16 + 20y$
$= 4(4 + 5y)$

You can factorise harder expressions involving powers.
A Venn diagram might help.

a Show all of the factors of $3x^2$ and $9xy$ in a Venn diagram.
b Use your answer to **a** to fully factorise $3x^2 + 9xy$.

a factors of $3x^2$ factors of $9xy$ **b** $3x^2 + 9xy = 3x(x + 3y)$

$3x$ is the highest common factor.

With practice, you will not need to show all this working.
▸ $8mp + 4n = 4(2mp + n)$ ▸ $16x - 8xy = 8x(2 - y)$
▸ $4mp + 8m^2 = 4m(p + 2m)$ ▸ $4k + 8k^2 = 4k(1 + 2k)$

You can factorise a formula.
$S = \frac{1}{2}vt - \frac{1}{2}ut$
$= \frac{1}{2}t(v - u)$

$\frac{1}{2}$ and t are both common factors.

66

Plenary

Discuss any difficulties students encountered in the exercise, and how they solved them.

Encourage students to demonstrate how they factorise more complex expressions.

For example $8\pi x^2 y + 12\pi^2 x + 16xy^3 - 20xyz$

66

Further activities

Students can develop question 7 into further investigations. They can begin to investigate factorising double brackets, using simple examples.

A2.5F is a worksheet containing problem-solving activities that involve factorising.

Differentiation

Extension questions:
▸ Questions 1–3 focus on finding common factors and factorising expressions.
▸ Question 4 focuses on devising and factorising formulae.
▸ Questions 5–7 focus on investigations, using factorisation in the context of justifications of maths statements.

Core tier: focuses on substituting negative values into an expression.

Exercise A2.5

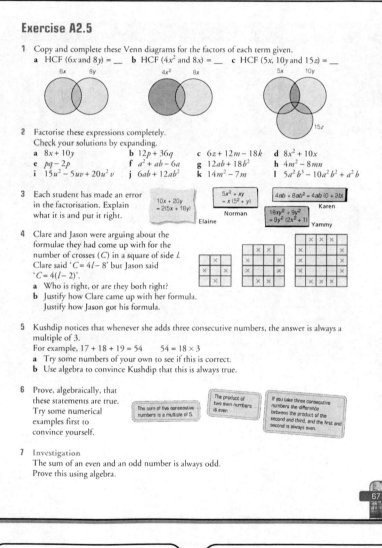

1 Copy and complete these Venn diagrams for the factors of each term given.
   a HCF ($6x$ and $8y$) = __   **b** HCF ($4x^2$ and $8x$) = __   **c** HCF ($5x$, $10y$ and $15z$) = __

2 Factorise these expressions completely.
   Check your solutions by expanding.
   a $8x + 10y$ **b** $12p + 36q$ **c** $6z + 12m - 18k$ **d** $8x^2 + 10x$
   e $pq - 2p$ **f** $a^2 + ab - 6a$ **g** $12ab + 18b^2$ **h** $4m^2 - 8mn$
   i $15u^2 - 5uv + 20u^2v$ **j** $6ab + 12ab^2$ **k** $14m^2 - 7m$ **l** $5a^2b^3 - 10a^2b^2 + a^2b$

3 Each student has made an error in the factorisation. Explain what it is and put it right.

   $10x + 20y = 2(5x + 18y)$ Elaine
   $5x^2 + xy = x(5^2 + y)$ Norman
   $4ab + 8ab^2 = 4ab(0 + 2b)$ Karen
   $18xy^2 + 9y^2 = 9y^2(2x^2 + 1)$ Yammy

4 Clare and Jason were arguing about the formulae they had come up with for the number of crosses (C) in a square of side l. Clare said '$C = 4l - 8$' but Jason said '$C = 4(l - 2)$'.
   a Who is right, or are they both right?
   b Justify how Clare came up with her formula. Justify how Jason got his formula.

5 Kushdip notices that whenever she adds three consecutive numbers, the answer is always a multiple of 3.
   For example, $17 + 18 + 19 = 54$ $54 = 18 \times 3$
   a Try some numbers of your own to see if this is correct.
   b Use algebra to convince Kushdip that this is always true.

6 Prove, algebraically, that these statements are true. Try some numerical examples first to convince yourself.

   The sum of five consecutive numbers is a multiple of 5.
   The product of two even numbers is even.
   If you take three consecutive numbers the difference between the product of the second and third, and the first and second is always even.

7 **Investigation**
   The sum of an even and an odd number is always odd. Prove this using algebra.

67

Exercise commentary

The questions assess the objectives on Framework Pages 117 and 123.

Problem solving
Questions 5–7 provide the opportunity to assess Framework Page 7.

Group work
Students should work in pairs on question 4 so they can discuss their justifications. Students often explain 'how' an expression works, rather than 'why' it works.

Misconceptions
Many students subtract terms as they remove them outside the brackets. Emphasise that factorisation involves **division**.
Students also have difficulty with the use of '1' as a place holder. Encourage students to check their solutions by expanding, which will show where a '1' is needed.
For example $3x + 9x^2 = 3x(1 + 3x)$, **not** $3x(0 + 3x)$

Links
The algebraic skills practised here are used in all areas of mathematics.

Homework

A2.5HW gives further practice in factorisation in various problem solving contexts.

Answers

2 **a** $2(4x + 5y)$ **b** $12(p + 3q)$ **c** $6(z + 2m - 3k)$ **d** $2x(4x + 5)$ **e** $p(q - 2)$
   f $a(a + b - 6)$ **g** $6b(2a + 3b)$ **h** $4m(m - 2n)$ **i** $5u(3u - v + 4uv)$
   j $6ab(1 + 2b)$ **k** $7m(2m - 1)$ **l** $a^2b(5b^2 - 10b + 1)$
3 Elaine $10(x + 2y)$, Norman $x(5x + y)$, Karen $4ab(1 + 2b)$, Yammy $9y^2(2x + 1)$
4 **a** Both right; the expressions are equivalent.
5 **b** $n + (n + 1) + (n + 2) = 3n + 3 = 3(n + 1)$ or equivalent
7 $2n + (2m + 1) = 2(n + m) + 1$ or equivalent

Mental starter

Draw a spider diagram on the board, with values of variables in the body and expressions in those variables in the 'legs'.

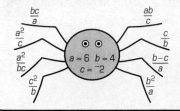

Challenge students to substitute the values into the expressions to find the value for each 'leg'. Using fraction and negative values for the variables will make the calculations more difficult.

Useful resources

Mini-whiteboards may be useful for the mental starter and the plenary activity.
Card

Introductory activity

Recap inverse operations $+/-$ and $\times/\div$.

Write these expressions on cards and stick them to the board (or write on the board):

$$= \quad 3x - 2 \quad 10 \quad 5x - 16 \quad 4(x+8) \quad \frac{8x+1}{2}$$

Students can come to the board and choose pairs of expressions to make equations to solve. Start with a simple one, for example:
$3x - 2 = 10$.

Encourage students to suggest strategies and methods for solving the equations they make. Go through the relevant example in the student book.

Summarise the equation-solving process:

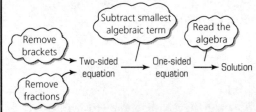

If students find the skills difficult, compare an equation to a balance scale – you have to 'do the same' to each side to keep the balance.

Emphasise that students should check their solutions by substituting back into the original equation.

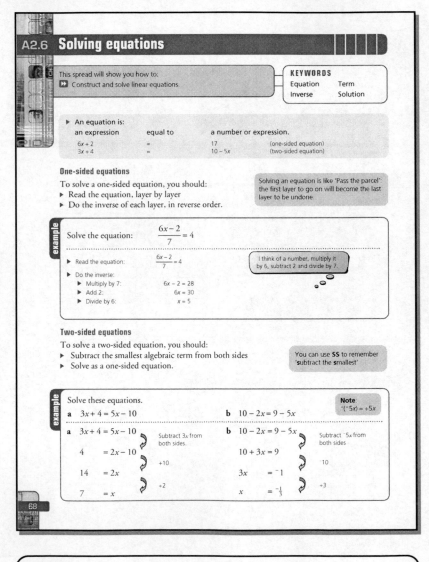

A2.6 Solving equations

This spread will show you how to:
▷ Construct and solve linear equations.

KEYWORDS
Equation   Term
Inverse   Solution

▷ An equation is:

an expression	equal to	a number or expression.
$6x + 2$	=	17	(one-sided equation)
$3x + 4$	=	$10 - 5x$	(two-sided equation)

One-sided equations

To solve a one-sided equation, you should:
▷ Read the equation, layer by layer
▷ Do the inverse of each layer, in reverse order.

Solving an equation is like 'Pass the parcel': the first layer to go on will become the last layer to be undone.

example

Solve the equation: $\dfrac{6x-2}{7} = 4$

▷ Read the equation: $\dfrac{6x-2}{7} = 4$

I think of a number, multiply it by 6, subtract 2 and divide by 7.

▷ Do the inverse:
  ▷ Multiply by 7: $6x - 2 = 28$
  ▷ Add 2: $6x = 30$
  ▷ Divide by 6: $x = 5$

Two-sided equations

To solve a two-sided equation, you should:
▷ Subtract the smallest algebraic term from both sides
▷ Solve as a one-sided equation.

You can use **SS** to remember 'subtract the smallest'

example

Solve these equations.

Note: $-(-5x) = +5x$

a $3x + 4 = 5x - 10$
b $10 - 2x = 9 - 5x$

a $3x + 4 = 5x - 10$ — Subtract $3x$ from both sides
  $4 = 2x - 10$ $+10$
  $14 = 2x$ $\div 2$
  $7 = x$

b $10 - 2x = 9 - 5x$ — Subtract $-5x$ from both sides
  $10 + 3x = 9$ -10
  $3x = -1$ $\div 3$
  $x = -\frac{1}{3}$

68

Plenary

Write four expressions and = on the board, for example:

$$= \quad 10 - 3x \quad 8x + 1 \quad 4(x + 3) \quad 3(2x + 4)$$

Challenge students to a race, working in pairs, to decide which two expressions make the equation with the largest solution.

Differentiation

Extension questions:
▶ Questions 1 allows students to practise solving equations.
▶ Questions 2–5 provide problem-solving activities centred on solving equations.
▶ Question 6 is a more demanding puzzle.

Core tier: focuses on working with formulae.

Exercise A2.6

1 Solve these equations.

a $\dfrac{3x-5}{2} = 10$ b $4(x+3) = 40$ c $2\left(\dfrac{x+1}{2} - 3\right) = 8$

d $5x - 5 = 3x + 5$ e $10 - 2x = 5x + 3$ f $3(x+1) = 9(x-4)$

2 a Arrange all of these expressions to make an equation and its solution.
  b Make a set of your own equation cards. Swap yours with a friend and solve each other's equation.

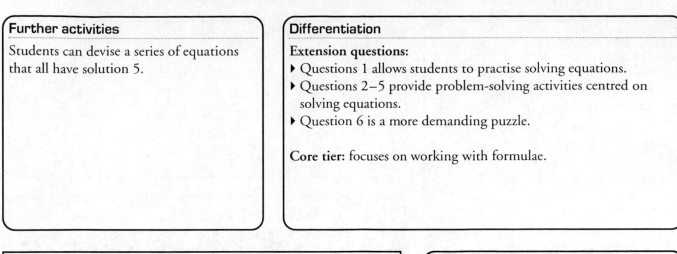

3 Copy and complete this crossword by solving the equations given in the clues.

Across
1 $2(x+2) = 30$
3 $2(x+10) = \frac{1}{2}(3x+80)$
4 $\dfrac{3x-11}{2} = 71$
6 $\dfrac{2x+50}{3} = \dfrac{x+175}{2}$

Down
1 $4(x-50) = 3x - 95$
2 $60 - 2x = 80 - 3x$
5 $x^2 = 196$

4 a This triangle is known to be isosceles. Use this information to find x.
  b Is the triangle equilateral?

$2x - 4$ $4x - 18$
$x + 2$

5 What number must the * represent in each case?

a $3x + * = 6x - 8$
  $* = 3x - 8$
  $12 = 3x$
  $* = x$

b $10 - *x = *x - 20$
  $x = 6$

c $*(10 - *x) = *x - 6$
  $* = x$

6 Puzzle
In each puzzle, the sum of the horizontal squares equals the sum of the vertical squares. What numbers should be in each square?

a
$4 - 3x$
$3x + 12$ $5 + 3x$ $7 - 8x$
$10x + 6$

b
$11a + 5$
$10a - ab$ $ab - 3$ $2ab + 6$
$4x - ab + 3a$

69

Exercise commentary

The questions assess the objectives on Framework Pages 123–125.

Problem solving
Questions 2–6 assess the objectives on Framework Page 9.

Group work
In question 2 students write their own equations and swap these with a partner. Small group discussion may be helpful for question 4.

Misconceptions
Students often have problems with negative terms. To 'undo' ^-3x students will often subtract $3x$.

$10 - 3x = 4x + 7$ wrongly becomes
$10 = x + 7$

Emphasise how far apart ^-3x and $4x$ are on the number line. They should remain at this distance:

$10 - 3x = 4x + 7 \rightarrow 10 = 7x + 7$

More difficult solutions including negatives and fractions may cause trouble. Remind students of previous work on converting fractions to mixed numbers.

Links
The algebraic skills practised here are used in all areas of mathematics, and also in science.

Homework

A2.6HW gives further practice in solving equations and opportunities for students to formulate their own equations.

Answers

1 a $x = 12\frac{1}{2}$ b $x = 7$ c $x = 13$ d $x = 5$ e $x = 1$ f $x = 6\frac{1}{2}$
2 a $7x - 12 = 10 - 4x$, $11x - 12 = 10$, $11x = 22$, $x = 2$
3 Across: 13, 40, 51, 425; Down: 105, 20, 14
4 a $x = 7$ b No 5 a 4 b $2\frac{1}{2}$ c 3 6 a $x = 1\frac{2}{7}$ b $x = \frac{1}{2}ab - a + \frac{1}{4}$

Summary

The key objectives for this unit are:

▸ Construct and solve linear equations with integer coefficients, using an appropriate method.

▸ Solve substantial problems by breaking them into simpler tasks.

▸ Present a concise, reasoned argument, using symbols, diagrams, graphs and related explanatory text.

Check out commentary

1   Students must be reminded to take extra care with negatives, writing out the question in full if stuck. $(^-3)^3 = {}^-3 \times {}^-3 \times {}^-3$.
Negative, fractional and zero indices are definitions that need to be learned and practised. The rules of indices are useful, but students should be encouraged to keep remembering why they work.

2   Students may find it hard to mix all simplification skills: only practice can make perfect. Part **i** will regularly be answered as $24x$ ... remind the students to think what they did in **c**. A number line should be available to recap concepts in adding/subtracting negative terms.

3   Students need to be reminded that factorising is dividing, not subtracting. $5(20y - 10)$ is a common response in **3b**.
'1' is required as a place holder in **3c** ... students should be reminded to expand their answer to check in each case.

4   In **a** part **i** students must take care with the ^-9x that is achieved on expanding: a common error is to subtract $9x$, not ^-9x, from both sides.
The importance of brackets may need to be mentioned in both **b** and **c** ... **all** the expression is being multiplied.

Plenary activity

Write these words on the board:
factorise simplify solve expand equation expression
Encourage students to explain what each word means.

Discuss the importance of the four skills listed. In pairs, students can choose one of these skills and devise a question to test it. These questions could be used for homework.

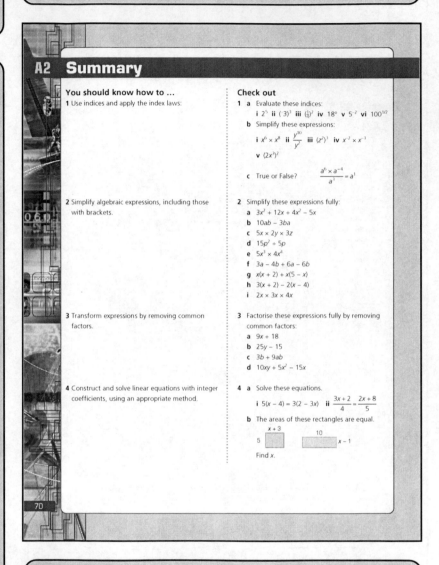

A2 Summary

You should know how to ...

1 Use indices and apply the index laws:

2 Simplify algebraic expressions, including those with brackets.

3 Transform expressions by removing common factors.

4 Construct and solve linear equations with integer coefficients, using an appropriate method.

Check out

1 a Evaluate these indices:
  i 2^5 ii $(^-3)^3$ iii $(\frac{1}{4})^2$ iv 18^0 v 5^{-2} vi $100^{1/2}$
 b Simplify these expressions:
  i $x^6 \times x^8$ ii $\frac{y^{10}}{y^4}$ iii $(z^2)^3$ iv $x^{-2} \times x^{-1}$
  v $(2x^3)^2$
 c True or False? $\frac{a^6 \times a^{-4}}{a^1} = a^1$

2 Simplify these expressions fully:
 a $3x^2 + 12x + 4x^2 - 5x$
 b $10ab - 3ba$
 c $5x \times 2y \times 3z$
 d $15p^2 + 5p$
 e $5x^3 \times 4x^4$
 f $3a - 4b + 6a - 6b$
 g $x(x + 2) + x(5 - x)$
 h $3(x + 2) - 2(x - 4)$
 i $2x \times 3x \times 4x$

3 Factorise these expressions fully by removing common factors:
 a $9x + 18$
 b $25y - 15$
 c $3b + 9ab$
 d $10xy + 5x^2 - 15x$

4 a Solve these equations.
  i $5(x - 4) = 3(2 - 3x)$ ii $\frac{3x + 2}{4} = \frac{2x + 8}{5}$
 b The areas of these rectangles are equal.

  5 | $x + 3$ 10 | $x - 1$

  Find x.

70

Development

The skills developed in this unit will be used and extended in further study of mathematics, including problem solving.

Links

Algebraic skills are used throughout mathematics, particularly in sequences and in problem solving.
Constructing and solving equations skills are also needed in science.

Mental starters

Objectives covered in this unit:
- Use metric units for calculations.
- Use metric units for estimation.
- Convert between metric units.
- Use partitioning to multiply.
- Know and use cubes and corresponding roots.

Resources needed

* means class set needed

Essential:
S2.1OHP – area conversion
S2.2OHP – volume conversion
S2.4OHP – circle
S2.5OHP – right prisms
S2.6OHP – nets
Compasses
String

Useful:
Mini-whiteboards
R1 – digit cards
R2 – decimal digit cards
Box of shapes
Cylindrical tube
S2.3ICT – areas

S2 Measures and measurements

This unit will show you how to:

- Use units of measurement to calculate, estimate, measure and solve problems in a variety of contexts.
- Convert between area measures and between volume measures.
- Know and use the formulae for the circumference and area of a circle.
- Calculate surface area and volume of prisms.
- Use formulae from mathematics and other subjects.
- Solve increasingly demanding problems and evaluate solutions.
- Solve substantial problems by breaking them into simpler tasks.
- Give solutions to problems to an appropriate degree of accuracy.

You've got a head start!

You get to run around a smaller track!

Running tracks contain semicircles at either end.

Before you start

You should know how to ...

1 Convert between metric measures of length.

2 Use of formula to calculate the area of a:
 – rectangle
 – triangle
 – parallelogram
 – trapezium.

3 Calculate the volume of a cuboid.

4 Round decimals to one or two decimal places.

Check in

1 Copy and complete.
 a 1 cm = ____ mm   **b** 56 cm = ____ m

2 Find the area of these shapes.
 a 3 m, 7 m
 b 7 m, 9 m
 c 1.2 m, 2.4 m
 d 6 m, 3 m, 9 m

3 Calculate the volume of this cuboid. 1.1 m, 3.4 m, 1.2 m

4 Round these numbers to the accuracy given.
 a 42.35 (1 d.p.)   **b** 27.93 (1 d.p.)
 c 4.255 (2 d.p.)   **d** 115.999 (1 d.p.)

71

Unit commentary

Aim of the unit

This unit aims to consolidate understanding of the metric system. It then builds on students' knowledge of metric measure to explore circles and prisms.

Introduction

Discuss the two main systems of measurement used in the UK: metric and imperial.
Relate the metric system to the decimal system, and discuss quantities that are measured in metric units.

Framework references

This unit focuses on:
- Teaching objectives pages: 229, 235–241
- Problem-solving objectives pages: 19, 21, 29, 31.

Check in activity

Ask students to estimate the length or height of objects or people in the classroom for example: the board, the door, windows, students. Focus on using metric units.

Progress to estimating the area of rectangular shapes within the classroom for example: the board, a desk, a window.

Differentiation

Core tier

Focuses on metric and imperial units, and the area and volume of rectilinear shapes.

Mental starter

Try some quick-fire metric conversions, including:
- g to kg
- mm to cm
- cm to m
- m to km
- ml to l   and vice versa for each.

Students can respond orally or using mini-whiteboards.
Discuss when you divide and when you multiply.

Useful resources

Mini-whiteboards may be useful for the starter activity.

S2.1OHP illustrates metric conversion of area.

Introductory activity

Discuss units of area.

Focus on metric units, specifically:
- Square millimetres
- Square centimetres
- Square metres
- Hectares
- Square kilometres

Ask for real-life examples of each, using the pictures in the student book for guidance.

Extend to conversion between two metric units of area.

Use a 1 cm square divided into a 10 by 10 grid to illustrate how 1 cm^2 is equivalent to 100 mm^2.

Progress to the relationship between m^2 and cm^2, ha and m^2, and also m^2 and km^2. These are all depicted on **S2.1OHP**.

Recap the formulae for the area of:
- A triangle
- A parallelogram
- A trapezium.

Describe the example in the student book, and emphasise the importance of an **estimate**.

As a further example, ask students to estimate and calculate the area of a trapezium with $a = 4.2$ cm, $b = 5.9$ cm and $h = 28$ mm.

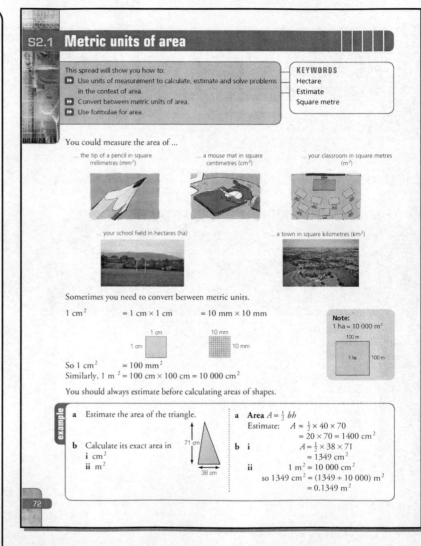

S2.1 Metric units of area

This spread will show you how to:
- Use units of measurement to calculate, estimate and solve problems in the context of area.
- Convert between metric units of area.
- Use formulae for area.

KEYWORDS
Hectare
Estimate
Square metre

You could measure the area of ...

... the tip of a pencil in square millimetres (mm^2)

... a mouse mat in square centimetres (cm^2)

... your classroom in square metres (m^2)

... your school field in hectares (ha)

... a town in square kilometres (km^2)

Sometimes you need to convert between metric units.

1 cm^2 $= 1 \text{ cm} \times 1 \text{ cm}$ $= 10 \text{ mm} \times 10 \text{ mm}$

Note:
1 ha = 10 000 m^2

So $1 \text{ cm}^2 = 100 \text{ mm}^2$
Similarly, $1 \text{ m}^2 = 100 \text{ cm} \times 100 \text{ cm} = 10\,000 \text{ cm}^2$

You should always estimate before calculating areas of shapes.

example

a Estimate the area of the triangle.

b Calculate its exact area in
 i cm^2
 ii m^2

71 cm
38 cm

a **Area** $A = \frac{1}{2}bh$
  Estimate: $A \approx \frac{1}{2} \times 40 \times 70$
  $= 20 \times 70 = 1400 \text{ cm}^2$

b i $A = \frac{1}{2} \times 38 \times 71$
  $= 1349 \text{ cm}^2$

 ii $1 \text{ m}^2 = 10\,000 \text{ cm}^2$
  so $1349 \text{ cm}^2 = (1349 \div 10\,000) \text{ m}^2$
  $= 0.1349 \text{ m}^2$

72

Plenary

Ask for the area of a rectangle measuring 2 cm by 3 cm.
Now ask for the possible dimensions of these shapes if they contain the same area (6 cm^2):
- A triangle
- A parallelogram
- A trapezium
- A square

Which of these shapes gives a unique answer (and cannot be given as an accurate measurement)?

Further activities

Students can work in pairs to produce a rough plan of the classroom and estimate its area.

S2.3ICT requires students to use an interactive geometry package to calculate areas of shapes.

Differentiation

Extension questions:
▶ Question 1 revises calculating the areas of standard shapes.
▶ Questions 2–5 focus on area, including compound shapes and metric conversion.
▶ Questions 6 and 7 are word problems involving area.

Core tier: focuses on metric units, including length, area, volume, capacity and mass.

Exercise S2.1

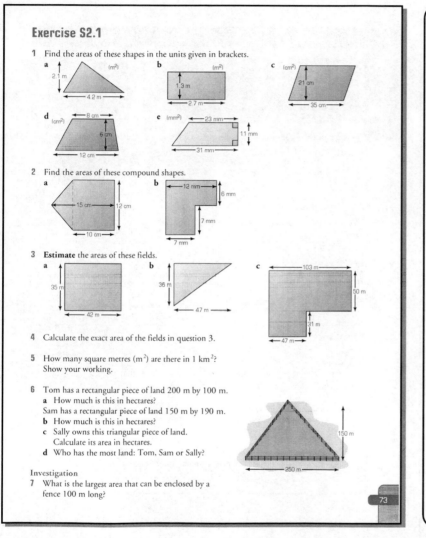

1 Find the areas of these shapes in the units given in brackets.

2 Find the areas of these compound shapes.

3 **Estimate** the areas of these fields.

4 Calculate the exact area of the fields in question 3.

5 How many square metres (m²) are there in 1 km²? Show your working.

6 Tom has a rectangular piece of land 200 m by 100 m.
 a How much is this in hectares?
 Sam has a rectangular piece of land 150 m by 190 m.
 b How much is this in hectares?
 c Sally owns this triangular piece of land. Calculate its area in hectares.
 d Who has the most land: Tom, Sam or Sally?

Investigation
7 What is the largest area that can be enclosed by a fence 100 m long?

73

Exercise commentary

The questions assess the objectives on Framework Page 229.

Problem solving
Questions 5 and 6 assess the objectives on Framework Page 19. Questions 5 and 7 assess Page 29.

Group work
Question 6 can be attempted in pairs.

Misconceptions
Students often have problems with mixed units, and may either ignore the fact that the units are different or incorrectly convert one to the other.
Encourage students to appreciate that the quantities are units of measure and not just numbers. Emphasise the importance of writing the correct units explicitly in the final answer.

Links
This topic links to the mental recall of measurement facts, which is often useful in number problems (Framework Page 91).

Homework

S2.1HW provides practice at calculating the area of compound shapes, and includes conversion between units.

Answers

1 a 4.41 m² b 3.51 m² c 735 cm² d 60 cm² e 297 mm²
2 a 150 cm² b 121 mm²
3 a About 1600 m² b About 1000 m² c About 6500 m²
4 a 1470 m² b 846 m² c 6607 m²
5 1 000 000 m²
6 a 2 ha b 2.85 ha c 1.875 ha d Sam
7 625 m²

Mental starter

Practise cubes and cube roots. Ask questions like:
What is the ...
▶ cube of 2
▶ cube root of 27
▶ cube of 4
▶ cube root of 125?
Students can respond with digit cards.

Useful resources

R1 digit cards may be useful for the mental starter.

S2.2OHP illustrates the conversion between metric units of volume.

Introductory activity

Discuss the meaning of volume.
Encourage the description of volume as the amount of space occupied by a 3-D shape.

Discuss metric units of volume: mm^3, cm^3, and m^3.
Demonstrate how to convert between one unit and another.
S2.2OHP illustrates the relationship between metric units of volume using a centimetre cube and a metre cube.

Recap the formula for the volume of a cuboid.
You can refer to the example in the student book.

Discuss capacity.
Relate capacity to volume, and emphasise that capacity is usually associated with liquids. Discuss the relationship between the units: ml, cl, l and cm^3.
Illustrate with the example in the student book.

Discuss mass.
Students should appreciate that, for everyday problems, mass is the same as weight.
Discuss the relationship between the units: g, kg and t.

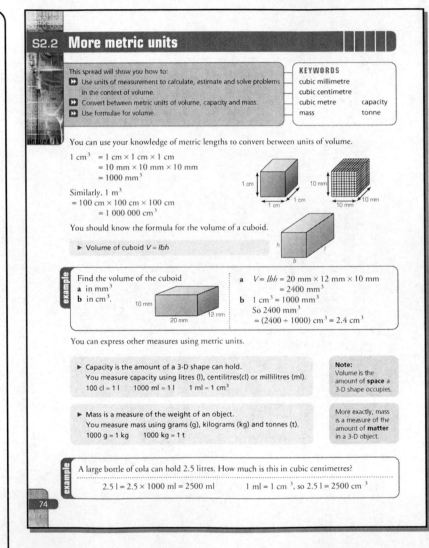

S2.2 More metric units

This spread will show you how to:
▶ Use units of measurement to calculate, estimate and solve problems in the context of volume.
▶ Convert between metric units of volume, capacity and mass.
▶ Use formulae for volume.

KEYWORDS
cubic millimetre
cubic centimetre
cubic metre   capacity
mass   tonne

You can use your knowledge of metric lengths to convert between units of volume.

$1 \, cm^3 = 1 \, cm \times 1 \, cm \times 1 \, cm$
$= 10 \, mm \times 10 \, mm \times 10 \, mm$
$= 1000 \, mm^3$

Similarly, $1 \, m^3$
$= 100 \, cm \times 100 \, cm \times 100 \, cm$
$= 1\,000\,000 \, cm^3$

You should know the formula for the volume of a cuboid.

▶ Volume of cuboid $V = lbh$

example

Find the volume of the cuboid
a in mm^3
b in cm^3.

a $V = lbh = 20 \, mm \times 12 \, mm \times 10 \, mm$
$= 2400 \, mm^3$
b $1 \, cm^3 = 1000 \, mm^3$
So $2400 \, mm^3$
$= (2400 \div 1000) \, cm^3 = 2.4 \, cm^3$

You can express other measures using metric units.

▶ Capacity is the amount of a 3-D shape can hold.
You measure capacity using litres (l), centilitres(cl) or millilitres (ml).
$100 \, cl = 1 \, l$ $1000 \, ml = 1 \, l$ $1 \, ml = 1 \, cm^3$

Note:
Volume is the amount of **space** a 3-D shape occupies.

▶ Mass is a measure of the weight of an object.
You measure mass using grams (g), kilograms (kg) and tonnes (t).
$1000 \, g = 1 \, kg$ $1000 \, kg = 1 \, t$

More exactly, mass is a measure of the amount of **matter** in a 3-D object.

example

A large bottle of cola can hold 2.5 litres. How much is this in cubic centimetres?
$2.5 \, l = 2.5 \times 1000 \, ml = 2500 \, ml$ $1 \, ml = 1 \, cm^3$, so $2.5 \, l = 2500 \, cm^3$

Plenary

Discuss imperial measures of volume, capacity and mass: cubic feet and cubic inches, pints and gallons, ounces and pounds.

Students should know that:
▶ $1 \, kg \cong 2.2 \, lb$, and $1 \, oz \cong 30 \, g$
▶ 1 litre is just less than 2 pints.

Further activities

Challenge students to solve this problem:
How many **different** cuboids can you make with volume 48 cm³?
Which has the greatest surface area?
Which has the least surface area?

Differentiation

Extension questions:
▶ Question 1 involves the dimension of a cube, and relates to the mental starter.
▶ Questions 2–4 focus on the volume of a cuboid.
▶ Questions 5–8 are word problems involving capacity and mass.

Core tier: focuses on imperial units of length, mass and capacity.

Exercise S2.2

1 The volume of a cube is 125 mm³.
What is the length of one side?

2 The volume of a cuboid is 240 cm³, its length is 30 cm, and its width is 2 cm.
 a What is its height?
 b Can you find a cuboid with this volume whose base is square?

3 A cuboid has a square top and bottom.
Its volume is 2880 mm³ and its height is 20 mm.
What are the dimension of its square faces?

4 Sani received this parcel. Find its volume in
 a mm³
 b cm³
 c m³.

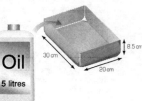

5 A tin is full of water.

 a Find its volume in **i** cm³ **ii** m³
 b What is the capacity of liquid it will hold?

6 Will the canister of oil fit in the oil tray?
Explain your answer.

7 A lift holds a maximum of 0.5 tonnes.
Will it hold these six people:
Zaina 90 kg, Ben 80 kg, Adebola 72 kg, Sarah 77 kg, Beth 64 kg, Tom 63 kg?
Explain your answer.

8 Packets of butter are cuboids with dimensions 4 cm by 5 cm by 3 cm.
How many packets will fit into a box measuring 80 cm by 1 m by 90 cm?

Exercise commentary

The questions assess the objectives on Framework Page 229.

Problem solving
Questions 4–8 assess the objectives on Framework Pages 19 and 21. Questions 6 and 8 assess Page 29 and questions 6 and 7 assess page 31.

Group work
Questions 4–8 may be attempted in pairs.

Misconceptions
When calculating volume, students may multiply too many numbers together. Emphasise that volume is a product of three lengths.
Students often use the wrong units, or none at all. Emphasise the importance of correct conversion, and of stating the correct units in the answer.

Links
This topic links to the mental recall of measurement facts (Framework page 91).

Homework

S2.2HW consolidates understanding of metric units.

Answers

1 5 mm   2 **a** 4 cm   **b** Yes   3 12 mm
4 **a** 7 500 000 mm³   **b** 7500 cm³   **c** 0.0075 m³
5 **a** **i** 5100 cm³   **ii** 0.0051 m³   **b** 5.1 litres
6 Yes, capacity of oil tray is 5.1 litres.
7 Yes, they weigh 0.446 tonnes.
8 12 000

75

Mental starter

Show various circles (rim of a cup, a coin, an angle indicator ...) and ask students to estimate:

▸ the diameter
▸ the circumference.

Ensure that students are familiar with these terms.

Useful resources

Compasses and **string** (lengths of about 30 cm) for the introductory activity.

Introductory activity

Discuss the similarity of circles.

Recap the terms: circumference (C) and diameter (D).

Ensure that students have compasses, and give out lengths of string.

Ask students to:

▸ Draw at least four circles of different sizes.
▸ Measure their circumferences (using string) and diameters.
▸ Calculate the ratio $\frac{C}{D}$ to 1 d.p.
▸ Describe what they notice.

Students should notice that the ratio $\frac{C}{D}$ is the same for all circles.

Discuss discrepancies, and highlight the inaccuracy in measuring a circumference.

Introduce the ratio $\frac{C}{D} = \pi$

Students should appreciate that π is a Greek letter pronounced 'pie' with a numerical value of approximately 3.14.

Recap the term 'radius' (r) and emphasise its relationship with D.

Rearrange the ratio $\frac{C}{D} = \pi$ into the common formulae:

$$C = \pi D \quad \text{and} \quad C = 2\pi r$$

Demonstrate the use of the formula to calculate missing dimensions in a circle.

Use the examples in the student book, and emphasise the importance of:

▸ Correct units
▸ Appropriate rounding.

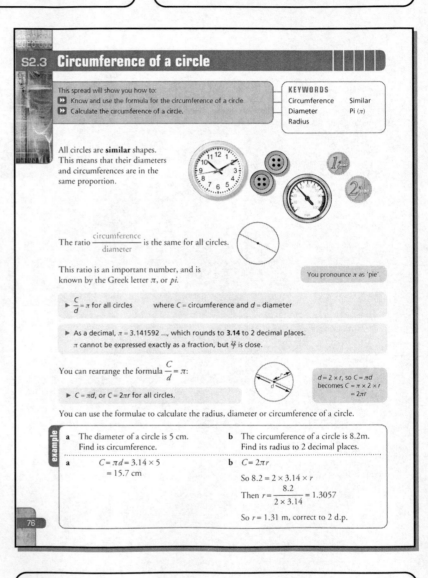

S2.3 Circumference of a circle

This spread will show you how to:
▸ Know and use the formula for the circumference of a circle.
▸ Calculate the circumference of a circle.

KEYWORDS
Circumference   Similar
Diameter   Pi (π)
Radius

All circles are **similar** shapes. This means that their diameters and circumferences are in the same proportion.

The ratio $\dfrac{\text{circumference}}{\text{diameter}}$ is the same for all circles.

This ratio is an important number, and is known by the Greek letter π, or *pi*.

You pronounce π as 'pie'

▸ $\dfrac{C}{d} = \pi$ for all circles   where C = circumference and d = diameter

▸ As a decimal, $\pi = 3.141592$ ..., which rounds to **3.14** to 2 decimal places. π cannot be expressed exactly as a fraction, but $\frac{22}{7}$ is close.

You can rearrange the formula $\dfrac{C}{d} = \pi$:

$d = 2 \times r$, so $C = \pi d$ becomes $C = \pi \times 2 \times r = 2\pi r$

▸ $C = \pi d$, or $C = 2\pi r$ for all circles.

You can use the formulae to calculate the radius, diameter or circumference of a circle.

example

a The diameter of a circle is 5 cm. Find its circumference.

b The circumference of a circle is 8.2 m. Find its radius to 2 decimal places.

a
$$C = \pi d = 3.14 \times 5$$
$$= 15.7 \text{ cm}$$

b
$$C = 2\pi r$$
So $8.2 = 2 \times 3.14 \times r$
Then $r = \dfrac{8.2}{2 \times 3.14} = 1.3057$
So $r = 1.31$ m, correct to 2 d.p.

76

Plenary

Discuss question 6.

Relate this problem to the school running track, and why the starting line for the lanes is staggered.

Further activities

Present this problem:
A circle has a circumference of 10 cm.
Give the dimensions of:
▶ a square
▶ a rectangle
▶ a triangle, with the same perimeter.
Which will have the largest area?
Investigate.

Differentiation

Extension questions:
▶ Question 1 involves measuring the diameter of a circle to calculate its circumference.
▶ Questions 2–5 focus on the circumferences of circles and shapes derived from circles.
▶ Question 6 is a substantial problem involving the distance around a running track.

Core tier: focuses on the area of a triangle and a parallelogram.

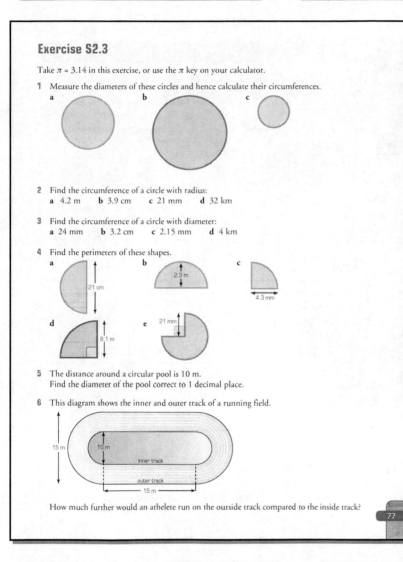

Exercise S2.3

Take $\pi = 3.14$ in this exercise, or use the π key on your calculator.

1   Measure the diameters of these circles and hence calculate their circumferences.
   a   b   c

2   Find the circumference of a circle with radius:
   a 4.2 m   b 3.9 cm   c 21 mm   d 32 km

3   Find the circumference of a circle with diameter:
   a 24 mm   b 3.2 cm   c 2.15 mm   d 4 km

4   Find the perimeters of these shapes.
   a   b   c
      21 cm   2.3 m
                                    4.3 mm

   d   e
      8.1 m   21 mm

5   The distance around a circular pool is 10 m.
   Find the diameter of the pool correct to 1 decimal place.

6   This diagram shows the inner and outer track of a running field.
   15 m   10 m
           inner track
           outer track
           15 m

   How much further would an athlete run on the outside track compared to the inside track?

77

Exercise commentary

The questions assess the objectives on Framework Page 235.

Problem solving
Question 6 assesses the objectives on Framework Page 19. Questions 5 and 6 also assess Page 29. The exercise as a whole assesses Page 31.

Group work
Question 6 can be attempted in pairs.

Misconceptions
In question 2, you may need to stress that the radius is given (not the diameter). Emphasise that the formulae $C = \pi D$ and $C = 2\pi r$ are identical, because $D = 2r$.
In question 4, remind students to allow for the straight edges.
Students often have problems in rounding the calculator display to the required number of decimal places. You may need to recap rounding.

Links
Calculating with π involves rounding (Framework Page 45).

Homework

S2.3HW contains word problems involving the circumference of a circle, and shapes derived from circles.

Answers

1 a 2.5 cm, 7.9 cm   b 3.5 cm, 11.0 cm   c 1.5 cm, 4.7 cm
2 a 26.4 m   b 24.5 m   c 132 mm   d 201 km
3 a 75 mm   b 10 cm   c 6.75 mm   d 12.6 km
4 a 54 cm   b 11.8 m   c 15.4 mm   d 28.9 m   e 141 mm
5 3.2 m   6 15.7 m

Mental starter

Give decimal values of the radius, and ask for the diameter.
Choose values like 5.4 cm, 2.9 cm, 28.7 cm, and 49.39 cm

Now ask for the radius given the diameter.
Choose values like 8.6 cm, 19.4 cm, 29.7 cm and 31.56 cm

Students can respond using decimal digit cards.

Useful resources

R2 decimal digit cards may be useful for the mental starter.

S2.4OHP shows a circle divided into equal sectors, to accompany the introductory activity.

Introductory activity

Ask students to imagine a circle divided into small sectors.
S2.4OHP shows the diagrams at the top of the page in the student book.

Show how you can rearrange the sectors into an approximate rectangle.
Discuss the dimensions of the 'rectangle', and encourage students to appreciate that the length will be πr.
Show that the area $= \pi r^2$.

Write the formula for the area of a circle: $A = \pi r^2$.
Illustrate the use of the formula with the first example in the student book.
Emphasise that:
▸ You need to round the calculator answer to the specified accuracy.
▸ The units will be squared (cm^2, m^2).
Highlight the π key on a scientific calculator.

Show how you can use the formula to solve problems.
Refer to the last example in the student book.
Emphasise that to find r you need to rearrange the equation by using inverse operations.
Recap that the inverse operation of 'square' is 'square root'.

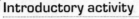

S2.4 Area of a circle

This spread will show you how to:
▸ Know and use the formula for the area of a circle.

KEYWORDS
Sector   Circumference
Diameter Radius

Imagine dividing a circle into tiny sectors ...   ... and rearranging the sectors into a rectangular shape.

The length of the 'rectangle' is πr because it is half of the circumference.

The smaller you make the sectors, the closer the shape becomes to a rectangle.

The area of the rectangle $\approx \pi r \times r = \pi r^2$.
This leads to the formula for the area of a circle.

▸ $A = \pi r^2$ for any circle.

example

Find to one decimal place the area of a circle:
a with radius 2.3 m   b with diameter 14 cm.

a $A = \pi r^2$
   $= 3.14 \times 2.3 \times 2.3$
   $= 16.6106$
   $A = 16.6$ m correct to 1 d.p.

b $d = 14$ cm, so $r = 14$ cm $\div 2 = 7$ cm
   $A = \pi r^2$
   $= 3.14 \times 7 \times 7$
   $= 153.9$ cm correct to 1 d.p.

You can use the formula $A = \pi r^2$ to solve problems.

example

A semicircular window has an area of 1.2 m^2.
Find the radius of the window to the nearest centimetre.

For the whole circle, $A = 2 \times 1.2$ m$^2 = 2.4$ m^2
$A = \pi r^2$, so $2.4 = 3.14 \times r^2$

$r^2 = \dfrac{2.4}{3.14} = 0.7643 \ldots$

$r = \sqrt{0.764331} = 0.8742 \ldots$

Remember:
The opposite of **square** is **square root**.

So the radius of the window is 0.8742 m, or 87 cm to the nearest centimetre.

78

Plenary

Discuss different approximations to π:
For a rough mental estimate, you can use $\pi \approx 3$.
Decimal approximations are 3.14 and 3.142.
The closest fraction approximation is $\frac{22}{7}$.

Invite students with a scientific calculator to quote π to 8 or 9 decimal places. You could use the memory aid: 'How I wish I could calculate pi' (count the letters in each word: 3, 1, 4, 1, 5, 9, 2)

Further activities

Set this problem as a challenge:
A circular pond of radius 5 m has a circular path of width 1 m running around its edge.
What is the area of the path?

Encourage students to sketch the pond first.

Differentiation

Extension questions:

▸ Question 1 involves calculating the area of a circle, given either the radius or the diameter.
▸ Question 2 involves calculating the area of shapes derived from circles.
▸ Questions 3–5 are word problems associated with the area of a circle.

Core tier: focuses on the area of compound shapes made up of triangles and rectangles.

Exercise S2.4

Take $\pi = 3.14$ in this exercise, or use the π key on your calculator.

1   Find the area of a circle with:

　a   radius　　**i** 4 cm　**ii** 5 mm　**iii** 10 m　**iv** 6.2 km

　b   diameter　**i** 5.3 mm　**ii** 4.1 cm　**iii** 27 m　**iv** 82 km

　Give your answers to an appropriate degree of accuracy.

2   Find the area of each shape.

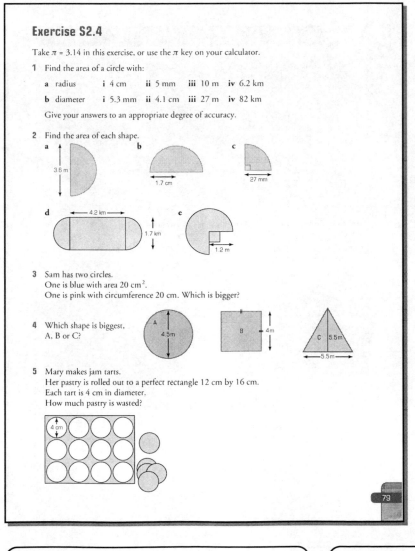

3   Sam has two circles.
　One is blue with area 20 cm².
　One is pink with circumference 20 cm. Which is bigger?

4   Which shape is biggest, A, B or C?

5   Mary makes jam tarts.
　Her pastry is rolled out to a perfect rectangle 12 cm by 16 cm.
　Each tart is 4 cm in diameter.
　How much pastry is wasted?

79

Exercise commentary

The questions assess the objectives on Framework Page 237.

Problem solving
Questions 3, 4 and 5 assess the objectives on Framework Page 19 and Page 29.

Group work
Students can work on question 5 and the further activity in pairs.

Misconceptions
Students often calculate πr^2 incorrectly. Emphasise the correct order of operations when dealing with powers.
Another common error is to use diameter instead of radius in the formula.
Encourage students to read the question carefully.

Links
Area of a circle links to using formulae (Framework page 141).

Homework

S2.4HW provides practice at using the formula for the area of a circle, and includes problem solving.

Answers

1 **a i** 50.2 cm²　**ii** 78.5 mm²　**iii** 314 m²　**iv** 121 km²
　b i 22.1 mm²　**ii** 13.2 cm²　**iii** 572 m²　**v** 5278 km²
2 **a** 5.1 m²　**b** 4.5 cm²　**c** 572 mm²　**d** 9.4 km²
　e 3.4 m²
3 Pink　4 A　5 41.3 cm²

Mental starter

Try some area conversions, using cm² and mm², for example convert:

▸ 561 mm² to cm²
▸ 2.85 cm² to mm²

Recap how to convert between cm³ and mm³, and try some volume conversions.

Students respond orally to keep the exercise brisk and lively.

Useful resources

S2.5OHP shows a variety of right prisms for class discussion.

Introductory activity

Define a prism as a solid with a uniform cross-section.

Ensure that students appreciate the meaning of the words uniform and cross-section, and illustrate with examples.

Use the analogy of a stick of seaside rock: wherever you break it along its length, you get the same lettering.

S2.5OHP shows a variety of **right prisms** (with rectangles making up the other faces), which include the ones in the student book.

Write the word formula for the volume of a prism.

Illustrate with the example of the cuboid and the triangular prism in the student book.

Emphasise that the units of volume are cubed (cm³ and mm³), and refer briefly to the mental starter.

Progress to compound 3-D shapes, made up of two or more prisms.

Discuss strategies for dealing with these shapes, and encourage different methods. Emphasise that there is often more than one way to split a shape up.

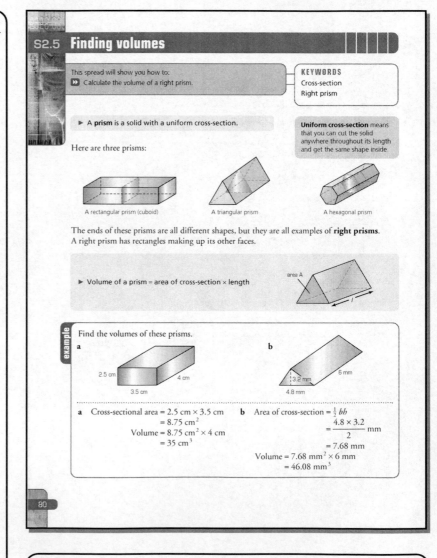

S2.5 Finding volumes

This spread will show you how to:
▶▶ Calculate the volume of a right prism.

KEYWORDS
Cross-section
Right prism

▶ A **prism** is a solid with a uniform cross-section.

Uniform cross-section means that you can cut the solid anywhere throughout its length and get the same shape inside.

Here are three prisms:

A rectangular prism (cuboid)   A triangular prism   A hexagonal prism

The ends of these prisms are all different shapes, but they are all examples of **right prisms**. A right prism has rectangles making up its other faces.

▶ Volume of a prism = area of cross-section × length

area A

l

example

Find the volumes of these prisms.

a

2.5 cm
3.5 cm
4 cm

b

3.2 mm
6 mm
4.8 mm

a Cross-sectional area = 2.5 cm × 3.5 cm
= 8.75 cm²
Volume = 8.75 cm² × 4 cm
= 35 cm³

b Area of cross-section = $\frac{1}{2}bh$
= $\frac{4.8 \times 3.2}{2}$ mm
= 7.68 mm
Volume = 7.68 mm² × 6 mm
= 46.08 mm³

Plenary

Discuss question 6, which involves a cylinder.

Although not strictly a prism because the cross-section is not a polygon, the volume of a cylinder follows the same formula.

Recap the area of a circle and discuss discrepancies within answers, particularly with accuracy.

Further activities

Students can work in pairs or small groups to discuss an estimate for the volume of water in their local swimming pool. Encourage students to think about the information they will need to make the calculation, and any assumptions they make.

Differentiation

Extension questions:
▸ Question 1 requires simple calculation of the volume of a cuboid.
▸ Questions 2–5 focus on the volume of more complex prisms, including compound shapes.
▸ Question 6 extends to the volume of a cylinder.

Core tier: focuses on the volume of a cuboid.

Exercise S2.5

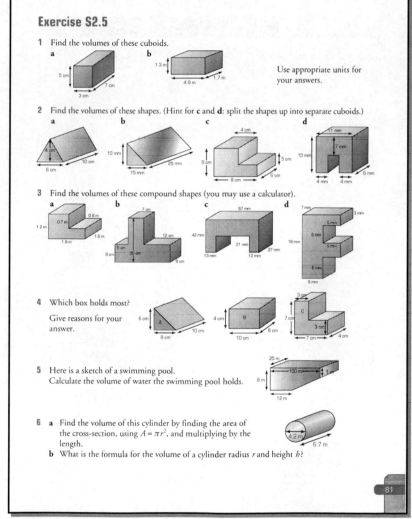

1 Find the volumes of these cuboids.

a   b

Use appropriate units for your answers.

2 Find the volumes of these shapes. (Hint for **c** and **d**: split the shapes up into separate cuboids.)

a   b   c   d

3 Find the volumes of these compound shapes (you may use a calculator).

a   b   c   d

4 Which box holds most?

Give reasons for your answer.

5 Here is a sketch of a swimming pool.
Calculate the volume of water the swimming pool holds.

6 a Find the volume of this cylinder by finding the area of the cross-section, using $A = \pi r^2$, and multiplying by the length.
   b What is the formula for the volume of a cylinder radius r and height h?

Exercise commentary

The questions assess the objectives on Framework Pages 239 and 241.

Problem solving

Questions 4 and 5 assess the objectives on Framework Page 19. Question 5 also assesses Pages 29 and 31.

Group work

Questions 5 and 6 can be attempted in pairs.

Misconceptions

Students will commonly either use area units or no units at all in stating a volume. Emphasise that volumes have cubed units (length × length × length).
For more complex prisms (triangular, trapezoidal) students may use an incorrect formula for the area of cross-section. You may wish to write all relevant formulae on the board.

Links

The volume of a prism links to analysis of 3-D shapes (Framework Pages 199 and 201).

Homework

S2.5HW provides practice at finding the volume of a prism.

Answers

1 a 105 cm³   b 10.8 m³
2 a 120 cm³   b 1875 mm³   c 336 cm³   d 606 mm²
3 a 2.56 m³   b 2272 cm³   c 52 164 mm³   d 1001 mm³
4 B
5 12 300 m³
6 a 92.8 m³   b $V = \pi r^2 h$

Mental starter

Practise multiplying by decimals mentally.
For example:

32 × 0.01   45 × 0.03   12 × 3.8   19 × 0.4

Students can respond using decimal digit cards.
Discuss strategies (for example, place value or partitioning).

Useful resources

R2 decimal digit cards for the mental starter.
Box of shapes for the introductory activity.
S2.6OHP shows two pairs of solids and their nets.
A **cylindrical tube** may be useful for the plenary.

Introductory activity

Recall the definition of area in the context of a 2-D (plane) shape.
Define **surface area** as a measure of the space occupied by the faces of a 3-D (solid) shape.
Illustrate by holding up a solid shape from a box of shapes, and asking the class to describe its faces.

Recap the net of a shape.
Show how you can use the net of a cuboid to find its surface area.
You can use the first example in the student book, which is illustrated on **S2.6OHP**.
Emphasise that the faces come in identical pairs, so you can just find the area of three rectangular faces and double them.

Extend to more complicated prisms.
The second example in the student book involves the surface area of a triangular prism. This example is illustrated on **S2.6OHP**.

Note that there are two identical triangles and three rectangles. Emphasise that in this case, the rectangles are different.

You may need to recap the area of a triangle.

Ask what shape the cross-section would be if all the rectangles were identical (equilateral triangle).

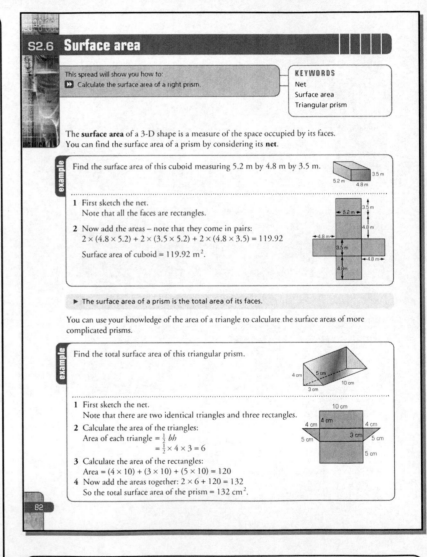

S2.6 Surface area

This spread will show you how to:
▶ Calculate the surface area of a right prism.

KEYWORDS
Net
Surface area
Triangular prism

The **surface area** of a 3-D shape is a measure of the space occupied by its faces. You can find the surface area of a prism by considering its **net**.

example

Find the surface area of this cuboid measuring 5.2 m by 4.8 m by 3.5 m.

1 First sketch the net.
   Note that all the faces are rectangles.
2 Now add the areas – note that they come in pairs:
   2 × (4.8 × 5.2) + 2 × (3.5 × 5.2) + 2 × (4.8 × 3.5) = 119.92

   Surface area of cuboid = 119.92 m².

▶ The surface area of a prism is the total area of its faces.

You can use your knowledge of the area of a triangle to calculate the surface areas of more complicated prisms.

example

Find the total surface area of this triangular prism.

1 First sketch the net.
   Note that there are two identical triangles and three rectangles.
2 Calculate the area of the triangles:
   Area of each triangle = ½ bh
                         = ½ × 4 × 3 = 6
3 Calculate the area of the rectangles:
   Area = (4 × 10) + (3 × 10) + (5 × 10) = 120
4 Now add the areas together: 2 × 6 + 120 = 132
   So the total surface area of the prism = 132 cm².

82

Plenary

Discuss question 5, which refers to the surface area of a cylinder. Ask a volunteer to describe the net of a cylinder to the class.

Encourage students to visualise the curved surface of a cylinder when it is flattened out, and pay particular attention to its dimensions.

Further activities

Students can extend question 4: how many cuboids can be found with a surface area of 52 cm²? What about other surface areas, for example 48 cm²?

Or:

Recall the swimming pool problem from S2.5.
Estimate the area that would need to be tiled in such a pool.

Differentiation

Extension questions:
▸ Question 1 provides simple practice at finding the surface area of a cuboid.
▸ Questions 2–4 focus on the surface area of a prism, including a triangular prism.
▸ Question 5 extends to the surface area of a cylinder.

Core tier: focuses on the surface area of a cuboid.

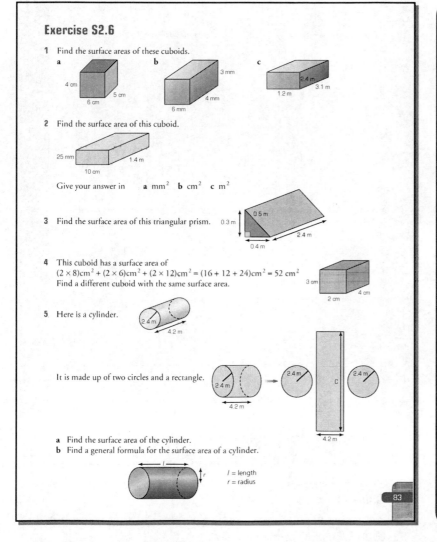

Exercise S2.6

1 Find the surface areas of these cuboids.
  a (4 cm, 6 cm, 5 cm) b (3 mm, 4 mm, 6 mm) c (2.4 m, 1.2 m, 3.1 m)

2 Find the surface area of this cuboid.
  (25 mm, 1.4 m, 10 cm)
  Give your answer in **a** mm² **b** cm² **c** m²

3 Find the surface area of this triangular prism. (0.3 m, 0.5 m, 2.4 m, 0.4 m)

4 This cuboid has a surface area of
  (2×8)cm² $+ (2 \times 6)$cm² $+ (2 \times 12)$cm² $= (16 + 12 + 24)$cm² $= 52$ cm²
  Find a different cuboid with the same surface area. (3 cm, 2 cm, 4 cm)

5 Here is a cylinder. (2.4 m, 4.2 m)

  It is made up of two circles and a rectangle. (2.4 m, 4.2 m, c)

  a Find the surface area of the cylinder.
  b Find a general formula for the surface area of a cylinder.
  l = length
  r = radius

Exercise commentary

The questions assess the objectives on Framework Pages 239 and 241.

Problem solving
Question 3 assesses the objectives on Framework Page 29, question 4 assesses Page 19 and question 5 assesses Page 31.

Group work
Questions 4 and 5 can be attempted in pairs. Students can try to find different solutions to question 4.

Misconceptions
Triangular prisms (such as in question 3) can lead to mistakes, particularly in calculating the cross-sectional area. Emphasise that the area of a triangle involves the **vertical** height, not the slant height.
Students should take particular care of question 2, as it contains mixed units.

Links
Surface area of a prism links to analysis of 3-D shapes (Framework Pages 199 and 201).

Homework

S2.6HW is a problem-solving activity that involves finding the amount of material needed to make a tent.

Answers

1 **a** 148 cm² **b** 108 mm² **c** 28.08 m²
2 **a** 355 000 mm² **b** 3550 cm² **c** 0.355 m²
3 3 m²
5 **a** 99.5 m² **b** $A = 2\pi rl + 2\pi r^2$

Summary

The key objectives for this unit are:
▸ Know and use the formula for the circumference and area of a circle.
▸ Solve substantial problems by breaking them into simpler tasks.
▸ Give solutions to problems to an appropriate degree of accuracy.

Plenary activity

Draw three circles on the board with different radii: 7 cm, 7.2 cm, 7.25 cm.
Ask students to work out the area and circumference of each circle.
Discuss the accuracy to which each answer should be given.
What about circles with radii 5 cm and 5.00 cm?

Check out commentary

1   Students often get the formulae for area and circumference of a circle mixed up. Remind students that the formula for the area of a circle has r^2 in it because area is measured in square units.
Encourage rounding to one decimal place for this question as the radius is given to one decimal place.

2   Surface area can be explained more easily using nets. Emphasise that for a cuboid or cube opposite faces are identical. This should save calculation.

Appropriate units often cause problems.
Encourage students to look carefully at the formula that they are going to use, and decide whether they are looking for a length, an area or a volume.

3   Students may be unsure of the order in which to multiply the lengths: emphasise that multiplication is commutative.
Encourage students to round the answer to the same number of decimal places as the original lengths (1 decimal place).

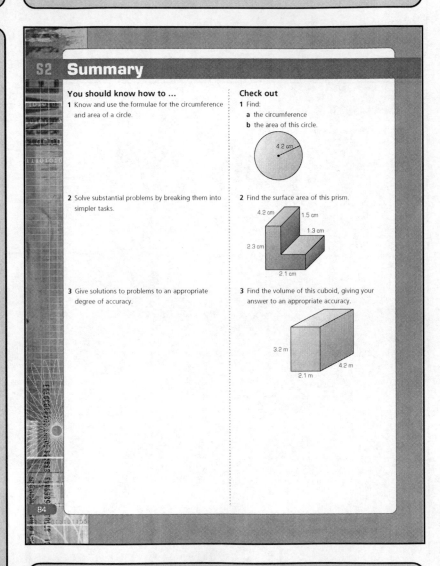

Development

The relationship between area and volume is further developed in the context of enlargement in S3.
The properties of cuboids are further explored in S4.

Links

Metric measure links to practical work across the curriculum, particularly in science and design and technology.

Mental starters

Objectives covered in this unit:
▸ Discuss and interpret graphs.
▸ Apply mental skills to solve simple problems.
▸ Convert between improper fractions and mixed numbers.

Resources needed

* means class set needed
Essential:
A3.2OHP – gradient of diagonal lines
A3.3OHP – equation of a straight line
A3.5OHP – curved graphs
A3.6OHP – distance-time graphs
Coordinate cards
Squared paper*
Graph paper*

Useful:
R8 – blank coordinate grid
A3.6F – real-life graphs
A3.3ICT – linear functions

A3 Functions and graphs

This unit will show you how to:

▸▸ Generate points and plot graphs of linear functions.
▸▸ Given values for *m* and *c*, find the gradient of lines given by equations of the form $y = mx + c$.
▸▸ Construct functions arising from real-life problems and plot their corresponding graphs.
▸▸ Interpret graphs arising from real situations, including distance-time graphs.

▸▸ Solve increasingly demanding problems and evaluate solutions.
▸▸ Represent problems and synthesise information in algebraic or graphical form.
▸▸ Solve substantial problems by breaking them into simpler tasks.

Road signs often tell you the steepness or gradient of a hill.

Before you start

You should know how to ...
1 Substitute values into an expression.

2 Plot points on a coordinate grid.

3 Rearrange a linear function to make *y* the subject.
For example, if $4y + 2x = 8$.
Then $4y = 8 - 2x$
And $y = 2 - \frac{1}{2}x$

Check in
1 If $a = 3$, $b = {}^-2$ and $c = 5$, place these expressions in ascending order:
b^2 $a - 2c$ c^2 abc $c - 3b$ $\frac{10-2c}{2}$ $2a - 3b$

2 a On a grid labelled ${}^-8$ to 8, plot and join these points in order.
$(6, 4), ({}^-2, 4), ({}^-2, 8), ({}^-8, 1), ({}^-2, {}^-2),$
$({}^-2, 1), (6, 1),(6, 4)$.
b What do you see? What is its mathematical name?

3 Make *y* the subject of these formulae:
a $y + x = 10$ b $y - 4 = 2x$ c $2y = 6x + 7$
d $2y - x = 14$ e $3y + 5x = 12$

Unit commentary

Aim of the unit
This unit aims to develop understanding of graphs. Linear graphs are explored in detail, with a particular emphasis on gradient. Distance-time graphs, and other contextual graphs, are also explored in this unit.

Introduction
Discuss the prevalence of graphs as a means of communication in the media, and use this to illustrate the importance of studying graphs.
Ask what these coordinate pairs have in common: (2, 3), (1, 4), (5, 0), (0, 5), (6, ⁻1), and (4, 1).
Encourage the response that each pair adds to 5, and link to the equation $x + y = 5$.

Framework references
This unit focuses on:
▸ Teaching objectives pages: 165–175
▸ Problem-solving objectives pages: 9, 33
The unit as a whole assesses objectives on page 27.

Check in activity

Using a coordinate grid on an OHT (you can use **R8**), plot 26 pairs of coordinates, each labelled with a letter of the alphabet. Read out sequences of coordinates that spell words.
Students should follow the coordinates and tell you the word.

Differentiation

Core tier
Focuses on constructing a linear graph from a function machine, and real-life graphs.

Mental starter

Have prepared a set of coordinate cards, from $(0, 0)$ to $(5, 5)$ – there should be 36 in total. Arrange the classroom in a square grid of chairs, with each student occupying a chair according to their coordinate card.

Ask questions such as:

▸ Stand up if your y value is 5   ($y = 5$)
▸ Stand up if your x value, doubled and then add 1, is your y value   ($y = 2x + 1$)

Useful resources

Coordinate cards for the mental starter.

R8 is a square grid for use in the introductory activity.

Introductory activity

Refer to the mental starter.

Give students an equation such as $y = 6$, and ask where it would be on the graph.

Vary the equations to cover horizontal, diagonal and vertical graphs.

Focus on diagonal line graphs.

Emphasise that the equation of a diagonal graph contains both an x and a y term.

Identify the main features of a straight-line graph:

▸ Direction
▸ Gradient
▸ 'height' (y-intercept)

Plot the graph of $y = 2x - 1$.

Go through the steps carefully, ensuring that students appreciate the importance of drawing a table first.

Emphasise that at least three points should be considered.

Spend time to discuss the scaling of the axes. This should really be done on a square grid, perhaps on an OHT (you can use **R8**).

Before plotting the graph, ask students to predict (from the table) what direction it will go in, how steep and how high it will be.

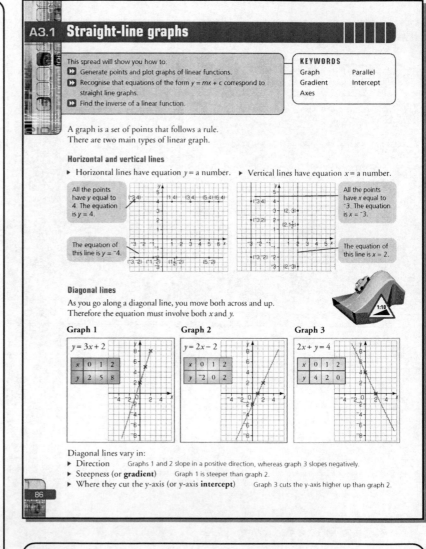

A3.1 Straight-line graphs

This spread will show you how to.
▶ Generate points and plot graphs of linear functions.
▶ Recognise that equations of the form $y = mx + c$ correspond to straight line graphs.
▶ Find the inverse of a linear function.

KEYWORDS
Graph   Parallel
Gradient   Intercept
Axes

A graph is a set of points that follows a rule.
There are two main types of linear graph.

Horizontal and vertical lines

▸ Horizontal lines have equation $y =$ a number.   ▸ Vertical lines have equation $x =$ a number.

All the points have y equal to 4. The equation is $y = 4$.

The equation of this line is $y = ^-4$.

All the points have x equal to $^-3$. The equation is $x = ^-3$.

The equation of this line is $x = 2$.

Diagonal lines

As you go along a diagonal line, you move both across and up. Therefore the equation must involve both x and y.

Graph 1
$y = 3x + 2$

x	0	1	2
y	2	5	8

Graph 2
$y = 2x - 2$

x	0	1	2
y	$^-2$	0	2

Graph 3
$2x + y = 4$

x	0	1	2
y	4	2	0

Diagonal lines vary in:
▸ Direction   Graphs 1 and 2 slope in a positive direction, whereas graph 3 slopes negatively.
▸ Steepness (or **gradient**)   Graph 1 is steeper than graph 2.
▸ Where they cut the y-axis (or y-axis **intercept**)   Graph 3 cuts the y-axis higher up than graph 2.

86

Plenary

Draw three unlabelled graphs on the board:

In pairs, students should discuss possible equations. Then pool the ideas from the class and identify the correct equations with reasons.

Further activities

Students can use graphical calculators to explore the nature of straight-line graphs.

Alternatively, students in pairs can devise a 'graph match game' whereby they create five graph and five equation cards – the object of the game is to match the graphs to the equations.

Differentiation

Extension questions:

▸ Questions 1 and 2 reinforce the relation between the slopes of line graphs and their equations and give simple cases of plotting line graphs.
▸ Questions 3 and 4 focus on plotting line graphs and investigating their characteristics.
▸ Question 5 requires a geometrical understanding of $y = mx + c$.

Core tier: focuses on linear function machines.

Exercise A3.1

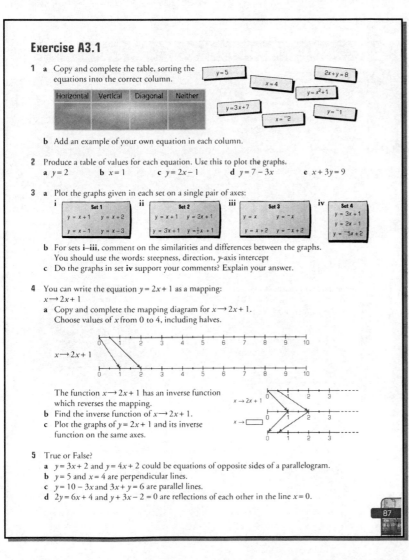

1 a Copy and complete the table, sorting the equations into the correct column.

Horizontal	Vertical	Diagonal	Neither

$y = 5$ $2x + y = 8$ $x = 4$ $y = x^2 + 1$ $y = 3x + 7$ $y = ^-1$ $x = ^-2$

   b Add an example of your own equation in each column.

2 Produce a table of values for each equation. Use this to plot the graphs.
   a $y = 2$ b $x = 1$ c $y = 2x - 1$ d $y = 7 - 3x$ e $x + 3y = 9$

3 a Plot the graphs given in each set on a single pair of axes:

   i Set 1
   $y = x + 1$ $y = x + 2$
   $y = x - 1$ $y = x - 3$

   ii Set 2
   $y = x + 1$ $y = 2x + 1$
   $y = 3x + 1$ $y = \frac{1}{2}x + 1$

   iii Set 3
   $y = x$ $y = ^-x$
   $y = x + 2$ $y = ^-x + 2$

   iv Set 4
   $y = 3x + 1$
   $y = 2x - 1$
   $y = ^-5x + 2$

   b For sets **i–iii**, comment on the similarities and differences between the graphs. You should use the words: steepness, direction, y-axis intercept
   c Do the graphs in set **iv** support your comments? Explain your answer.

4 You can write the equation $y = 2x + 1$ as a mapping:
   $x \rightarrow 2x + 1$
   a Copy and complete the mapping diagram for $x \rightarrow 2x + 1$. Choose values of x from 0 to 4, including halves.

   $x \rightarrow 2x + 1$

   The function $x \rightarrow 2x + 1$ has an inverse function which reverses the mapping.
   b Find the inverse function of $x \rightarrow 2x + 1$.
   c Plot the graphs of $y = 2x + 1$ and its inverse function on the same axes.

   $x \rightarrow 2x + 1$

   $x \rightarrow \boxed{}$

5 True or False?
   a $y = 3x + 2$ and $y = 4x + 2$ could be equations of opposite sides of a parallelogram.
   b $y = 5$ and $x = 4$ are perpendicular lines.
   c $y = 10 - 3x$ and $3x + y = 6$ are parallel lines.
   d $2y = 6x + 4$ and $y + 3x - 2 = 0$ are reflections of each other in the line $x = 0$.

87

Exercise commentary

The questions assess the objectives on Framework Pages 165 and 167.

Problem solving
Question 4 assesses the objectives on Framework Page 9. Questions 3 and 5 assess Page 33.

Group work
Students can work in pairs or small groups on question 3, perhaps with each group focusing on a particular set.

Misconceptions
Incorrect plotting of coordinates is common, especially in cases involving negatives. Aids to memory may be helpful, (for example x comes first alphabetically, and when you write an x you write a cross ('across')).
Students may create a short line segment with their three coordinates. Encourage students to extend the line, and emphasise that it is infinite.

Links
This topic links to linear sequences (Framework Page 149) and also proportionality (Page 79).

Homework

A3.1HW gives further practice in graph plotting and equation matching.
It provides a link to sequences, in showing students how to establish an equation.

Answers

1 a Horizontal: $y = 5$, $y = ^-1$; Vertical: $x = 4$, $x = ^-2$; Diagonal: $y = 3x + 7$, $2x + y = 8$; Neither: $y = x^2 + 1$
4 a $0 \rightarrow 1, \frac{1}{2} \rightarrow 2, 1 \rightarrow 3, 1\frac{1}{2} \rightarrow 4, 2 \rightarrow 5, 2\frac{1}{2} \rightarrow 6, 3 \rightarrow 7, 3\frac{1}{2} \rightarrow 8, 4 \rightarrow 9$
   b $x \rightarrow \frac{1}{2}(x - 1)$
5 a False b True c True d True

Mental starter

Present a pair of numbers – one mixed and one improper, for example $2\frac{1}{2}$ and $\frac{36}{11}$. Ask students to decide which is the larger.

You could introduce a simple response, like 'Raise your right hand for the mixed number and your left hand for the improper fraction'.

You could turn it into a game – eliminate students at each round until a winner is found.

Useful resources

A3.2OHP shows the graphs in the student book.

R8 is a blank coordinate grid for use in demonstrating gradients.

Introductory activity

Discuss the meaning of the word 'gradient'.

Encourage students to discuss real-life occurrences, such as road signs. Emphasise that the bigger the gradient, the steeper the slope.

Show the three graphs in the student book.

A3.2OHP shows the graphs of $y = 2x + 1$, $y = 3x + 1$ and $y = \frac{1}{2}x + 1$.

Focus on the steepest graph, $y = 3x + 1$. Show that by counting squares along and up, you can say: 'For every one square that you go along, you go three up'.

Write the gradient as $\frac{3}{1} = 3$. Reinforce the technique for measuring gradient with the other two graphs.

Invite volunteers to draw lines with varying gradients on a square grid (you can use R8), for example 4, $^-3$, $\frac{2}{3}$, $2\frac{1}{2}$ (= $\frac{5}{2}$).

Generalise the result for any two points (x_1, y_1) and (x_2, y_2).

Derive the formula for gradient:

$$m = \frac{y_2 - y_1}{x_2 - x_1}$$

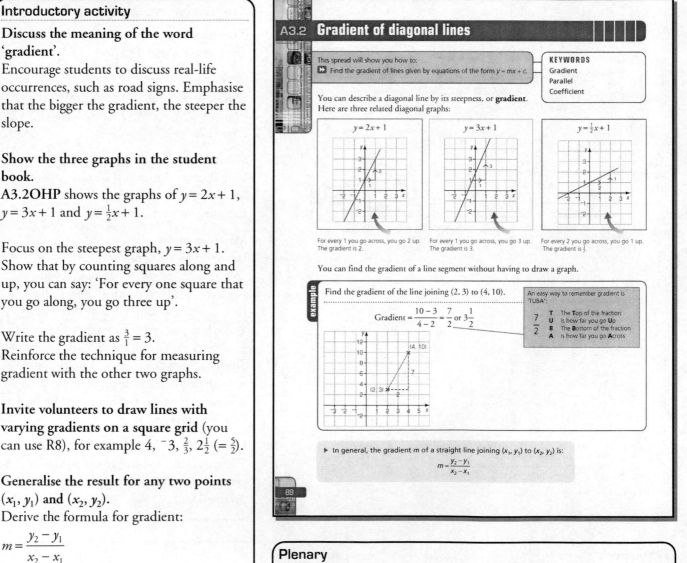

A3.2 Gradient of diagonal lines

This spread will show you how to:
▶ Find the gradient of lines given by equations of the form $y = mx + c$.

KEYWORDS
Gradient
Parallel
Coefficient

You can describe a diagonal line by its steepness, or **gradient**. Here are three related diagonal graphs:

$y = 2x + 1$

$y = 3x + 1$

$y = \frac{1}{2}x + 1$

For every 1 you go across, you go up 2 up. The gradient is 2.

For every 1 you go across, you go 3 up. The gradient is 3.

For every 2 you go across, you go 1 up. The gradient is $\frac{1}{2}$.

You can find the gradient of a line segment without having to draw a graph.

example

Find the gradient of the line joining (2, 3) to (4, 10).

$$\text{Gradient} = \frac{10 - 3}{4 - 2} = \frac{7}{2} \text{ or } 3\frac{1}{2}$$

An easy way to remember gradient is 'TUBA':

$\frac{7}{2}$

T The **T**op of the fraction
U is how far you go **U**p
B The **B**ottom of the fraction
A is how far you go **A**cross

▶ In general, the gradient m of a straight line joining (x_1, y_1) to (x_2, y_2) is:

$$m = \frac{y_2 - y_1}{x_2 - x_1}$$

Plenary

Draw a parallelogram ABCD on an OHT containing a square grid.
Ask students to count squares to find the gradient of AB.
What will the gradient of CD be?
What about the gradients of BC and DA?
Repeat for a parallelogram without a square grid, given the coordinates.

Further activities

Students can work in pairs to investigate the gradients of perpendicular lines on a square grid.
Encourage generalisation, using algebra if possible.

Differentiation

Extension questions:
▸ Question 1 requires students to think of gradient as a fraction.
▸ Questions 2 and 3 focus on the link between gradient and equations, and include geometrical reasoning.
▸ Questions 4–6 involve use of the general formula for gradient.

Core tier: focuses on the graph of a linear function.

Exercise A3.2

1 On a squared grid, draw lines with these gradients:
 a Line A = $\frac{1}{3}$ **b** Line B = 4 **c** Line C = $^-\frac{1}{2}$ **d** Line D = $\frac{3}{5}$
 e Line E = $1\frac{1}{2}$ **f** Line F = $^-2\frac{1}{4}$

2 For each straight line on the graph, complete the row in the table:

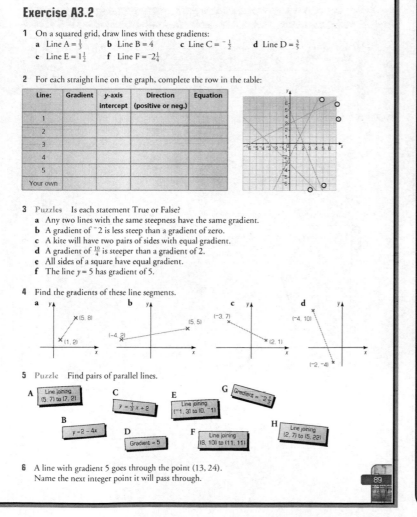

Line:	Gradient	y-axis intercept	Direction (positive or neg.)	Equation
1				
2				
3				
4				
5				
Your own				

3 Puzzles Is each statement True or False?
 a Any two lines with the same steepness have the same gradient.
 b A gradient of $^-2$ is less steep than a gradient of zero.
 c A kite will have two pairs of sides with equal gradient.
 d A gradient of $\frac{10}{4}$ is steeper than a gradient of 2.
 e All sides of a square have equal gradient.
 f The line $y = 5$ has gradient of 5.

4 Find the gradients of these line segments.
 a (5, 8) (1, 2) **b** $^-4$, 2) (5, 5) **c** ($^-3$, 7) (2, 1) **d** ($^-4$, 10) ($^-2$, $^-4$)

5 Puzzle Find pairs of parallel lines.
 A Line joining (5, 7) to (7, 2)
 B $y = 2 - 4x$
 C $y = \frac{1}{3}x + 2$
 D Gradient = 5
 E Line joining ($^-1$, 3) to (0, $^-1$)
 F Line joining (8, 10) to (11, 11)
 G Gradient = $^-2\frac{1}{2}$
 H Line joining (2, 7) to (5, 22)

6 A line with gradient 5 goes through the point (13, 24). Name the next integer point it will pass through.

89

Exercise commentary

The questions assess the objectives on Framework Page 167.

Problem solving
Question 5 assesses the objectives on Framework Page 9. Question 3 assesses Page 33.

Group work
Students should be encouraged to discuss the statements in question 3 in a small group.
The activity suggested in the 'Further resources' box should be attempted in pairs.

Misconceptions
Students often write the fraction the wrong way around, putting the change in x on top. A useful memory aid is TUBA ('top up, bottom across).
When using the general formula, encourage students to label the coordinates (x_1, y_1) and (x_2, y_2) first.

Links
The concept of gradient will be used to interpret the equation of a line of best fit in later statistical work.

Homework

A3.2HW provides further practice at finding gradients, both by counting squares and using the general formula.

Answers

2 2, $^-2$, pos, $y = 2x - 2$; $\frac{1}{2}$, 3, pos, $y = \frac{1}{2}x + 3$; $^-1$, $^-3$, neg, $y = ^-x - 3$; $2\frac{1}{2}$, 5, neg, $y = ^-2\frac{1}{2}x + 5$; 0, 4, neither, $y = 4$
3 **a** True **b** False **c** False **d** True **e** False **f** False
4 **a** $1\frac{1}{2}$ **b** $\frac{1}{3}$ **c** $^-1\frac{1}{5}$ **d** $^-7$
5 A and G, B and E, C and F, D and H
6 (14, 29)

Mental starter

Students stand up.
Read out the equation of a graph.
Students hold out their arms to make the shape of the graph.
For example:

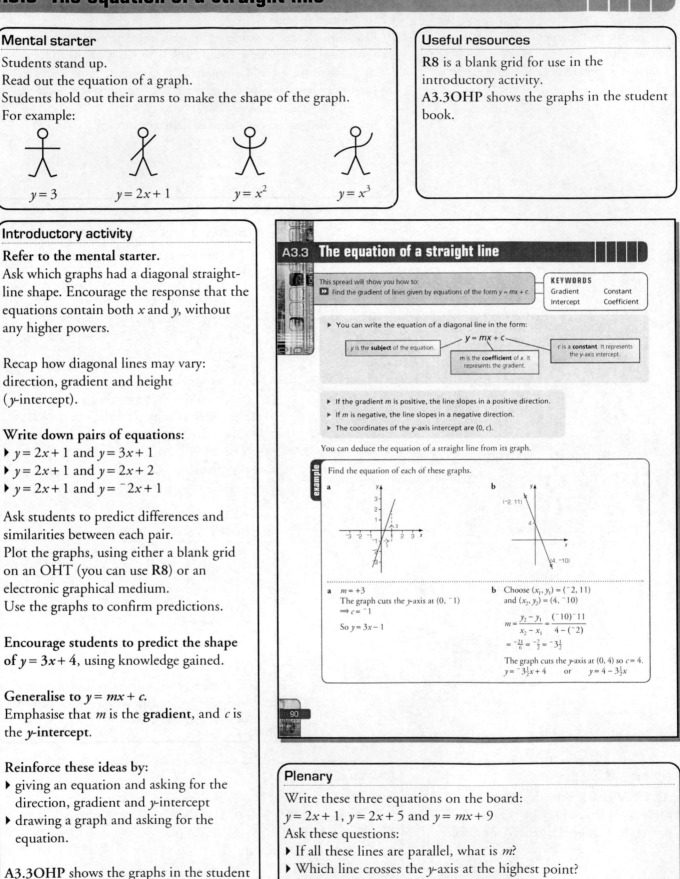

$y = 3$ $y = 2x + 1$ $y = x^2$ $y = x^3$

Useful resources

R8 is a blank grid for use in the introductory activity.
A3.3OHP shows the graphs in the student book.

Introductory activity

Refer to the mental starter.
Ask which graphs had a diagonal straight-line shape. Encourage the response that the equations contain both x and y, without any higher powers.

Recap how diagonal lines may vary: direction, gradient and height (y-intercept).

Write down pairs of equations:
▸ $y = 2x + 1$ and $y = 3x + 1$
▸ $y = 2x + 1$ and $y = 2x + 2$
▸ $y = 2x + 1$ and $y = {}^-2x + 1$

Ask students to predict differences and similarities between each pair.
Plot the graphs, using either a blank grid on an OHT (you can use R8) or an electronic graphical medium.
Use the graphs to confirm predictions.

Encourage students to predict the shape of $y = 3x + 4$, using knowledge gained.

Generalise to $y = mx + c$.
Emphasise that m is the **gradient**, and c is the **y-intercept**.

Reinforce these ideas by:
▸ giving an equation and asking for the direction, gradient and y-intercept
▸ drawing a graph and asking for the equation.

A3.3OHP shows the graphs in the student book.

A3.3 The equation of a straight line

This spread will show you how to:
▶▶ Find the gradient of lines given by equations of the form $y = mx + c$.

KEYWORDS
Gradient   Constant
Intercept   Coefficient

▸ You can write the equation of a diagonal line in the form:

$$y = mx + c$$

y is the **subject** of the equation.

m is the **coefficient** of x. It represents the gradient.

c is a **constant**. It represents the y-axis intercept.

▸ If the gradient m is positive, the line slopes in a positive direction.
▸ If m is negative, the line slopes in a negative direction.
▸ The coordinates of the y-axis intercept are $(0, c)$.

You can deduce the equation of a straight line from its graph.

example

Find the equation of each of these graphs.

a

b
$(^-2, 11)$

$(4, ^-10)$

a $m = +3$
The graph cuts the y-axis at $(0, ^-1)$
$\Rightarrow c = ^-1$
So $y = 3x - 1$

b   Choose $(x_1, y_1) = (^-2, 11)$
and $(x_2, y_2) = (4, ^-10)$
$$m = \frac{y_2 - y_1}{x_2 - x_1} = \frac{(^-10)^-11}{4 - (^-2)}$$
$$= \frac{^-21}{6} = \frac{^-7}{2} = ^-3\tfrac{1}{2}$$
The graph cuts the y-axis at $(0, 4)$ so $c = 4$.
$y = ^-3\tfrac{1}{2}x + 4$ or $y = 4 - 3\tfrac{1}{2}x$

90

Plenary

Write these three equations on the board:
$y = 2x + 1$, $y = 2x + 5$ and $y = mx + 9$
Ask these questions:
▸ If all these lines are parallel, what is m?
▸ Which line crosses the y-axis at the highest point?
▸ Does the first graph pass through $(0, 1)$? $(5, 10)$?
▸ If the last graph passes through $(10, 10)$, what is m?

Further activities

Students make up their own questions to try on others, for example:

a What is the equation of the x-axis?

b What is the equation of the reflection of the line $y = 2x + 5$

i in the x-axis **ii** in the y-axis?`

A3.3ICT explores linear functions, graphs, and points of intersection using a spreadsheet program.

Differentiation

Extension questions:

▸ Questions 1 and 2 can be tackled by comparing given equations to $y = mx + c$.

▸ Questions 3 and 4 require an understanding of $y = mx + c$ in order to match equations and graphs.

▸ Questions 5 and 6 focus on finding an equation given certain clues, and require a good knowledge of gradient.

Core tier: focuses on sketching straight-line graphs from an equation.

Exercise A3.3

1 True or False?

a $y = 5x + 3$ is the equation of the line with gradient 5 that cuts the y-axis at (0, 3).

b $y = 10 - 2x$ is the equation of the line with gradient 10 that cuts the y-axis at $(0, {}^-2)$.

c $3y = 6x - 18$ is the equation of the line with gradient 6 that cuts the y-axis at $(0, {}^-18)$.

d $y = 10x + 1$ and $y = 10x - 2$ slope in opposite directions to each other.

2
a $y = 4x + 3$ 　**b** $y = {}^-2x + 5$
c $y = 10x - 2$ 　**d** $y = 7 - 3x$
e $2y = 5x + 4$ 　**f** $y + 2x = 9$
g $3y + 2x = 7$ 　**h** $y + px = q$

Fill in the required information for each equation.

EQUATION	GRADIENT	DIRECTION	Y-AXIS INTERCEPT
$y = 5x + 1$	5	Positive	(0, 1)

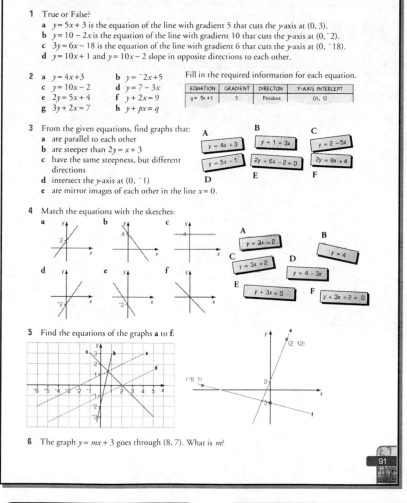

3 From the given equations, find graphs that:

a are parallel to each other

b are steeper than $2y = x + 3$

c have the same steepness, but different directions

d intersect the y-axis at $(0, {}^-1)$

e are mirror images of each other in the line $x = 0$.

A $y = 4x + 3$　**B** $y + 1 = 3x$　**C** $y = 2 - 5x$
D $y = 5x - 1$　**E** $2y + 6x - 2 = 0$　**F** $2y = 8x + 4$

4 Match the equations with the sketches:

a b c d e f

A $y = 3x - 2$　**B** $y = 4$
C $y = 3x + 2$　**D** $y = 4 - 3x$
E $y + 3x = 0$　**F** $y + 3x + 2 = 0$

5 Find the equations of the graphs **a** to **f**:

(2, 12)　(⁻8, 1)

6 The graph $y = mx + 3$ goes through (8, 7). What is m?

Exercise commentary

The questions assess the objectives on Framework Page 167.

Problem solving

Question 1 assesses the objectives on Framework Page 33. Questions 3, 4 and 6 assess Page 9.

Group work

The questions in this exercise are best tackled and discussed in pairs.

Misconceptions

Students may assume that a graph travels in a negative direction if there is a negative present, such as in $y = 2x - 5$. Reinforce that the **gradient** must be negative. Errors often occur where an equation needs rearranging, such as $2y = 5x + 1$. Students may assume that the gradient is 5. Reinforce the general form $y = mx + c$, with y as the subject.

Links

The equation of a straight line may be used in later experimental work in science.

Homework

A3.3HW practises $y = mx + c$, including sketching graphs and matching equations.

It provides more SAT-style questions, as in the plenary.

Answers

1 a True 　**b** False 　**c** False 　**d** False

2 a 4, pos, (0, 3) 　**b** $^-2$, neg, (0, 5) 　**c** 10, pos, $(0, {}^-2)$ 　**d** $^-3$, neg, (0, 7)
　e $2\frac{1}{2}$, pos, (0, 2) 　**f** $^-2$, neg, (0, 9) 　**g** $-\frac{2}{3}$, neg, $(0, 2\frac{1}{3})$ 　**h** ^-p, neg, (0, q)

3 a A and F 　**b** C and D 　**c** B and E, C and D 　**d** B, D 　**e** None

4 a C 　**b** D 　**c** B 　**d** A 　**e** F 　**f** E

5 a $y = \frac{1}{2}x + 1$ 　**b** $y = \frac{1}{2}x - 1$ 　**c** $y = 2 - x$ 　**d** $y = 5x - 2$ 　**e** $y = 5x + 2$
　f $y = {}^-\frac{1}{2}x - 3$

6 $\frac{1}{2}$

Mental starter

Give students a piece of squared paper.

Instruct students to draw lines with these gradients:
$\frac{2}{3}$, 5, $-\frac{1}{2}$, $1\frac{1}{5}$

Discuss strategies for drawing a line with a decimal gradient, such as 1.4.

Useful resources

Squared paper for the mental starter.

Introductory activity

Refer to the mental starter.
Ask students to:
▸ draw a line parallel to their first line (the one with gradient $\frac{2}{3}$)
▸ measure the gradient of the new line.

Ask students to comment on what they notice. Ensure by asking questions that students understand what 'parallel' means.

Generalise: Two lines are parallel if they have the same gradient.
Discuss whether $y = 2x + 4$ and $2y = 10 + 4x$ are parallel.

For the other three lines in the mental starter, ask students to:
▸ draw a line at 90° (use the term 'perpendicular'). Students should use a protractor for this.
▸ measure the gradient of the perpendicular line, and give the answer as a fraction.
Encourage students to comment on what they notice.

Generalise: Two lines are perpendicular if their gradients are the negative reciprocals of each other.
Recap the meaning of a reciprocal, and discuss the consequences of this result:
If one line has a positive gradient, the other line will be negative.

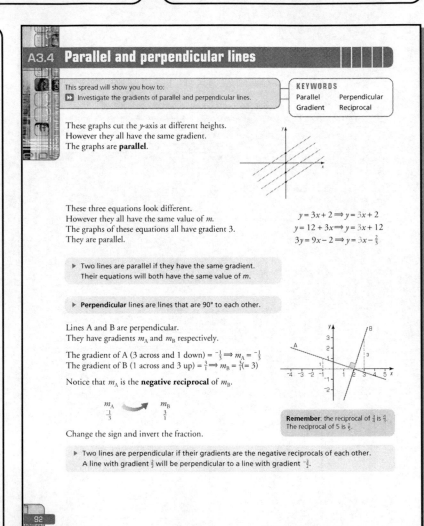

A3.4 Parallel and perpendicular lines

This spread will show you how to:
▶ Investigate the gradients of parallel and perpendicular lines.

KEYWORDS
Parallel   Perpendicular
Gradient   Reciprocal

These graphs cut the y-axis at different heights.
However they all have the same gradient.
The graphs are **parallel**.

These three equations look different.
However they all have the same value of m.
The graphs of these equations all have gradient 3.
They are parallel.

$y = 3x + 2 \Rightarrow y = 3x + 2$
$y = 12 + 3x \Rightarrow y = 3x + 12$
$3y = 9x - 2 \Rightarrow y = 3x - \frac{2}{3}$

▸ Two lines are parallel if they have the same gradient.
Their equations will both have the same value of m.

▸ **Perpendicular** lines are lines that are 90° to each other.

Lines A and B are perpendicular.
They have gradients m_A and m_B respectively.

The gradient of A (3 across and 1 down) = $-\frac{1}{3} \Rightarrow m_A = -\frac{1}{3}$
The gradient of B (1 across and 3 up) = $\frac{3}{1} \Rightarrow m_B = \frac{3}{1} (= 3)$

Notice that m_A is the **negative reciprocal** of m_B.

m_A m_B
$-\frac{1}{3}$ $\frac{3}{1}$

Change the sign and invert the fraction.

Remember: the reciprocal of $\frac{3}{4}$ is $\frac{4}{3}$.
The reciprocal of 5 is $\frac{1}{5}$.

▸ Two lines are perpendicular if their gradients are the negative reciprocals of each other.
A line with gradient $\frac{2}{3}$ will be perpendicular to a line with gradient $-\frac{3}{2}$.

92

Plenary

Extend to horizontal and vertical graphs, to introduce the idea of an infinite and a zero gradient.
For example, ask students to write down the equation of a graph
i parallel to $y = 5$ **ii** perpendicular to $y = 5$.
▸ What is the gradient of $y = 5$?
▸ So what is the gradient of the line perpendicular to it?

Students can find the gradients of curves at various points by drawing tangents.

Extension questions:
▸ Questions 1 and 2 provide practice at identifying parallel and perpendicular lines.
▸ Questions 3 and 4 focus on the link to $y = mx + c$.
▸ Questions 5–8 link to geometrical reasoning.

Core tier: focuses on conversion graphs.

Exercise A3.4

1 For each value of m, find the gradient of a line that is
   i parallel   **ii** perpendicular.
   a 4   **b** $\frac{1}{5}$ **c** $\frac{2}{3}$ **d** $1\frac{1}{2}$ **e** $^-1\frac{1}{4}$ **f** m

2 Copy this line on a square grid.

On the same grid, draw a line that is perpendicular to it.

3 True or False?
   a Two parallel lines never meet each other.
   b Two perpendicular lines always meet each other.
   c The lines $y = 3x + 2$
      and $y = \frac{1}{3}x + 2$ are perpendicular.
   d The lines $y = 5x - 1$
      and $2y = 10(x + 2)$ are parallel.

4 Copy and complete this table:

Equation of a parallel line	Equation	Equation of a perpendicular line
	$y = 4x + 1$	
	$y = 2\frac{1}{2}x - 3$	
	$2y = 7x + 4$	
	$2x + 3y = 9$	
	your own	

5 Under what circumstances could a pair of perpendicular lines be mirror images of each other in the y-axis?

6 Give the equation of each line used to construct these diagrams.

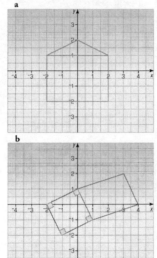

7 Give possible equations that could be plotted to construct:
   a a parallelogram
   b a rhombus
   c a right-angled triangle with no horizontal or vertical lines
   d a kite

8 Construct your own coordinate picture with instructions on which equations to plot.

93

The questions assess the objectives on Framework Pages 167 and 169.

Problem solving
Questions 3 and 5 assess the objectives on Framework Page 33.

Group work
Questions 3–8 can be attempted with discussion in pairs.

Misconceptions
Students often express gradients inappropriately, either as a decimal or as a mixed number.
Emphasise that the gradient should be expressed as an improper fraction.
A common error is to interpret a gradient of $\frac{2}{3}$ as 'across 2 and up 3', probably due to prior knowledge of coordinates. Emphasise that the top number is the change in y.

Links
This topic links to geometrical properties of quadrilaterals (Framework Page 187).

A3.4HW reinforces the key ideas of the lesson with a focus on geometrical properties.

1 **a i** 4 **ii** $^-\frac{1}{4}$ **b i** $\frac{1}{5}$ **ii** $^-5$ **c i** $\frac{2}{3}$ **ii** $^-1\frac{1}{2}$
  d i $1\frac{1}{2}$ **ii** $^-\frac{2}{3}$ **e i** $1\frac{1}{4}$ **ii** $^-\frac{4}{5}$ **f i** m **ii** $^-\frac{1}{m}$
3 **a** True **b** True **c** False **d** True
5 Only if gradients are 1 and $^-1$
6 **a** $y = \frac{1}{2}x + 2$, $y = ^-\frac{1}{2}x + 2$, $y = 1$, $x = ^-2$, $x = 2$, $y = ^-2$
  b $y = \frac{1}{2}x + 1$, $y = \frac{1}{2}x - 1\frac{1}{2}$, $y = ^-2x - 4$, $y = ^-2x + 1$, $y = \frac{1}{3}x + 1$,
     $y = \frac{1}{3}x - 1\frac{1}{3}$, $y = ^-2x + 8$

Mental starter

Students play 'Bingo'.

They enter nine numbers, between ⁻10 and 10 inclusive, in a 3 × 3 grid.

Ask random questions, for example $(^-3)^2$. Students cross out the solution if it is on their grid. Solve equations outside the range, for example $(^-4)^2$, $(5)^3$ can be included.

Useful resources

A3.5OHP shows the graphs of $y = x^2$ and $y = x^3$.

Graph paper

Introductory activity

The purpose of the lesson is to plot quadratic and cubic graphs.

Firstly, it is important that students know the slope of a graph from its equation in order to test that it is correctly plotted. Ask who can remember the slopes of $y = x^2$ and $y = x^3$ from previous starters.

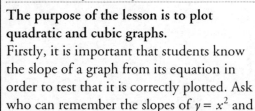

For x^2, discuss: shape, vocabulary → quadratic, parabola, why it is a curve (not a line), why it comes down, turns around and goes back up again.

For x^3, discuss: shape, vocabulary → cubic, ∫-shaped, why it is a curve (not a line), why it goes up, rests and carries on again, unlike x^2.

The key point to bring out is that, when you square a negative it is positive but when you cube a negative it is still negative.

Ask students to remind you how many points they plotted when drawing a line graph. Is this enough for a curve? You need enough values to get the full shape. A table is useful.

Try $y = x^2 - 4x + 3$. Demonstrate it is easiest to have rows for x^2, $- 4x$, and $+ 3$ when compiling a table.

Show students that the final graph is useful for solving equations or working out substitutions. For example $x^2 - 4x + 3 = 9$, $x^2 - 4x + 3$ when $x = 1.6$.

Ask students what rows they would choose for $y = x^3 + 2x^2 - 1$ and to remind you of this graph's shape.

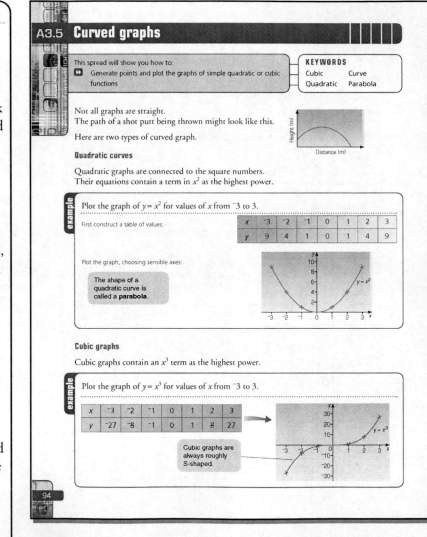

A3.5 Curved graphs

This spread will show you how to:
▶ Generate points and plot the graphs of simple quadratic or cubic functions

KEYWORDS
Cubic   Curve
Quadratic   Parabola

Not all graphs are straight.
The path of a shot putt being thrown might look like this.

Here are two types of curved graph.

Quadratic curves

Quadratic graphs are connected to the square numbers.
Their equations contain a term in x^2 as the highest power.

example

Plot the graph of $y = x^2$ for values of x from ⁻3 to 3.

First construct a table of values:

x	⁻3	⁻2	⁻1	0	1	2	3
y	9	4	1	0	1	4	9

Plot the graph, choosing sensible axes:

The shape of a quadratic curve is called a **parabola**.

Cubic graphs

Cubic graphs contain an x^3 term as the highest power.

example

Plot the graph of $y = x^3$ for values of x from ⁻3 to 3.

x	⁻3	⁻2	⁻1	0	1	2	3
y	⁻27	⁻8	⁻1	0	1	8	27

Cubic graphs are always roughly S-shaped.

94

Plenary

Ask students to think about the solutions of the equation $x^3 = x^2$.
A number cubed gives the same answer as that number squared.
(Encourage the responses $x = 0$ and $x = 1$). Ask how you can use their graphs to show that there are only two solutions.
Predict the number of solutions of $x^2 = 2x + 1$ in this way.

Further activities

Students could extend the ideas introduced in later questions: that is, use graphs to solve complex equations for example $x^2 = 2x + 1$.

Further graph slopes could be investigated:

$\left.\begin{array}{l} y = \frac{1}{x} \\ y = 2^x \end{array}\right\}$ and their shapes explained.

Differentiation

Extension questions:

▸ Questions 1 and 2 provide practice in drawing curved graphs.
▸ Questions 3 and 4 focus on using and recognising curved graphs.
▸ Questions 5 and 6 extend to sketching curved graphs.

Core tier: focuses on distance–time graphs.

Exercise A3.5

1 For each equation given:
   i   predict the shape of the graph
   ii complete the table of values
   iii plot the graph on appropriately scaled axes.

a $y = x^2 + 3$

x	-4	-3	-2	-1	0	1	2	3	4
x^2	16	9					4		
+3	+3	+3	+3	+3	+3	+3	+3	+3	+3
y	19	12					7		

b $y = x^3 - x$

x	-4	-3	-2	-1	0	1	2	3	4
x^3	-64	-27					8		
$-x$	+4	+3					-2		
y	-60	-24					6		

c $y = x^2 + x - 6$

y	-5	-4	-3	-2	-1	0	1	2	3
x^2	25							4	
x	-5							+2	
-6	-6							-6	
y	14							0	

d $y = 3x^2 + 1$

y	-3	-2	-1	0	1	2	3	4
x^2								
$3x^2$								
+1								
y								

2 Devise your own table and plot the graphs over the range of values given.
   a $y = 2x^2 - 3$ $-3 \leqslant x \leqslant 3$
   b $y = 10 - x^2$ $-2 \leqslant x \leqslant 3$
   c $y = x^3 - x^2 + x$ $-2 \leqslant x \leqslant 3$

3 Use your graph in question 2a to find
   a the value of y when $x = 2.5$
   b the values of x when $y = 6$.

4 Match these shape and equation cards:

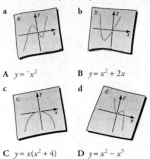

a   b

c   d

A $y = {}^-x^2$ **B** $y = x^2 + 2x$

C $y = x(x^2 + 4)$ **D** $y = x^2 - x^3$

5 a Use one set of axes, taking x from $^-5$ to 5, to draw each graph group. Comment on how the change in equation causes the graph to change.

> **Group 1**
> $y = x^2$; $y = x^2 + 1$
> $y = x^2 + 3$; $y = x^2 - 2$

> **Group 2**
> $y = x^2$; $y = (x + 1)^2$
> $y = (x + 2)^2$; $y = (x + 2)^2$

b Use your findings to sketch:
   i $y = x^2 + 10$ ii $y = (x - 4)^2$
   iii $y = (x + 5)^2 - 2$

6 Use your findings from question 5 to sketch the graph $y = (x - 2)^2 + 4$. What is the equation of its line of symmetry?

95

Exercise commentary

The questions assess the objectives on Framework Page 171.

Problem solving
Question 4 assesses the objectives on Framework Page 9.

Group work
Solutions to question 1 should be compared in pairs and students should check their values in question 2 with one another before plotting the graphs.

Misconceptions
Students should be encouraged to work mentally when substituting negatives, but they can check their answers on a calculator.

You may need to suggest sensible axes, as students may be unfamiliar with using graph paper.

Links
This work links to later work on real-life graphs, and also subsequent work on quadratic equations.

Homework

A3.5HW gives further practice in plotting graphs, extending to other curves such as $y = x^4$ and $y = 2^x$.

Answers

3 a 9.5   b 2.1, $^-$2.1
4 a C   b B   c A   d D
6 $x = 2$

Mental starter

Draw some real-life graphs on the board, without any axes labelled.

Ask students, in small groups, to identify a fitting story for each one.

For example, the left-hand graph could represent currency conversion and the right-hand graph could represent pulse rate during exercise.

Useful resources

A3.6OHP shows the distance-time graph in the student book.

Introductory activity

Refer to the mental starter.

Use the examples to illustrate that a graph can be useful to show what is happening in many situations.

Introduce distance-time graphs.

You can show the graph in the student book, which is also illustrated on **A3.6OHP**.

Discuss each stage of the journey, and focus in particular on the gradient of each stage. Encourage students to describe how the gradient relates to the story.

Bring out the key points:

▸ The labelled axes tell you what the graph shows.

▸ Flat sections represent a break in the journey.

▸ Steeper sections represent a faster speed.

▸ Lines with negative gradient represent a homeward journey.

Use the example to show that the gradient of a distance-time graph represents speed.

Ask students to draw four sets of distance-time axes (no units needed).

Explain that you are going to make four different journeys across the front of the room (walk then run, across to the door, then back again ...)

Students should draw a graph to represent each journey.

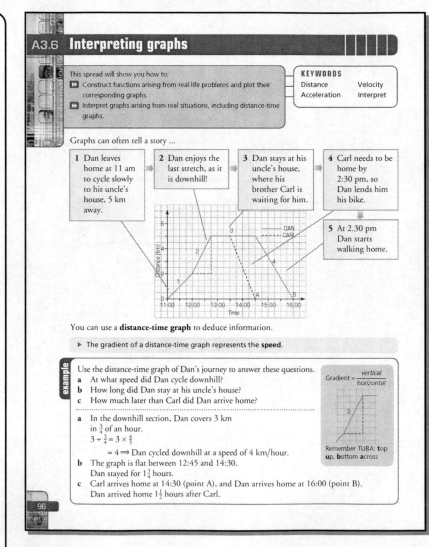

A3.6 Interpreting graphs

This spread will show you how to:
▶ Construct functions arising from real-life problems and plot their corresponding graphs.
▶ Interpret graphs arising from real situations, including distance-time graphs.

KEYWORDS
Distance   Velocity
Acceleration   Interpret

Graphs can often tell a story ...

1 Dan leaves home at 11 am to cycle slowly to his uncle's house, 5 km away.

2 Dan enjoys the last stretch, as it is downhill!

3 Dan stays at his uncle's house, where his brother Carl is waiting for him.

4 Carl needs to be home by 2:30 pm, so Dan lends him his bike.

5 At 2.30 pm Dan starts walking home.

You can use a **distance-time graph** to deduce information.

▸ The gradient of a distance-time graph represents the **speed**.

Use the distance-time graph of Dan's journey to answer these questions.

a   At what speed did Dan cycle downhill?
b   How long did Dan stay at his uncle's house?
c   How much later than Carl did Dan arrive home?

Gradient = $\dfrac{vertical}{horizontal}$

a   In the downhill section, Dan covers 3 km in $\frac{3}{4}$ of an hour.
$3 \div \frac{3}{4} = 3 \times \frac{4}{3}$
$= 4 \Rightarrow$ Dan cycled downhill at a speed of 4 km/hour.

b   The graph is flat between 12:45 and 14:30. Dan stayed for $1\frac{3}{4}$ hours.

c   Carl arrives home at 14:30 (point A), and Dan arrives home at 16:00 (point B). Dan arrived home $1\frac{1}{2}$ hours after Carl.

Remember TUBA: **t**op **u**p, **b**ottom **a**cross

96

Plenary

Refer again to Dan and Carl's journey, which is illustrated in the student book and on **A3.6OHP**.

Discuss how to draw a speed-time graph of their journey.

Further activities

Challenge students to draw a speed-time graph for Jo and Vicky's journey in question 3.

A3.6F contains extra practice at interpreting real-life graphs.

Differentiation

Extension questions:
▸ Question 1 is based on interpreting a distance-time graph without units labelled.
▸ Questions 2 and 3 require students to interpret and construct distance-time graphs in detail.
▸ Question 4 explores another type of real-life graph.

Core tier: focuses on interpreting real-life graphs.

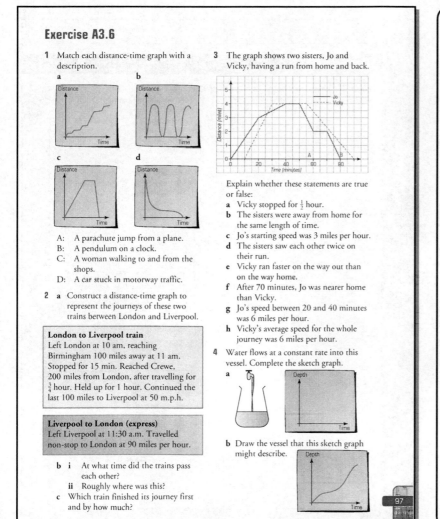

Exercise A3.6

1 Match each distance-time graph with a description.

a

b

c

d

A: A parachute jump from a plane.
B: A pendulum on a clock.
C: A woman walking to and from the shops.
D: A car stuck in motorway traffic.

2 a Construct a distance-time graph to represent the journeys of these two trains between London and Liverpool.

London to Liverpool train
Left London at 10 am, reaching Birmingham 100 miles away at 11 am. Stopped for 15 min. Reached Crewe, 200 miles from London, after travelling for $\frac{3}{4}$ hour. Held up for 1 hour. Continued the last 100 miles to Liverpool at 50 m.p.h.

Liverpool to London (express)
Left Liverpool at 11:30 a.m. Travelled non-stop to London at 90 miles per hour.

b i At what time did the trains pass each other?
   ii Roughly where was this?
c Which train finished its journey first and by how much?

3 The graph shows two sisters, Jo and Vicky, having a run from home and back.

Explain whether these statements are true or false:

a Vicky stopped for $\frac{1}{2}$ hour.
b The sisters were away from home for the same length of time.
c Jo's starting speed was 3 miles per hour.
d The sisters saw each other twice on their run.
e Vicky ran faster on the way out than on the way home.
f After 70 minutes, Jo was nearer home than Vicky.
g Jo's speed between 20 and 40 minutes was 6 miles per hour.
h Vicky's average speed for the whole journey was 6 miles per hour.

4 Water flows at a constant rate into this vessel. Complete the sketch graph.

a

b Draw the vessel that this sketch graph might describe.

97

Exercise commentary

The questions assess the objectives on Framework Pages 173 and 175.

Problem solving
Questions 1 and 4 assess the objectives on Framework Page 9.

Group work
Students should discuss question 1 in pairs.
Encourage students to compare their constructed graphs when they have completed question 2.

Misconceptions
Students often draw a homeward journey as one that goes backwards to the origin. Emphasise that this would imply going back in time.
Students may not interpret speed correctly. Emphasise that 60 mph means that you travel 60 miles in 1 hour (so how far do you go in $\frac{3}{4}$ of the time ... ?).

Links
This topic links to the use of line graphs in statistics (Framework Page 265).

Homework

A3.6HW focuses on distance-time graphs, and also explores other real-life graphs.

Answers

1 a D b B c C d A
2 b i About 12:40 ii Crewe
  c Liverpool to London (express) by 10 min
3 a True b True c False d True e True
  f True g False h True

Summary

The key objectives for this unit are:
▸ Given values for m and c, find the gradients of lines given by equations of the form $y = mx + c$.
▸ Construct functions arising from real-life problems and plot their corresponding graphs.
▸ Interpret graphs arising from real situations.

Check out commentary

1 Students should have learned, from starters and lessons, how to tell the slope of a graph from its equation so part **a** should be accessible to all and useful as a reminder if forgotten.

In part **b**, students may need to check their calculations, mentally or using a calculator as negatives cause difficulty.

2 **a** Students should recall that m gives the gradient, but may need to be reminded that y should be the subject first.
In part **ii**, TUBA may need to be revisited as an aid to memory.

b Students tend to be able to interpret each part of the graph, but speed, rather than distance, may need to be emphasised in **iii**.

Students' descriptions often tend to lack detail and evidence; they should be encouraged to read each other's work to ensure they make sense.

Plenary activity

Sketch this graph on the board:

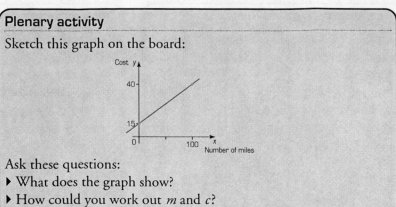

Ask these questions:
▸ What does the graph show?
▸ How could you work out m and c?
▸ What do m and c actually mean in this context?

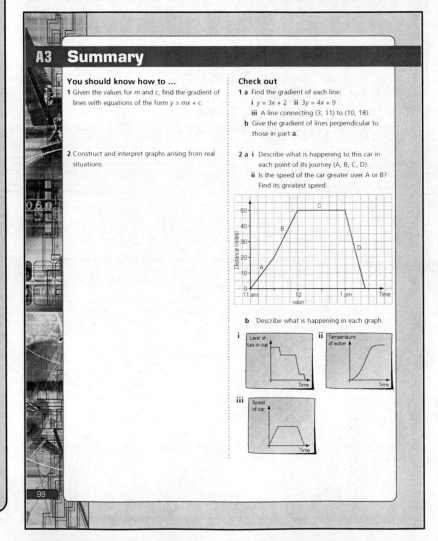

A3 Summary

You should know how to ...
1 Given the values for m and c, find the gradient of lines with equations of the form $y = mx + c$.

2 Construct and interpret graphs arising from real situations.

Check out
1 **a** Find the gradient of each line:
   i $y = 3x + 2$ **ii** $3y = 4x + 9$
   iii A line connecting (3, 11) to (10, 18).
   b Give the gradient of lines perpendicular to those in part **a**.

2 **a** **i** Describe what is happening to this car in each point of its journey (A, B, C, D).
   ii Is the speed of the car greater over A or B? Find its greatest speed.

b Describe what is happening in each graph.

98

Development

Graphs are revisited in A5, where $y = mx + c$ is interpreted in various real-life situations and graph work is linked to proportion.

Links

Graph work underpins many area of the maths curriculum, both in algebra and in data handling. Furthermore, the skills involved in constructing and interpreting graphs are transferable to other subjects such as geography and science.

Mental starters

Objectives covered in this unit:

▸ Know and use squares, cubes, roots and index notation.
▸ Convert between m, cm and mm, km and m, kg and g, litres and ml, cm^2 and mm^2.
▸ Add and subtract several small numbers of several multiples of 10.
▸ Use factors to multiply and divide mentally.
▸ Use partitioning to multiply and divide mentally.
▸ Calculate using knowledge of multiplication and division facts and place value.
▸ Use approximations to estimate the answers to calculations.
▸ Convert between fractions, decimals and percentages.

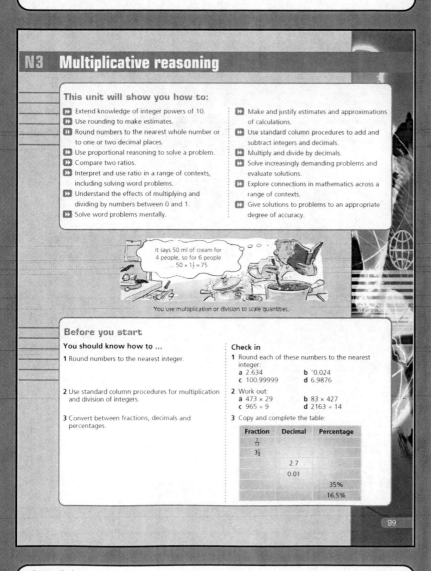

N3 Multiplicative reasoning

This unit will show you how to:

▸ Extend knowledge of integer powers of 10.
▸ Use rounding to make estimates.
▸ Round numbers to the nearest whole number or to one or two decimal places.
▸ Use proportional reasoning to solve a problem.
▸ Compare two ratios.
▸ Interpret and use ratio in a range of contexts, including solving word problems.
▸ Understand the effects of multiplying and dividing by numbers between 0 and 1.
▸ Solve word problems mentally.

▸ Make and justify estimates and approximations of calculations.
▸ Use standard column procedures to add and subtract integers and decimals.
▸ Multiply and divide by decimals.
▸ Solve increasingly demanding problems and evaluate solutions.
▸ Explore connections in mathematics across a range of contexts.
▸ Give solutions to problems to an appropriate degree of accuracy.

It says 50 ml of cream for 4 people, so for 6 people ... $50 \times 1\frac{1}{2} = 75$

You use multiplication or division to scale quantities.

Before you start

You should know how to ...

1 Round numbers to the nearest integer.

2 Use standard column procedures for multiplication and division of integers.

3 Convert between fractions, decimals and percentages.

Check in

1 Round each of these numbers to the nearest integer:
a 2.634   b ⁻0.024
c 100.99999   d 6.9876

2 Work out:
a 473 × 29   b 83 × 427
c 965 ÷ 9   d 2163 ÷ 14

3 Copy and complete the table:

Fraction	Decimal	Percentage
$\frac{1}{7}$		
$3\frac{1}{4}$		
	2.7	
	0.01	
		35%
		16.5%

99

Check in activity

Show a number, for example 900, and ask for different ways in which it can be expressed.
$900 = 90 \times 10$, or $9 \times 10 \times 10$, or 30^2 ...

Resources needed

* means class set needed
Essential:
Graph or squared paper
N3.1OHP – place value table
N3.3OHP – number lines
N3.7OHP – comparing ages using ratio
N3.8OHP – scale factor as a decimal and percentage

Useful:
Bag of counters or beads
Mini whiteboards
R4 – place value tables*
R6 – number lines*
Mail order catalogues
N3.7F – ratio problems
N3.8F – percentage problems
N3.5ICT – multiplying decimals

Unit commentary

Aim of the unit

The first sections cover multiplication by powers of 10 and rounding – the following sections consolidate mental and written methods using decimals. Ratio and proportion are developed through looking at multiplicative relationships.

Introduction

Discuss the use of indices in expressing large numbers. Limit the discussion to powers of 10, for example:
$$10\ 000\ 000 = 10 \times 10 \times 10 \times 10 \times 10 \times 10 \times 10 = 10^7$$
Ask students to describe this quantity in words (10 million).

Framework references

This unit focuses on:
Teaching objectives pages:
37–39, 43–45, 57, 79–81, 93, 97, 105–107, 229.
Problem solving objectives pages 3, 5, 7, 19, 31, 33.

Differentiation

Core tier

Focuses on the same topics as the extension tier, but uses simpler numbers and contexts.

Mental starter

Write on the board questions like these:

I have the power that makes 4^{power} into 4096. What is my power?

I have the power that makes 5^{power} into 3125. What is my power?

I have the power that makes 2^{power} into 1024. What is my power?

I have the power that makes 3^{power} into 243. What is my power?

I have the power that makes 7^{power} into 343. What is my power?

Useful resources

Mini-whiteboards may be useful for the mental starter.

N3.1OHP – place value table

R4 – place value tables

R6 – number lines

Introductory activity

Recap basic ideas introduced in Year 7:

▸ Ordering a set of numbers that includes simple negative decimals,

for example 3.2, 7, ⁻1.8. 0.84, ⁻2.3

▸ Multiplying and dividing by 10, 100, 1000,

for example $34 \div 1000$, 0.0027×100

▸ Multiply by $\frac{1}{10}$,

for example what is $\frac{1}{10}$ of £300?

Remind students that the decimal system is based on powers of 10.

Emphasise that each 'column heading' can be written as a power of 10.

Encourage students to extend the table, which is shown on **N3.1OHP**.

Discuss the examples in the student book.

You will need to recap the relationship between multiplication and division:

▸ $\times \frac{1}{10} = \div 10$

▸ $\div \frac{1}{10} = \times 10$

Encourage the use of correct units during workings, particularly where a question requires conversion of units.

In the second example, encourage the use of a number line.

You could use **R6** as an OHT.

This spread will show you how to:

▶▶ Extend your knowledge of integer powers of 10.

▶▶ Multiply and divide by any integer power of 10.

▶▶ Understand the effects of multiplying and dividing by numbers between 0 and 1.

KEYWORDS
Power of 10
Place value

You can write numbers like 0.01 and 1000 as powers of 10.

Number		Power of 10
1000	$10 \times 10 \times 10$	10^3
100	10×10	10^2
10	10	10^1
1	1	10^0
0.1	$\frac{1}{10}$	10^{-1}
0.01	$\frac{1}{10 \times 10}$	10^{-2}

Numbers divided by 10

Powers decreasing by 1

On a calculator, you could input 10^3 like this.

`1 0 xʸ 3`

You can use powers of 10 in a place value table.

10^2(Th)	10^2(H)	10^1(T)	10^0(U)	.	10^{-1}(t)	10^{-2}(h)	10^{-3}(th)
3	9	2	7	.	4	7	3

The digit 9 stands for 9 hundreds, or 9×10^2.

The digit 7 stands for 7 hundredths, or 7×10^{-2}.

Place value helps you to multiply or divide by powers of 10 mentally.

example

Calculate mentally: **a** $500 \text{ kg} \times 0.1$ **b** $30 \text{ m} \div 0.01$

a $\times 0.1$ is the same as $\times \frac{1}{10}$ ... $\Rightarrow 500 \text{ kg} \times 0.1 = 500 \text{ kg} \times \frac{1}{10}$
... is the same as $\div 10$: $= 500 \text{ kg} \div 10$
$= 50 \text{ kg}$

b $\div 0.01$ is the same as $\div \frac{1}{100}$ ... $\Rightarrow 30 \text{ m} \div 0.01 = 30 \text{ m} \div \frac{1}{100}$
... is the same as $\times 100$: $= 30 \text{ m} \times 100$
$= 3000 \text{ m} (= 3 \text{ km})$

Remember (from page 48):
▶ $\times \frac{1}{10} = \div 10$
▶ $\div \frac{1}{10} = \times 10$

To put decimals in order, you need to compare digits working from left to right.

example

Place these numbers in ascending order: 1.76, ⁻2.46, ⁻0.4535, 1.756, ⁻0.5

Compare the first digits: ⁻2.46, ⁻0.4535, ⁻0.5, 1.76, 1.756
Now compare the second digits: ⁻2.46, ⁻0.5, ⁻0.4535, 1.76, 1.756
(be careful with the negative sign)
Finally compare the third digits: ⁻2.46, ⁻0.5, ⁻0.4535, 1.756, 1.76

Plenary

Write on the board:

0.000 001   10 000 000

Ask students to write these as powers of 10.

Conversely, give numbers as powers of 10, such as:

0.5×10^7 45×10^{-9}

Ask students to express them as normal numbers.

Students can work in pairs to explore how a scientific calculator deals with very large and very small numbers.
This will lead into standard form.

Differentiation

Extension questions:
▸ Questions 1–3 focus on mental methods for multiplying and dividing by powers of 10.
▸ Questions 4–8 extend to metric units and use of a calculator.
▸ Question 9 provides a link to later work on standard form.

Core tier: focuses on multiplying by powers of 10.

Exercise N3.1

1 Calculate, using a mental method:
 a $46 \div 100$ b 3.2×1000
 c $29 \div 1000$ d $0.2 \div 10$
 e $0.031 \times 10\,000$ f 1.3×10^2

2 Calculate, using a mental method:
 a 43×0.1 b $4.5 \div 0.01$
 c 234.6×0.01 d $0.06 \div 0.1$
 e 5.7×0.001 f $345 \div 0.01$
 g 0.006×0.1 h $7.44 \div 0.001$

3 Work these out without using a calculator.
 a 5×10^1 b $3 \div 10^3$
 c 98×10^5 d $^-7 \div 0.1$
 e $^-6 \times 10^{-2}$ f $19 \div 1000$
 g 217×10^2 h $^-287 \div 10^2$

4 On a map the scale is 1 : 100 000.
 a A field shown on the map has a real length of 2 km.
 What is its length on the map in centimetres?
 b What would be the actual length of an object that was 3 mm on the map?
 Give you answer in metres.

5 Use the facts in the shaded box to convert these areas into the units shown in brackets.
 $1\,m^2 = 10\,000\,cm^2$
 $1\,cm^2 = 100\,mm^2$
 a $2.74\,m^2$ (cm^2)
 b $4000\,cm^2$ (m^2)
 c $800\,000\,mm^2$ (cm^2)
 d $0.006\,m^2$ (cm^2)
 e $0.56\,cm^2$ (mm^2)

6 Arrange each of these sets of numbers from smallest to largest.
 a $^-10.49, 10.49, ^-0.49, 1.049, 14.09$
 b $^-0.6, 0.6, ^-0.062, 0.162, ^-0.006$
 c $5.6472, 5.648, 0.98276, ^-5.2, 5.7, ^-60$
 d $\frac{7}{8}, 0.72, 0.8171, \frac{2}{3}, \frac{11}{14}, ^-0.63417$

7 Investigation
Ivor Sum uses his calculator to divide the number 10 by each of the numbers, 0.1, 0.2, 0.3, ... 0.9, 1 and records his results

$10 \div 0.1$	$10 \div 0.2$	$10 \div 0.3$	etc.	$10 \div 1$
			etc.	10

 a Copy and complete Ivor's table of results and comment on the answers in the table.
 b Ivor continues the investigation by multiplying 10 by 0, 0.1, 0.2 etc. Set up another table similar to the one above and comment on the answers you get.
 c Copy and complete these sentences.
  ▸ *Multiplying by a number between 0 and 1 ...*
  ▸ *Dividing by a number between 0 and 1 ...*

8 Use your calculator to decide whether these are true or false.
 a $9.0 \times 10^4 = 90 \times 10^2$
 b $0.8 \times 10^{-3} = 800 \times 10^{-6}$
 c $4.0 \times 10^7 = 0.04 \times 10^9$
 d $0.006 = 0.6 \times 10^2$
 e $0.721 \times 10^3 = 721$
 f $0.84 = 0.840 \times 10^0$

9 The three numbers in each row are equivalent. Fill in the boxes in the table.

a	6×10^2	$\Box \times 10^3$	0.06×10^4
b	$\Box \times 10^4$	$0.1 \times \Box^5$	0.01×10^6
c	$12 \times \Box^6$	1.2×10^7	$0.12 \times 10^\Box$
d	0.5	$\Box \times 10^{-1}$	$0.05 \times 10^\Box$
e	1.63	$163 \times 10^\Box$	$16300 \times 10^\Box$
f	$300 \times 10^\Box$	0.03	$0.3 \times \Box^{-1}$
g	10	$10^\Box$	$0.01 \times 10^\Box$
h	$10^\Box$	1	$1000 \times 10^\Box$

101

Exercise commentary

The questions assess the objectives on Framework Pages 37–39 and 57.

Problem solving
Question 7 requires generalisation, assessing the objective on Framework Page 33.

Group work
Question 7 can be attempted in pairs or in small groups.

Misconceptions
Students may add or remove zeros from decimal numbers as a strategy for × and ÷ 10.
Encourage students to concentrate on how the digits move relative to the (fixed) decimal point.

Links
Using scales and converting units links to science and geography.

Homework

N3.1HW is based on multiplying and dividing by powers of 10.

Answers

1 a 0.46 b 3200 c 0.029 d 0.02 e 310 f 130 2 a 4.3 b 450 c 2.346
  d 0.6 e 0.0057 f 34 500 g 0.0006 h 7440 3 a 50 b 0.003 c 9 800 000
  d $^-70$ e $^-0.06$ f 0.009 g 21 700 h $^-2.87$ 4 a 2 cm b 300 m
5 a 27 400 b 0.4 c 8000 d 60 e 56 6 a $^-10.49, ^-0.49, 1.049, 10.49, 14.09$
  b $^-0.6, ^-0.062, ^-0.006, 0.162, 0.6$ c $^-60, ^-5.2, 0.982\,76, 5.6472, 5.648, 5.7$
  d $^-0.634\,17, \frac{2}{3}, 0.72, \frac{11}{14}, 0.8171, \frac{7}{8}$ 7 a 100, 50, 33.333, 25, 20, 16.667, 14.286, 12.5,
  11.111, 10 b 1, 2, 3, 4, 5, 6, 7, 8, 9, 10 c ... makes it smaller; ... makes it bigger
8 a F b T c T d F e T f F 9 a 0.6 b 1, 10 c 10, 8 d 5, 1 e $^-2, ^-4$
  f $^-4, 10$ g 1, 3 h 0, $^-3$

Mental starter

State that a length of 2.1 cm has been measured to the nearest mm. We know this because 0.1 cm = 1 mm.

Write these measurements on the board:
    5.2 litres   6.78 kg   5.4 cm   2.000 kg   7.02 tonnes
To what level of accuracy is each piece of data given?

Useful resources

Mini-whiteboards may be useful for the mental starter and the plenary activity.
R6 number lines may be useful for discussing rounding.
R4 is a place value table.

Introductory activity

Discuss with the students why it is important to know what level of accuracy a measurement is given to.
For example, Ben says the height of his kitchen is 2 m. Is this accurate enough?

Focus on the purpose of the measurement. Is it just an estimate to help Ben judge how tall something else is? Or is it a measurement for a kitchen fitter designing a new kitchen?

Recap work on decimal places from Year 7.

In rounding, some students may find it useful to 'cut off' the decimal with a dotted line and then circle the 'decider'.

Emphasise that in calculations you only round your final answer.

Discuss the examples in the student book.
In the first example, discuss a reasonable approximation of 182.7 kg. Some students may prefer to round it to 200 kg. Recap the conversion from kg to tonnes.
Discuss rounding in the second and third examples. You may wish to use a number line to help you. (You could use R6.)
R4 is a place value table that may also be helpful.

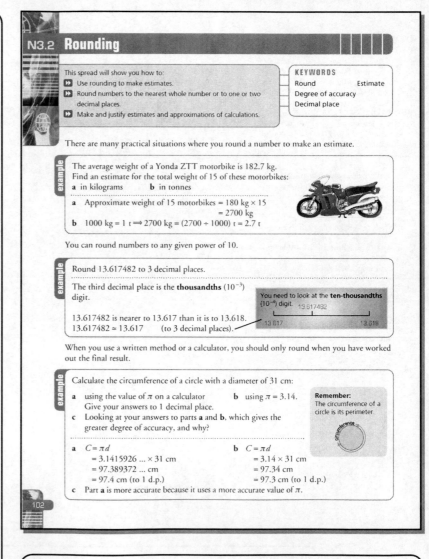

N3.2 Rounding

This spread will show you how to:
▶ Use rounding to make estimates.
▶ Round numbers to the nearest whole number or to one or two decimal places.
▶ Make and justify estimates and approximations of calculations.

KEYWORDS
Round   Estimate
Degree of accuracy
Decimal place

There are many practical situations where you round a number to make an estimate.

example
The average weight of a Yonda ZTT motorbike is 182.7 kg.
Find an estimate for the total weight of 15 of these motorbikes:
a in kilograms   b in tonnes

a   Approximate weight of 15 motorbikes = 180 kg × 15
                                        = 2700 kg
b   1000 kg = 1 t ⟹ 2700 kg = (2700 ÷ 1000) t = 2.7 t

You can round numbers to any given power of 10.

example
Round 13.617482 to 3 decimal places.

The third decimal place is the **thousandths** (10^{-3}) digit.

13.617482 is nearer to 13.617 than it is to 13.618.
13.617482 ≈ 13.617   (to 3 decimal places).

You need to look at the **ten-thousandths** (10^{-4}) digit. 13.617482

13.617   13.618

When you use a written method or a calculator, you should only round when you have worked out the final result.

example
Calculate the circumference of a circle with a diameter of 31 cm:
a   using the value of π on a calculator   b   using π = 3.14.
    Give your answers to 1 decimal place.
c   Looking at your answers to parts a and b, which gives the greater degree of accuracy, and why?

Remember:
The circumference of a circle is its perimeter.

a $C = \pi d$ b $C = \pi d$
    = 3.1415926 ... × 31 cm   = 3.14 × 31 cm
    = 97.389372 ... cm   = 97.34 cm
    = 97.4 cm (to 1 d.p.)   = 97.3 cm (to 1 d.p.)
c   Part a is more accurate because it uses a more accurate value of π.

102

Plenary

Ask students questions such as:
▶ What is the level of accuracy of the measurement 3.87 kg?
▶ What is 2.399 999 rounded to 3 decimal places?
▶ A measurement is given as 4 cm correct to the nearest centimetre. What is the largest and smallest the length could actually be?

Students could consider upper and lower bounds for values and measurements in questions such as:

▸ The population of a country is given as 68 million to the nearest million. What is the least the population could be? What is the greatest the population could be?

Differentiation

Extension questions:
▸ Questions 1, 2, 4, 5, 6 focus on rounding decimals.
▸ Questions 3 and 7 focus on accuracy of measurement.
▸ Questions 8 and 9 focus on rounding the final answer to calculations.

Core tier: focuses on ordering and rounding decimals to one or two decimal places.

Exercise N3.2

1 Round each of these decimals to the number of decimal places indicated in brackets.
  a 3.142 (2 d.p.)   **b** 0.9876 (3 d.p.)
  c 0.09 (1 d.p.)   **d** 0.003 (2 d.p.)
  e 5.8696 (2 d.p.)   **f** 6.997 (1 d.p.)
  g 7.936281 (4 d.p.) **h** 762.199 (2 d.p.)

2 Convert each fraction to a decimal, giving your answer
  i to 2 d.p.   **ii** to 1 d.p.
  iii to the nearest whole number.
  a $11\frac{4}{9}$ **b** $3\frac{5}{11}$ **c** $39\frac{7}{13}$ **d** $46\frac{3}{8}$

3 Ahmed has £1628 in his bank account. Ahmed says that he has nearly £2000. His father says that Ahmed has over £1500.
  a Explain how Ahmed and his father have different answers.
  b Who is the more accurate, Ahmed or his father?

4 Work these out using a calculator. Give your answer correct to 3 decimal places.
  a $\dfrac{34 \times 78}{16}$ **b** $\dfrac{14.6 \times 9.81}{14.8 \times 7.4}$

5 **a** Work out an approximate answer to:
    $\dfrac{45.611 \times 12.845}{3.923}$
  b Now use a calculator to work out the answer, giving your answer:
    i to 3 d.p.
    ii to the nearest whole number.

6 The weight of a footballer is given as 80 kg to the nearest 10 kg. Which one of these five weights could he **not** possibly be?
  78.5 kg, 84.3 kg, 74.9 kg, 80 kg, 81.053 kg

7 Puzzle
Box A contains a set of exact measurements.
Box B contains the same measurements, rounded to 1 decimal place.

Box A	
7.302 cm	14.90 m
14.99 m	25.06 ℓ
17.009 mm	182.33 s

Box B	
7.3 cm	15.0 m
	17.0 mm
182.3 s	25.1 ℓ

Pair each length in Box A with its rounded equivalent in Box B. Hence find the odd one out.

8 Investigation
Harry is trying to work out the circumference of a circle.
He uses the formula $C = \pi \times$ diameter.
He uses three different values for π: 3.1, 3.14 and $\frac{22}{7}$.

Investigate how the different values for π would affect Harry's final answer for different sized circles.

9 Calculate the value of each of these amounts to the accuracy stated in brackets:
  a $\frac{4}{9}$ of $\frac{1}{2}$ of €100 (to the nearest cent)
  b $\frac{2}{3}$ of $\frac{1}{3}$ of $\frac{3}{8}$ of 1 m (to the nearest cm)
  c 40% of $\frac{8}{11}$ of $\frac{1}{2}$ of 500 (to 3 d.p.)
  d $\frac{7}{13}$ of $\frac{6}{12}$ of 32% of 620 (to 2 d.p.)
  e The answers to **c** and **d** multiplied together and rounded to 1 d.p.

103

Exercise commentary

The questions assess the objectives on Framework Pages 43–45.

Problem solving
Question 8 provides an opportunity to assess Framework Page 19. This exercise as a whole assesses Page 31.

Group work
Question 8 is suitable for work in pairs.

Misconceptions
When rounding, students may forget to look at the next digit to see whether they should round up or not.
Problems often arise with 9s, such as in 1h. Emphasise that when you round up from 9, you get 10 and this has a knock-on effect on the previous digit.

Links
Accuracy in measurement: Framework Page 231.
Significant figures, upper and lower bounds: Framework Page 47.

Homework

N3.2HW provides practice at rounding to a given number of decimal places.

Answers

1 **a** 3.14 **b** 0.988 **c** 0.1 **d** 0.00 **e** 5.87 **f** 7.0 **g** 7.9363 **h** 762.20
2 **a i** 11.44 **ii** 11.4 **iii** 11 **b i** 3.45 **ii** 3.5 **iii** 3
  c i 39.54 **ii** 39.5 **iii** 40 **d i** 46.38 **ii** 46.4 **iii** 46
3 **b** Father 4 **a** 165.750 **b** 1.308
5 **a** About 150 **b i** 149.343 **ii** 149
6 74.9 kg 14.90m is the odd one out.
9 **a** €22.22 **b** 0.08 m **c** 72.727 **d** 53.42 **e** 3885.1

N3.3 Adding and subtracting

Mental starter

Write on the board   □ + □ = 10

Ask students to suggest two numbers to go in the boxes that have:

▸ 3 decimal places

▸ a different number of decimal places from each other

▸ 2 decimal places, use all the digits 1 to 8, and the number in the first box must be negative.

Useful resources

N3.3OHP shows the number lines in the student book, to illustrate partitioning and compensation.

Introductory activity

Discuss the addition strategies students used for the starter activity.

Recap on strategies students have previously learned for addition. Encourage students to use vocabulary such as *partitioning, compensating, pairing.*

Discuss partitioning.
Emphasise that you break an number into simpler parts.
Refer to the example in the student book, which is on **N3.3OHP**.

Discuss compensation.
The example in the student book is illustrated on **N3.3OHP**.

Demonstrate the standard column method of addition.
Emphasise that you need to:

▸ line up decimal points

▸ put the numbers you are adding into one group and the numbers you are subtracting into another group, to make the calculation more efficient.

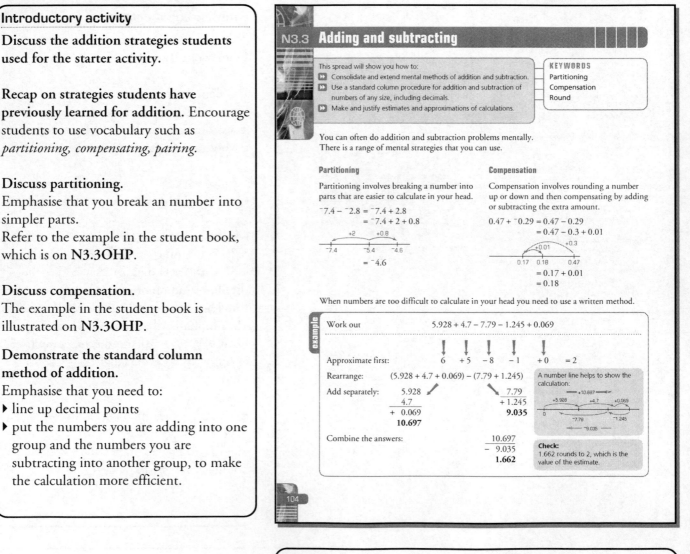

This spread will show you how to:

▸ Consolidate and extend mental methods of addition and subtraction.

▸ Use a standard column procedure for addition and subtraction of numbers of any size, including decimals.

▸ Make and justify estimates and approximations of calculations.

KEYWORDS
Partitioning
Compensation
Round

You can often do addition and subtraction problems mentally. There is a range of mental strategies that you can use.

Partitioning

Partitioning involves breaking a number into parts that are easier to calculate in your head.

$^-7.4 - ^-2.8 = ^-7.4 + 2.8$
$= ^-7.4 + 2 + 0.8$

$= ^-4.6$

Compensation

Compensation involves rounding a number up or down and then compensating by adding or subtracting the extra amount.

$0.47 + ^-0.29 = 0.47 - 0.29$
$= 0.47 - 0.3 + 0.01$

$= 0.17 + 0.01$
$= 0.18$

When numbers are too difficult to calculate in your head you need to use a written method.

Work out $5.928 + 4.7 - 7.79 - 1.245 + 0.069$

Approximate first: 6   + 5   − 8   − 1   + 0   = 2

Rearrange: $(5.928 + 4.7 + 0.069) - (7.79 + 1.245)$

Add separately:
```
    5.928              7.79
    4.7              + 1.245
 +  0.069             9.035
   10.697
```

A number line helps to show the calculation:

Combine the answers:
```
   10.697
 −  9.035
    1.662
```

Check:
1.662 rounds to 2, which is the value of the estimate.

104

Plenary

Discuss the strategies students used for the calculations in the exercise. Encourage students to demonstrate strategies that they find useful.

Highlight how looking for mental strategies as a 'first' resort can reduce the need for written work.

Further activities

Students can devise problems similar to those in question 4 for a partner to solve.

Differentiation

Extension questions:
▸ Question 1 provides practice at adding and subtracting mentally.
▸ Questions 2–6 are problem-solving activities focusing on addition and subtraction.
▸ Question 7 is a harder version of question 3.

Core tier: focuses on mental addition and subtraction with decimals up to 2 decimal places.

Exercise N3.3

1 Work out these using a mental method.
   a $43 + {}^-100$ **b** ${}^-5.0 - {}^-20 + 2.3$ **c** ${}^-8.9 + 79 - 69$ **d** $60 + 7.2 - 80 - 3$
   e $100 - 256 + 7.3$ **f** $62 - 71 + 3$ **g** ${}^-4.7 + 2 - 89$ **h** ${}^-8.2 - 9.6 + 1.7$

2 **a** Work out the missing value for each of the four people's bank accounts:

	Start balance	Transactions	Final balance
Bill	£1321.47	${}^-$£14.40, $+$£10.60, ${}^-$£80.24	
Jenny	£96.28	£14.87, [ ] , ${}^-$£24.00	${}^-$£20.36
Harry	£364.90	[ ] , ${}^-$£700.23, $+$£84.00	£500.60
Louise	£941.40	£26.41, ${}^-$£836.18, [ ]	£37.10

   b Work out the total of the final balances.

3 Puzzle
   Two numbers have a sum of ${}^-5.86$ and a difference of 3.74 when the smaller is subtracted from the larger.
   Work out the two numbers.

4 Place the digits in the clouds in the correct boxes.

5 In this number pyramid, each number is the sum of the two numbers directly above it.
   Copy and complete the pyramid.

6 Puzzle
   Work out the next three numbers in each sequence.
   Explain the pattern in each case.
   a 0.1, 0.2, 0.15, 0.175, 0.1625, 0.16875
   b 0.1, 0.2, ${}^-0.05$, 0.125, ${}^-0.0875$, 0.10625

7 Puzzle
   The difference between two numbers is 7.055.
   The number exactly halfway between the same two numbers is ${}^-1.1875$.
   a What are the two numbers?
   b Explain clearly the method you have used to solve the problem.

105

Exercise commentary

The questions assess the objectives on Framework Pages 93 and 105.

Problem solving
Question 3 provides an opportunity to assess the objective on Framework Page 7.

Group work
Question 5–7 can be attempted in pairs.

Misconceptions
Where a calculation involves both addition and subtraction students can easily get confused.
The strategy of separating the calculation into two group, one for addition and one for subtraction, can help here.
Estimating the answer first also provides a useful check.

Links
Laws of arithmetic: Framework Page 85.

Homework

N3.3HW requires students to demonstrate mental strategies for addition.

Answers

1 **a** ${}^-57$ **b** 17.3 **c** 1.1 **d** ${}^-15.8$ **e** ${}^-148.7$ **f** ${}^-6$
   g ${}^-91.7$ **h** ${}^-16.1$
2 **a** Bill £1237.43, Jenny ${}^-$£107.49, Harry ${}^+$£751.93, Louise ${}^-$£94.53
   b £1754.77
3 ${}^-1.06$, ${}^-4.8$
5 9.104, 4.477, 10.677; 13.268, 13.581, 15.154; 26.849, 28.735
6 **a** 0.165625, 0.1671875, 0.16640625   7 **a** 2.34, ${}^-4.715$

Mental starter

Write on the board:

33×0.03	$360 \div 15$	$252 \div 14$
17×1.6	5.6×31	4.9×0.01
$7.5 \div 0.1$	3.5×4.5	$^-8.4 \times 5$
46×11	39×75	$^-18 \times 6$

Put students into groups. Students volunteer to choose a calculation and describe a mental strategy for solving it. Each suitable strategy gains a point for the group, and the group with most points wins.

Useful resources

Mini-whiteboards may be useful for the mental starter.

Introductory activity

Recap mental strategies for multiplication and division from Year 7. The mental starter calculations cover all these strategies.

Select one of the questions from the starter and discuss the strategies used. Emphasise the variety of strategies available. Discuss which was the most efficient for the chosen question.

Emphasise that there are often several ways of finding an answer.
Students should take note of suggestions from the teacher or other students.

Draw a cuboid on the board and list its dimensions:
$L = 8$ m
$W = 3.5$ m
$H = ?$
$V = 70$ m^3
Give students two minutes in small groups to decide on strategies for calculating the height of this cuboid mentally.
Encourage students to come to the board and describe their strategy.

Discuss the methods referred to in the student book:
‣ factors
‣ partitioning
‣ near 10s
‣ equivalence.

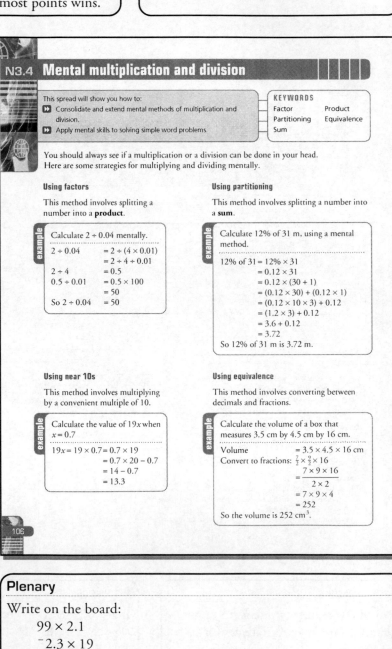

N3.4 Mental multiplication and division

This spread will show you how to:
▷ Consolidate and extend mental methods of multiplication and division.
▷ Apply mental skills to solving simple word problems.

KEYWORDS
Factor   Product
Partitioning   Equivalence
Sum

You should always see if a multiplication or a division can be done in your head. Here are some strategies for multiplying and dividing mentally.

Using factors

This method involves splitting a number into a **product**.

example
Calculate $2 \div 0.04$ mentally.
$2 \div 0.04 = 2 \div (4 \times 0.01)$
$\qquad\quad = 2 \div 4 \div 0.01$
$2 \div 4 \quad = 0.5$
$0.5 \div 0.01 = 0.5 \times 100$
$\qquad\quad = 50$
So $2 \div 0.04 = 50$

Using partitioning

This method involves splitting a number into a **sum**.

example
Calculate 12% of 31 m, using a mental method.
$12\% \text{ of } 31 = 12\% \times 31$
$\qquad\qquad = 0.12 \times 31$
$\qquad\qquad = 0.12 \times (30 + 1)$
$\qquad\qquad = (0.12 \times 30) + (0.12 \times 1)$
$\qquad\qquad = (0.12 \times 10 \times 3) + 0.12$
$\qquad\qquad = (1.2 \times 3) + 0.12$
$\qquad\qquad = 3.6 + 0.12$
$\qquad\qquad = 3.72$
So 12% of 31 m is 3.72 m.

Using near 10s

This method involves multiplying by a convenient multiple of 10.

example
Calculate the value of $19x$ when $x = 0.7$
$19x = 19 \times 0.7 = 0.7 \times 19$
$\qquad\qquad\quad = 0.7 \times 20 - 0.7$
$\qquad\qquad\quad = 14 - 0.7$
$\qquad\qquad\quad = 13.3$

Using equivalence

This method involves converting between decimals and fractions.

example
Calculate the volume of a box that measures 3.5 cm by 4.5 cm by 16 cm.
Volume $= 3.5 \times 4.5 \times 16$ cm
Convert to fractions: $\frac{7}{2} \times \frac{9}{2} \times 16$
$\qquad\qquad\qquad = \frac{7 \times 9 \times 16}{2 \times 2}$
$\qquad\qquad\qquad = 7 \times 9 \times 4$
$\qquad\qquad\qquad = 252$
So the volume is 252 cm^3.

106

Plenary

Write on the board:
99×2.1
$^-2.3 \times 19$
15×3.1
$^-3.4 \div 25$
$2.6 \times \frac{1}{1000}$

Students can find the answers using the mental strategies they have practised in the lesson.

Students can develop question 8 in pairs, by devising their own multiplication trails for a partner to solve.

Differentiation

Extension questions:

▸ Questions 1 and 2 allow students to practise mental strategies without context.

▸ Questions 3–5, are problems in context.

▸ Question 6–8 are harder problems, and include algebra.

Core tier: focuses on mental multiplication and division of simpler numbers.

Exercise N3.4

1 Calculate using an appropriate mental method:
 a 23×11 b $\frac{1}{8} \times 96$
 c 0.1×50 d $12.4 \div 0.1$
 e $0.36 \div 0.01$ f 16×0.01
 g 0.1×0.1 h $12 \times \frac{1}{6}$

2 Calculate using an appropriate mental method:
 a 19×2.1 b $^-2.3 \times 11$
 c 6.7×31 d $^-3.4 \div 0.2$
 e 2.6×2.5 f 29×6.7

3 a Work out the cost of a 'Giant Sweet Bag' weighing 3.6 kg, if sweets cost £1.20 per kg.
 b Work out the cost of a week's supply for Kevin if he eats 5% of a bag per day. (Give your answer to the nearest penny.)

4 Puzzle
 The diameter of a 2 pence piece is 2.6 cm. Use this information to answer these questions.
 a Henry receives a 2p coin from his grandmother each day. He lays them side by side in the hope of making a '1 km line of copper'.
 How many weeks will it take him?
 b Henry has a friend Gary who receives one 2p coin from his grandmother in week 1, two 2p coins in week 2, four in week 3, eight in week 4 and so on. Will Gary make the '1 km line of copper' before Henry or not? Show your working to justify your answer.

5 Calculate the area of the rectangles with the dimensions given. Use a mental method.
 a $2.3 \text{ m} \times 1.5 \text{ m}$ b $250 \text{ cm} \times 30 \text{ cm}$
 c $0.06 \text{ m} \times 0.14 \text{ m}$

6 Work out the value of the expression $\frac{x^2}{2y}$ when x and y take these values:
 a $x = 0.02$ $y = ^-0.1$
 b $x = ^-50$ $y = \frac{1}{100}$
 c $x = 3.2$ $y = (^-0.2)^2$

7 Puzzle
 Work out the value of each letter in the grid using this rule:

4	A	18
B	7.2	C
5	D	E

 Multiply the first two numbers in each row or column to produce the third number, for example $4 \times A = 18$, $A \times 7.2 = D$.

8 In each multiplication trail you must choose numbers, one from each row, which multiply together to give the target number.

3.2	3.4	3.5
3	5	2
4	6	5

96

Example $3.2 \times 5 = 16$
$16 \times 6 = 96$

a

1.3	1.4	1.5
5	10	15
9	8	7

49

b

2.4	1.2	0.6
10	15	30
6	8	7

84

Exercise commentary

The questions assess the objectives on Framework Page 97.

Problem solving

Questions 3 and 4 provide an opportunity to assess the objectives on Framework Page 7.

Group work

For the further activities, students need to work in pairs.

Misconceptions

Students may think that there is only one way to solve a problem, and so may not know where to start.
Emphasise the variety of strategies suitable for given problems to show that there are usually several possible strategies to use.

Links

Laws of arithmetic: Framework Page 85.

Homework

N3.4HW requires students to demonstrate mental strategies for multiplication and division.

Answers

1 a 253 b 12 c 5 d 124 e 36 f 0.16 g 0.01 h 2
2 a 39.9 b $^-$25.3 c 207.7 d $^-$17 e 6.5 f 194.3
3 a £0.72 b £0.25 4 a 5495 weeks b Yes, after 16 weeks.
5 a 3.45 m^2 b 75 cm^2 c 0.0084 m^2
6 a $^-$0.002 b 125 000 c 128
7 A = 4.5, B = 1.25, C = 9, D = 32.4, E = 162
8 a $1.4 \times 5 \times 7$ b $1.2 \times 10 \times 7$

Mental starter

Each student needs a number line from 1 to 10.

Ask students to estimate these square roots to 1 decimal place and mark them on their number lines.

$\sqrt{23}$ $\sqrt{40}$ $\sqrt{78}$ $\sqrt{37}$ $\sqrt{89}$

Students score 1 point for placing the square root between the correct two integers, and a further point if they are within ±0.1 of the exact answer.

Useful resources

Mini-whiteboards may be useful for the mental starter, or **R6** – number lines
Graph or squared paper

Introductory activity

Discuss the strategies students used for the starter, encouraging them to explain their own approaches. Emphasise that $\sqrt{40}$ lies between $\sqrt{36} = 6$ and $\sqrt{49} = 7$.

Write on the board:
3×7.5 61×2.5 48.6×0.0001
3.92×6.2 6.3×6.97

Ask students which of these they would do mentally, and which by a written method.
Encourage the different responses. The method students prefer depends on how confident they are with mental strategies.

Emphasise the importance of looking for mental methods first, to reduce the amount of writing. If a mental method is not possible, you need to know the written method.

Go over the first example in the student book. If necessary, work through a second example to demonstrate the written method, such as 3.92×6.2

Emphasise the need to:
▶ Estimate an answer first, so you can check your final answer for accuracy.
▶ Keep the numbers lined up in columns.
▶ Think carefully about the accuracy required for the final answer.

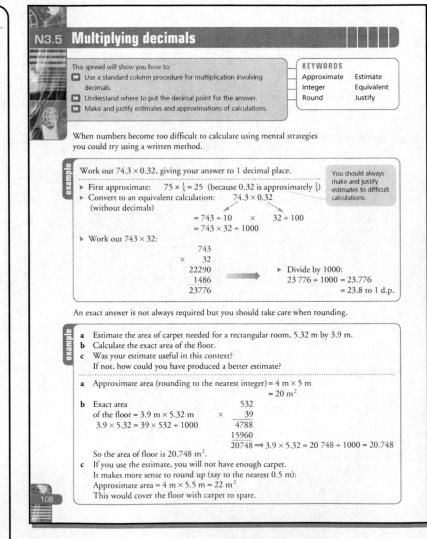

N3.5 Multiplying decimals

This spread will show you how to:
▶ Use a standard column procedure for multiplication involving decimals.
▶ Understand where to put the decimal point for the answer.
▶ Make and justify estimates and approximations of calculations.

KEYWORDS
Approximate   Estimate
Integer   Equivalent
Round   Justify

When numbers become too difficult to calculate using mental strategies you could try using a written method.

example
Work out 74.3×0.32, giving your answer to 1 decimal place.
▶ First approximate: $75 \times \frac{1}{3} = 25$ (because 0.32 is approximately $\frac{1}{3}$)
▶ Convert to an equivalent calculation: 74.3×0.32
(without decimals)
$$= 743 \div 10 \quad \times \quad 32 \div 100$$
$$= 743 \times 32 \div 1000$$
▶ Work out 743×32:

```
     743
  ×   32
   22290
    1486
   23776
```
▶ Divide by 1000:
$23\,776 \div 1000 = 23.776$
$= 23.8$ to 1 d.p.

You should always make and justify estimates to difficult calculations.

An exact answer is not always required but you should take care when rounding.

example
a   Estimate the area of carpet needed for a rectangular room, 5.32 m by 3.9 m.
b   Calculate the exact area of the floor.
c   Was your estimate useful in this context?
If not, how could you have produced a better estimate?

a   Approximate area (rounding to the nearest integer) = 4 m × 5 m
$$= 20 \text{ m}^2$$
b   Exact area
of the floor = 3.9 m × 5.32 m
$3.9 \times 5.32 = 39 \times 532 \div 1000$

```
      532
  ×    39
     4788
    15960
    20748
```
$\Rightarrow 3.9 \times 5.32 = 20\,748 \div 1000 = 20.748$

So the area of floor is 20.748 m².
c   If you use the estimate, you will not have enough carpet.
It makes more sense to round up (say to the nearest 0.5 m):
Approximate area = 4 m × 5.5 m = 22 m²
This would cover the floor with carpet to spare.

108

Plenary

Explore some of the 'oddities' that occur when making estimates and approximating answers. For example:
In what circumstances might you not need an exact answer?
When is it better to always round up the numbers in the calculation, even if this is against the normal rules for rounding?
What approximations are commonly used for 'pi' in mathematics?
How can you tell instantly that $0.08 \times 0.3 \neq 0.24$?

Further activities

Students can invent context questions, like in questions 2 and 3, for a partner to solve.

N3.5ICT involves using a spreadsheet to multiply decimals.

Differentiation

Extension questions:

▸ Question 1 provides simple products, mostly integers, to calculate.
▸ Questions 2–6 focus on multiplying decimals, using estimation and problems in context.
▸ Question 7 links to possibility space diagrams.

Core tier: focuses on multiplying simpler decimals.

Exercise N3.5

1 Calculate using a mental or written method:
   a 39×54 **b** 48×25
   c 397×47 **d** 860×21
   e 109×708 **f** 16.8×53
   g 859×3.8 **h** 12.3×4.6
   i 47.1×2.8

2 **a** Which is larger: 64% of £0.81 or 81% of £0.64?
      Show working to justify your answer and comment on your results.
   b A runner estimates that on average it will take him 72.3 seconds to run each lap of a 400 m track in a 5 km race. Calculate
      i how long it will take him to run the race
      ii his time if he improves his average lap time to 68.3 seconds.
   c A tin of paint states that the contents will cover 8.3 m². Mary buys 14 of these tins and estimates that she will lose 2.5% of the paint washing out the brush each evening. What area will she be able to cover?
   d Work out the volume in m³ of a cuboid that measures
      $0.03 \text{ m} \times 0.12 \text{ m} \times 0.036 \text{ m}$.

3 The manual for Janet's car contains this table for fuel consumption:

30 mph = 9.3 miles per litre
56 mph = 12.5 miles per litre
70 mph = 6.7 miles per litre

   The tank will hold 47 litres of fuel. The current price of fuel for her car is 73p per litre.
   a How much would it cost her to fill the tank?
   b How far could she travel on a full tank at each of the three speeds listed?

4 Calculate using a written method:
   a 84.2×6.3 **b** 113×2.7
   c 15.6×8.2 **d** 56×4.3
   e 6.3×882 **f** 0.57×345

5 Estimate each answer by rounding each number to an appropriate degree of accuracy. You do not need to work out the exact answer.
   a $(3075 \times 498) + 289$ **b** $32^2 \times 469$
   c $\dfrac{97 \times 0.53}{\sqrt{390}}$
   d $1.31 \times (6.4 + 2.76)$

6 Calculate these products using a written or calculator method.
   Estimate the answer to each question before you begin.
   a $^-0.2 \times 0.3214$ **b** $^-0.505 \times 0.120$
   c $2.3 \times (0.12)^2$ **d** $^-16.1 \times 0.032$
   e $(^-0.4)^3 \times 1.7$ **f** $0.00024 \times ^-0.0316$
   g $570\,000 \times 1\frac{1}{5}$ **h** $20.4 \times 0.4 \times ^-0.006$

7 **a** The spinner shown is spun three times.

   The three numbers are then multiplied together.

Number 1	Number 2	Number 3	Product

   Copy and complete the table to show all the possible products that could be achieved. The order of the numbers does not matter.
   b The numbers on the spinner use the digits 1, 2, 3, 4, 5 and 6.
      Place these digits in the boxes to achieve
      i the highest possible product
      ii the lowest possible product.
      $0.\Box\Box \times 0.\Box\Box \times 0.\Box\Box$

Exercise commentary

The questions assess the objectives on Framework Pages 103 and 105.

Problem solving

Question 2(b) and 3 provide an opportunity to assess Framework Page 7. Questions 2(c) and (d) provide an opportunity to assess Framework Page 19. This exercise as a whole assesses Page 31.

Group work

Questions 2 and 3 can be discussed in pairs.

Misconceptions

Remind students that in the written method they need to convert their answer back to a decimal at the end. Encourage the use of estimation whenever calculating with decimals.

Links

Measures: Framework Page 229
Area of 2-D shapes: Framework Page 235

Homework

N3.5HW provides practice at multiplying decimals.

Answers

1 **a** 2106 **b** 1200 **c** 18 659 **d** 18 060 **e** 77 172 **f** 890.4 **g** 3264.2 **h** 56.58
   i 131.88 **2 a** Same **b i** 15 min 3.75 sec **ii** 14 min 13.75 sec **c** 113.3 m²
   d 0.0001296 m³ **3 a** £34.31 **b** 437.1 miles, 587.5 miles, 314.9 miles **4 a** 530.46
   b 305.1 **c** 127.92 **d** 240.8 **e** 5556.6 **f** 196.65 **5 a** 1 500 000 **b** 450 000
   c 2.5 **d** 10 **6 a** $^-0.06428$ **b** $^-0.0606$ **c** 0.03312 **d** $^-0.5152$ **e** $^-0.1088$
   f $^-0.000007584$ **g** 684 000 **h** $^-0.04896$ **7 a** 0.001728, 0.004896, 0.008064,
      0.022848, 0.039304, 0.064736, 0.106624, 0.175616, 0.013872, 0.037632
   b i $0.61 \times 0.52 \times 0.43$ **ii** $0.14 \times 0.25 \times 0.36$

Mental starter

Give students calculations and ask them to show you the remainder.
For example:

354 divided by 7
1000 divided by 19
679 divided by 13
528 divided by 23
2000 divided by 11

Useful resources

Mini-whiteboards may be useful for the mental starter.

Introductory activity

Encourage students to demonstrate the methods they used to find the remainder in the starter.
Discuss the efficiency of these methods and how easy they are to use.
Encourage students to try new methods.

Emphasise that it is much easier to divide by an integer than by a decimal.
Use an OHP of a calculator to show students how division problems can be transformed in this way to make the problem easier, and still produce the correct answer.
For example: 20 ÷ 0.4 is the same as 200 ÷ 4
325 ÷ 0.05 is the same as 32 500 ÷ 5
0.006 ÷ 0.2 is the same as 0.06 ÷ 2

Recap the standard method for division used in Year 7.
Discuss the method used in the examples in the student book. This method is based on repeated subtraction.
The second example requires working to 2 decimal places, as this gives a final answer correct to 1 decimal place.

Encourage students to find an approximate answer first, to check their calculated answer against.
Once students have an exact answer, students can work backwards through the problem as a multiplication 'estimate', for a further check.

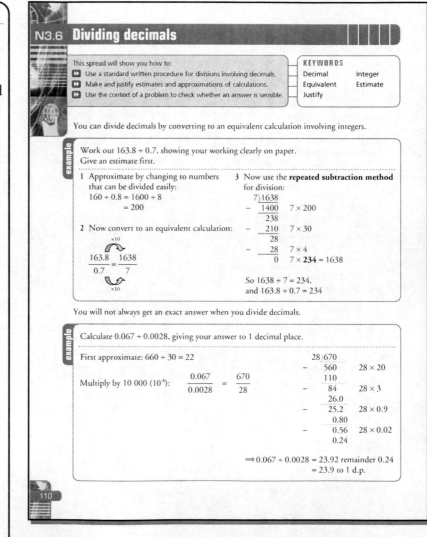

N3.6 Dividing decimals

This spread will show you how to:
▶ Use a standard written procedure for divisions involving decimals.
▶ Make and justify estimates and approximations of calculations.
▶ Use the context of a problem to check whether an answer is sensible.

KEYWORDS
Decimal   Integer
Equivalent Estimate
Justify

You can divide decimals by converting to an equivalent calculation involving integers.

example

Work out 163.8 ÷ 0.7, showing your working clearly on paper.
Give an estimate first.

1 Approximate by changing to numbers that can be divided easily:
160 ÷ 0.8 = 1600 ÷ 8
= 200

2 Now convert to an equivalent calculation:
$$\frac{163.8}{0.7} = \frac{1638}{7}$$
(×10)

3 Now use the **repeated subtraction method** for division:

```
  7)1638
-  1400    7 × 200
    238
-   210    7 × 30
     28
-    28    7 × 4
      0    7 × 234 = 1638
```

So 1638 ÷ 7 = 234,
and 163.8 ÷ 0.7 = 234

You will not always get an exact answer when you divide decimals.

example

Calculate 0.067 ÷ 0.0028, giving your answer to 1 decimal place.

First approximate: 660 ÷ 30 = 22

Multiply by 10 000 (10^4): $\frac{0.067}{0.0028} = \frac{670}{28}$

```
  28)670
-    560    28 × 20
     110
-     84    28 × 3
      26.0
-     25.2   28 × 0.9
       0.80
-      0.56   28 × 0.02
       0.24
```

⟹ 0.067 ÷ 0.0028 = 23.92 remainder 0.24
= 23.9 to 1 d.p.

110

Plenary

Use the following question to assess whether students can find an approximate answer to a division problem and use a standard method:
A piece of paper is 0.086 mm thick. A pile of this paper is 18.92 mm high.
How many sheets of paper are there in the pile?
Calculate an estimate, and then an exact answer.

Further activities

Students can develop question 6 by finding population and area figures for other countries, and calculating the population densities.

Differentiation

Extension questions:
▶ Question 1 involves division of integers.
▶ Questions 2–6 focus on harder divisions, including context.
▶ Questions 7 requires students to give their answer to 4 decimal places.

Core tier: focuses on dividing simpler decimals, using simpler contexts.

Exercise N3.6

1 Work out each of thse divisions using either a mental or a written method.
  a 2710 ÷ 5   b 232 ÷ 4
  c 2583 ÷ 3   d 644 ÷ 14
  e 442 ÷ 17   f 4256 ÷ 28
  g 1089 ÷ 99   h 10 197 ÷ 11

2 Answer each of these problems, showing your working out.
  a A cinema has 26 rows of seats with an equal number of seats in each row. The total number of seats in the cinema is 1274. How many seats are there in each row?
  b Akil's football team has 18 players in its squad. Their total weight is 1422 kg. What is the mean weight of the players?
  c Jenny's car travelled 255 miles and used 17 litres of fuel. How far could she travel with 1 litre?
  d George wanted to share his £37 728 lottery winnings equally between the 24 families in the street. How much should he give to each family?

3 Work out the remainder from each of these divisions.
  a 190 ÷ 4.3   b 70.6 ÷ 2.2
  c 4.83 ÷ 0.4   d 35.26 ÷ 1.3
  e 40 568 ÷ 0.13   f 0.0236 ÷ 0.00018
  g What length of string would be left over if lengths of 1.21 m were cut from a 12 m ball of string?

4 Work out each of these divisions using the repeated subtraction method. Give your answers to 2 decimal places.
  a 140 ÷ 9   b 270 ÷ 17
  c 354 ÷ 11   d 1552 ÷ 3
  e 6410 ÷ 6   f 7302 ÷ 7

5 Work out the missing side of each of these shapes.
Give your answers to 1 d.p.

  a rectangle, area = 32.5 cm², width = 3.4 cm
  b rectangle, area = 106 cm², length = 9.1 cm
  c triangle, area = 50.3 cm², base = 8.4 cm
  d triangle, area = 210.5 cm², vertical height = 12.7 cm

6 The table shows the areas of six countries and their population.

Country	Area (thousand km²)	Population (millions)
Czech Republic	78.866	10.264
Greenland	2175.6	0.05635
Bangladesh	144.0	131.270
UK	244.82	59.648
Australia	7686.85	19.358
Jamaica	10.99	2.666

Calculate the average number of people per square kilometre for each country.

7 Work out each of these divisions, giving your answer to 4 decimal places.
  a ⁻0.056 ÷ 0.11   b 0.00003 ÷ ⁻0.07
  c 0.01 ÷ 9   d ⁻0.0004 ÷ ⁻0.03
  e ⁻1.19 ÷ 0.0019   f 0.1 ÷ ⁻72

111

Exercise commentary

The questions assess the objectives on Framework Pages 103 and 107.

Problem solving
Question 5 provides an opportunity to assess Framework Page 7.
Question 6 provides an opportunity to assess Framework Page 3.
This exercise as a whole assesses Page 31.

Group work
Students could work together on the further activity.

Misconceptions
Students may not know where to stop in long division.
Encourage them to read the question and stop their working when they have the required number of decimal places. This will involve working to one decimal place more.

Links
Area of a rectangle and trapezium: Framework Page 235.

Homework

N3.6HW provides practice at dividing a decimal by a decimal.

Answers

1 a 542   b 58   c 861   d 46   e 26   f 152   g 11   h 927
2 a 49   b 79 kg   c 15 miles   d £1572
3 a 0.8   b 0.2   c 0.03   d 0.16   e 0.07000006   f 0.0002   g 1 m   h 14 g
4 a 15.56   b 15.88   c 32.18   d 517.33   e 1068.33   f 1043.14
5 a 9.6 cm   b 11.6 cm   c 12.0 cm   d 33.1 cm
6 130.14, 0.03, 911.60, 243.64, 2.52, 242.58
7 a ⁻0.5091   b ⁻0.0004   c 0.0011   d 0.0133   e ⁻626.3158   f ⁻0.0014

Mental starter

Two students each take a handful of beads or counters from a bag, and count them.
Write the two numbers on the board as a ratio.
Ask the students to express the ratio in its simplest form.
Repeat with different pairs of students.

Useful resources

Mini-whiteboards may be useful for the mental starter.
Bag of beads or counters
N3.7OHP – comparing ages using ratio (Leon and Naomi)

Introductory activity

Choose ratios from the starter activity and ask students questions about them:
How many times smaller is the first amount than the second?
How many times larger is the second amount than the first?

Write some of the ratios from the starter activity in the form 1 : *m* or *m* : 1.
Emphasise how making one of the numbers equal to '1' makes it easier to see how one quantity compares to the other.

Discuss how to deal with ratios where the numbers are not integers,
For example 3.4 : 2.
Emphasise that you should finish by cancelling the ratio into its simplest form.

Discuss the relationship between the ages of Leon and Naomi in the student book.
This is shown on **N3.7OHP.**
Use the example to emphasise the equivalence of decimals, fractions and percentages.

Discuss the two examples in the student book.
Use them to show how ratio can be used in context.

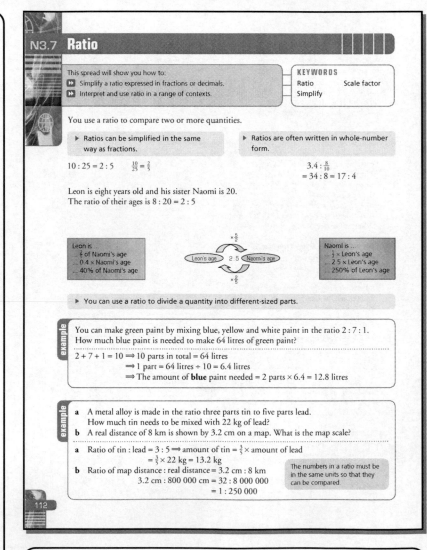

Plenary

Write on the board:

$$a \qquad : b \qquad : c \qquad = 1 : 2 : 5$$
red paint   white paint   pink paint (litres)

Ask questions linked to these.
For example:
▸ How many litres of red paint are needed to make 15 litres of pink paint?
▸ How many times smaller than *b* is *a*?

Further activities

Students can work in pairs to extend question 10. What scale would you choose to design a poster showing a map of the world?

N3.7F contains more ratio problems to solve.

Differentiation

Extension questions:

▶ Questions 1, 6 and 9 focus on simplifying ratios.
▶ Questions 2, 3, 4, 7 and 8 are word problems in context.
▶ Questions 5 and 10 are investigations using ratio.

Core tier: focuses on dividing a quantity in a given ratio.

Exercise N3.7

1 Write each of these ratios both in its whole number form and in its simplest form.
   a 0.5 : 6 **b** $\frac{1}{4}$: 3 **c** 4.2 : 0.7
   d 130 : 17.5 **e** 5 : 6.2 **f** 6 : 3.21
   g 7 : $\frac{1}{8}$ **h** 2$\frac{4}{5}$: 5.7

2 Maximum growth of a particular type of sunflower is achieved by mixing three plant food chemicals A, B and C in the ratio 2 : 6 : 7.
   Use this information to solve these problems:
   a How many grams of chemical A and chemical B would be needed to mix with 30 g of chemical C?
   b What is the weight of each chemical in a tub of plant food containing 80 g of the mixture?

3 The angles of a quadrilateral are in the ratio 3 : 7 : 4 : 10.
   Work out the size of each angle of the quadrilateral.

4 Puzzle
   Dion, Esther and Francis earn £18 000, £25 000 and £32 000 respectively.
   They each receive a 30% pay rise.
   Are their salaries in the same ratios as they were before the pay rise?

5 Investigation
   A map has a scale of 1 : 30 000.
   a What would be the real area corresponding to a square on the map measuring 2.5 cm by 2.5 cm?
   b Repeat part **a** for a map scale of:
   i 1 : 10 000 **ii** 1 : 200
   c Investigate the actual area of a map square measuring 2.5 cm × 2.5 cm for different scales. Write down what you notice.

6 Change these ratios to the forms 1 : m or m : 1.
   a 25 : 40 **b** 8 : 7 **c** 4 cm : 2 m
   d 340 g : 2.5 kg **e** $\frac{1}{6}$: 3 **f** 5.2 : 0.5
   g 56 : 0.7 **h** $\frac{1}{10}$: 3.2 **i** 1$\frac{4}{5}$: 0.6

7 A copper-lead alloy is in the ratio of four parts copper to seven parts lead.
   a If a block of this alloy contains 21 kg of copper, how much lead is there?
   Another block of the alloy weighs 2.64 kg.
   b How much lead is in this block?

8 William and Wilma both like blackcurrant cordial drink. William likes the blackcurrant and water to be mixed in a ratio 3 : 20 but Wilma likes the mixture to be in the ratio 4 : 25.
   a Who prefers the 'stronger' mix (a higher proportion of blackcurrant)?
   b If both children are given a litre of water how much blackcurrant can they make?

9 By converting each ratio in the cloud to the form 1 : m decide which ratio is
   a the smallest
   b the largest
   c the closest to 13 : 120

   2 : 17
   3 : 28 5 : 49 7 : 88
   4 : 38 6 : 59

10 Investigation
   Here are some common map scales,
   1 : 2 000 000; 1 : 200 000; 1 : 25 000;
   1 : 500 and 1 : 200.
   a What distance does 4 cm on the map represent in real life using each of these scales?
   b Investigate some other distances. What kind of map should each scale be used for? Explain and justify your answer.

113

Exercise commentary

The questions assess the objectives on Framework Page 81.

Problem solving
Questions 2, 3, 7 and 8 provide an opportunity to assess Framework Page 5.

Group work
Students could work in pairs or small groups on the investigations in questions 5 and 10.

Misconceptions
Students may confuse the term 'ratio' with 'proportion'.
Reinforce ratio as 'part to part', whereas proportion is 'part to whole'.
You may need to guide students through question 5, particularly in the link between map scale and area.

Links
Enlargement and scale: Framework Pages 212–217.

Answers

1 **a** 1 : 12 **b** 1 : 12 **c** 6 : 1 **d** 52 : 7 **e** 25 : 31 **f** 200 : 107 **g** 56 : 1 **h** 28 : 57
2 **a** 8.6 g, 25.7 g **b** 10.7 g, 32 g, 37.3 g **3** 45°, 105°, 60°, 150° **4** Yes
5 **a** 0.5625 km^2 **b i** 62 500 m^2 **ii** 25 m^2 **c** The area ratio is the length ratio squared.
6 **a** 1 : 1.6 **b** 1$\frac{1}{7}$: 1 **c** 1 : 50 **d** 1 : 7$\frac{6}{17}$ **e** 1 : 18 **f** 10.4 : 1 **g** 80 : 1 **h** 1 : 32
   i 3 : 1
7 **a** 36.75 kg **b** 1.68 kg **8 a** Wilma **b** 1.15 litres, 1.16 litres
9 **a** 2 : 17 **b** 6 : 59 **c** 3 : 28 **10 a** 80 km, 8 km, 1 km, 20 m, 8 m

N3.8 Multiplicative relationships

Mental starter

Are you my equal?
Write on the board:

0.52	0.24	0.15	0.08	0.3
$\frac{6}{25}$	$\frac{3}{20}$	$\frac{2}{25}$	$\frac{3}{10}$	$\frac{13}{25}$
15%	8%	24%	52%	30%

Each student chooses one value and writes it on a mini-whiteboard or piece of paper.
On the word 'equivalence', students get into groups of equivalent values.

Useful resources

Mini-whiteboards are useful for the mental starter.
N3.8OHP – scale factor as a fraction, decimal or percentage
Mail order catalogues

Introductory activity

Discuss ways of tackling this problem:
You score 45 out of 60 in test A and 32 out of 42 in test B.
Which test did you do better in?

Discuss how to deal with fractions that do not cancel, for example $\dfrac{32}{42}$.

Encourage students to demonstrate converting fractions to decimals by division, and decimals to percentages, for example $32 \div 42 = 0.762 = 76.2\%$

Recap on proportion from Year 7, emphasising that proportion expresses the relationship of the part to the whole.

Emphasise that proportion can be expressed in fractions, decimals or percentages.
For test A, 45 out of 60 is $\frac{3}{4}$ or 0.75 or 75%.
Ask students for the fraction, decimal and percentage relating to test B.

Emphasise that when proportions are expressed in the same form they can be compared as ratios.
Ratio of test scores
    A : B
    = 75 : 76

N3.8OHP illustrates the multiplicative relationship between Clare's and Jenny's money.

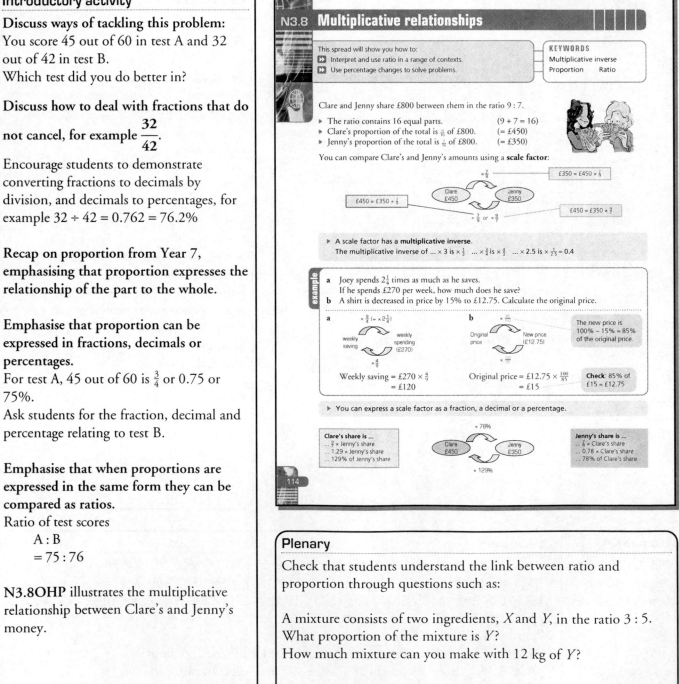

Plenary

Check that students understand the link between ratio and proportion through questions such as:

A mixture consists of two ingredients, X and Y, in the ratio 3 : 5. What proportion of the mixture is Y?
How much mixture can you make with 12 kg of Y?

Further activities

Students can calculate sale prices of 15% or 20% off, for goods in a mail order catalogue.

N3.8F contains further problems involving percentage change.

Differentiation

Extension questions:
Questions 1, 2 and 3 focus on finding percentages and fractions.
Questions 4 and 5 focus on calculating percentage increase and decrease.
Questions 6 and 7 focus on proportion and ratio.

Core tier: focuses on scaling quantities up and down.

Exercise N3.8

1 Calculate
   a $\frac{2}{7}$ of 420 kg   b 23% of £110
   c $1\frac{1}{3} \times$ €520   d $5\frac{1}{8}$ lots of 72 g
   e 16% of 120 litres   f $53 \times \frac{9}{16}$
   g 20 cm $\times 8\frac{2}{5}$ h 81% of 2.4 cm

2 Express the first quantity in each pair:
   i as a fraction of the second quantity, in its simplest form
   ii as a percentage.
   a 180, 150   b 150, 180
   c 2 hours, $1\frac{1}{4}$ hours   d 2.5 kg, 120 g
   e 38 cm, 0.912 m

3 At Mixenmatch High school there are 320 boys and 280 girls.
   Copy and complete this spider diagram.

The number of girls is – – – % of the number of boys
The ratio of boys to girls is – – – :7
Boys : Girls 320 : 280
$\frac{\Box}{\Box}$ of the students are boys
– – – % of the students are girls
There are – – – times as many boys than girls

   Make up some statements of your own.

4 Callum's wage was £8000 per annum until his boss gave him a £3000 per year increase.
   Calculate the percentage increase that Callum received.

5 The table shows the original and sale prices of three Play Station games.

Game	Original price	Sale price
Scooby Doo	£24.99	£18.99
Jet Racer	£19.99	£16.99
FIFA Soccer	£23.99	£19.99

   For each game calculate the percentage decrease (give your answers to the nearest whole numbers).

6 The table shows information about students in Motley High School.

Year	Boys	Girls
7	87	68
8	70	87
9	78	69
10	60	69
11	72	69

   a What proportion of Motley High's population are:
      i girls   ii boys
      iii Year 8 pupils?
   b What is the ratio of boys to girls in each year group?
   c In which year group is the proportion of girls the highest?

7 A window blind comes in three different sizes. 'Super size' and 'Supermax' size are enlargements of the 'Original' size (this means that the length and width have been increased by the same scale factor).

ORIGINAL   SUPER   SUPERMAX

	Original	Super	Supermax
Length	120 cm	150 cm	195 cm
Width		112.5 cm	

   a Copy and complete this table.
   b What is the ratio of the length of the 'Supermax' : length of the 'Original'?
   c What percentage of the width of the 'Supermax' is the width of the 'Original'?

115

Exercise commentary

The questions assess the objectives on Framework Pages 75–79.

Problem solving
Questions 4 and 5 provide an opportunity to assess Framework Page 3. Question 7 provides an opportunity to assess Framework Page 5.

Group work
Students can compare their spider diagrams for question 3.

Misconceptions
Students confuse ratio and proportion. Emphasise that ratio only shows the parts, but proportion compares the parts with the total.

Links
Enlargement and scale: Framework Page 212–217
Scale factors links to work on maps in geography.

Homework

N3.8HW explores the relationship between ratio and proportion in the context of a table.

Answers

1 a 120 kg   b £25.30   c €693.33   d 369 g   e 19.2 litres   f $29\frac{13}{16}$ g 168 cm   h 1.944 cm

2 a i $1\frac{1}{5}$ ii 120%   b i $\frac{5}{6}$ ii 83.3%   c i $1\frac{3}{5}$ ii 160%
   d i $20\frac{5}{6}$ ii 2083.3%   e i $\frac{5}{12}$ ii 41.7%

3 Clockwise from top-right: 8, $\frac{8}{15}$, $1\frac{1}{7}$, 46.7%, 87.5%   4 37.5%   5 24%, 15%, 17%

6 a i $\frac{362}{729}$ ii $\frac{367}{729}$ iii $\frac{157}{729}$ b 87 : 68, 70 : 87, 26 : 23, 20 : 23, 24 : 23   c Year 8

7 a 90 cm, 146.25 cm   b 13 : 8   c 61.5%

Mental starter

Write on the board:

Magic potion – cures students who cannot hand homework in on time.
400 cl olive oil
250 cl bat's blood
0.5 g garlic   Enough to cure 30 students.

Ask students questions such as:

▸ How much of the potion could you make with 100 cl of olive oil?

▸ How much garlic would you need for seven students?

Useful resources

Mini-whiteboards may be useful for the mental starter.

Graph or squared paper

Introductory activity

Write this table on the board:

Hours worked	Pay (£) to nearest penny
1	8.25
0.5	4.13
6	49.50
3.4	28.05
7	57.75
2.8	23.10
0.7	5.78

Emphasise that in this example, pay is in **direct proportion** to hours worked. Encourage students to explain what they think direct proportion means.

Introduce the term 'multiplying factor'. What is the multiplying factor that links pay and wages in the table above?

Encourage students to give other examples of quantities that are in direct proportion, and explain their reasoning.

Remind students that the area of a circle is directly proportional to the square of its radius.

Introduce the symbol for direct proportion
          Area ∝ radius²
This relationship can be written as
          Area = multiplying factor × radius²
Recap the formula for the area of a circle:
$A = \pi r^2$
In this case the multiplying factor is π.

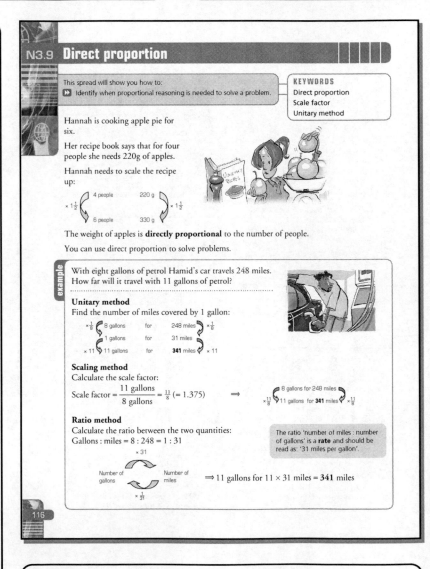

N3.9 Direct proportion

This spread will show you how to:
▸ Identify when proportional reasoning is needed to solve a problem.

KEYWORDS
Direct proportion
Scale factor
Unitary method

Hannah is cooking apple pie for six.

Her recipe book says that for four people she needs 220g of apples.

Hannah needs to scale the recipe up:

$\times 1\frac{1}{2}$ 4 people   220 g $\times 1\frac{1}{2}$
        6 people   330 g

The weight of apples is **directly proportional** to the number of people.

You can use direct proportion to solve problems.

example

With eight gallons of petrol Hamid's car travels 248 miles. How far will it travel with 11 gallons of petrol?

Unitary method
Find the number of miles covered by 1 gallon:

$\times \frac{1}{8}$ 8 gallons   for   248 miles $\times \frac{1}{8}$
      1 gallons   for   31 miles
$\times 11$ 11 gallons for   **341** miles $\times 11$

Scaling method
Calculate the scale factor:
Scale factor $= \dfrac{11 \text{ gallons}}{8 \text{ gallons}} = \frac{11}{8}\ (= 1.375)$ ⟹ 8 gallons for 248 miles
                                                                                        11 gallons for **341** miles

Ratio method
Calculate the ratio between the two quantities:
Gallons : miles = 8 : 248 = 1 : 31
            ×31

The ratio 'number of miles : number of gallons' is a **rate** and should be read as: '31 miles per gallon'.

Number of gallons   Number of miles   ⟹ 11 gallons for 11 × 31 miles = **341** miles
            × $\frac{1}{31}$

116

Plenary

Assess students' understanding of the topic through questions such as:

          W and *F* are directly proportional.
          W = 15 when *F* = 3.
          Write an equation connecting *W* and *F*.
          What is the multiplying factor that links *W* and *F*?

Further activities

Investigate the ratio of kilometres to miles. Calculate the number of miles in 0 km and 80 km. Use these two points to draw a conversion graph for kilometres to miles. Translate speed limits into km/h, for example 30 mph, 40 mph, 60 mph.

Differentiation

Extension questions:
▶ Questions 1, 3, 4 and 5 focus on word problems in context.
▶ Question 6 focuses on scale factor and similarity.
▶ Questions 2 and 7 focus on calculations using direct proportion.

Core tier: focuses on conversion problems.

Exercise N3.9

1 Solve these direct proportion problems.
  a If 12 tickets cost £9 what do five tickets cost?
  b If a car travels 60 miles in 5 hours how far does it travel in $3\frac{1}{2}$ hours?
  c If ribbon costs £0.80 for 2.5 m what does it cost for 32 m?
  d If a meal for nine people costs £62.91 what would a meal for seven people cost?

2 A fruit stall sells produce at these prices.

Calculate the price of
  a 8 kg of pears
  b 2 kg of bananas and 5 kg of pears
  c A bag of fruit with 6 kg of apples containing apples, bananas and pears in the ratio 1 : 2 : 3.

3 Fred says 'I swam 10 lengths in 350 seconds. I must have swum each length in 35 seconds'.
  a Comment on Fred's statement.
  b Is the number of lengths that Fred swims proportional to the time taken? Explain your answer.

4 Kofi pays £16.99 for five reams of A4 paper.
  a How many reams could he buy for £75 and how much change would he receive?
  b If the price was reduced to $\frac{2}{3}$ of its original price what would the answers to **a** be?

5 A car hire firm charges £30 fixed fee and £42 per day for a family saloon car. A family is charged £366 for the hire period.
  a How many days did the family hire the car for?
  Another family hires the same car for half the number of days, but the cost is not half of £366.
  b Have the company made a mistake? Explain and justify your answer.

6 Triangles ABC and ADE are similar. This means their sides are in proportion to each other.

  a What is the scale factor that links the sides of triangle ABC to those of triangle ADE?
  b Write the ratio of the side AB : side AD in its simplest form.
  c Work out the lengths of AC and DE.

7 The table shows the conversions for gallons to litres.

Gallons	Litres
3.?	17
1.5	7
0.6	3
2.2	10

  a Express as a ratio in the form 1 : *n*
    i gallons : litres
    ii litres : gallons
  b Calculate the number of litres in 200 gallons.
  c Construct a conversion graph for gallons to litres
    (use *x*-axis for litres 0 → 50
    y-axis for gallons 0 → 12)
  d Use the graph to work out the number of litres in nine gallons.

117

Exercise commentary

The questions assess the objectives on Framework Pages 79 and 229.

Problem solving
Questions 4 and 5 provide an opportunity to assess Framework Page 5.

Group work
Questions 5–7 can be tackled and discussed in pairs.

Misconceptions
Students may find it difficult to move from the unitary method to using a scale factor. You could illustrate with simple cases:
If 5 miles = 8 km, then $\times \frac{8}{5}$ to convert from miles to km, and $\times \frac{5}{8}$ to convert from km to miles.

Links
Conversion graphs: Framework Page 173
Similarity: Framework Page 193

Homework

N3.9HW consists of contextual problems based on direct proportion.

Answers

1 a £3.33 b 42 miles c £10.24 d £48.93
2 a £2.93 b 72p c £13.32
3 b No
4 a 22 reams, 24p change b 33 reams, 24p change
5 a 8 days b No
6 a 3 b 1 : 3 c 3 cm, 30 cm
7 a i 1 : 5 ii 1 : 0.2 b 900 litres d 40 litres

Summary

The key objectives for this unit are:
▸ Use proportional reasoning to solve a problem, choosing the correct numbers to take as 100% or as a whole.
▸ Make and justify estimates and approximations of calculations.
▸ Give solutions to problems to an appropriate degree of accuracy.

Check out commentary

1 Remind students to read the question carefully and not to assume that any question containing a percentage sign requires them to simply calculate a percentage of something.

Many students will still find the method of unity the most obvious way of solving this problem, although some may begin to realise that problems of this kind can be solved using multiplying factors.

2 Questions 2 and 3 could be used to promote a class discussion revisiting the idea that when working out estimates to problems you need to think carefully about what the result might be used for. The calculations in question 2 have been provided without a context so that a variety of contexts may be suggested.

In question 3, try to encourage real-life examples. For example, medical and engineering contexts require high levels of accuracy, but measurements are rounded even in these cases.
Calculations involving two small integers would not normally be rounded.

Plenary activity

The table shows a sample of 10% of students in a school who were asked their favourite colour of trainer.

Red	Blue	Green
40	50	60

Ask how many students might prefer each colour trainer in a school of 1500 students.

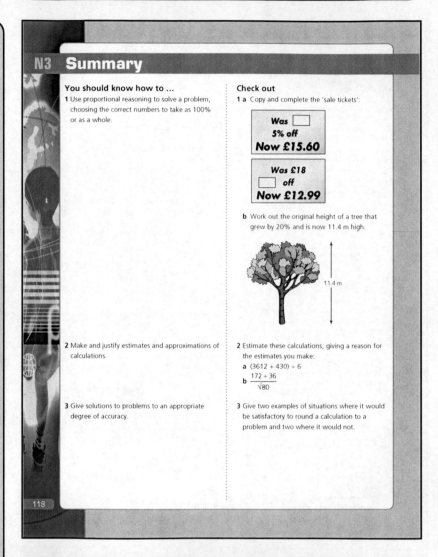

N3 Summary

You should know how to ...
1 Use proportional reasoning to solve a problem, choosing the correct numbers to take as 100% or as a whole.

2 Make and justify estimates and approximations of calculations.

3 Give solutions to problems to an appropriate degree of accuracy.

Check out
1 a Copy and complete the 'sale tickets':

Was ☐
5% off
Now £15.60

Was £18
☐ off
Now £12.99

b Work out the original height of a tree that grew by 20% and is now 11.4 m high.

11.4 m

2 Estimate these calculations, giving a reason for the estimates you make:
a $(3612 + 430) \div 6$
b $\dfrac{172 \div 36}{\sqrt{80}}$

3 Give two examples of situations where it would be satisfactory to round a calculation to a problem and two where it would not.

118

Development

Addition, subtraction, multiplication and division are used throughout mathematics.
A5 addresses proportion problems through algebra.
Ratio and proportion are used in P1.

Links

Addition, subtraction, multiplication and division, and proportional reasoning are used throughout mathematics.
Ratio and proportion are also used in science and geography.

Mental starters

Objectives covered in this unit:
▸ Visualise, describe and sketch 2-D shapes.
▸ Use factors to divide mentally.

Resources needed

* means class set needed
Essential:
S3.1OHP – congruent shapes
S3.2OHP – tessellations
S3.3OHP – combined transformations
S3.4OHP – planes of symmetry
S3.5OHP – negative enlargement
Useful:
R1 – digit cards
R8 – coordinate grid
Mini-whiteboards
S3.6ICT – enlargements

S3　Transformations and congruence

This unit will show you how to:
▸ Understand congruence.
▸ Transform 2-D shapes by combinations of translations, rotations and reflections.
▸ Know that translations, rotations and reflections preserve length and angle.
▸ Identify reflection symmetry in 3-D shapes.
▸ Enlarge 2-D shapes, given a centre of enlargement and a whole-number scale factor.
▸ Recognise that enlargements preserve angle but not length.
▸ Use ratio and proportion to solve problems.
▸ Solve increasingly demanding problems and evaluate solutions.

Reflections can combine to produce fascinating effects.

Before you start

You should know how to ...
1 Transform 2-D shapes.

2 Recognise straight-line graphs parallel to the axes.

3 Recognise 3-D shapes.

4 Simplify ratios.

Check in
1 On a square grid, draw the triangle with coordinates (1, 1), (2, 1) and (2, 4).
  a Reflect it in the y-axis.
  b Rotate it through ⁻90°, centre the origin.
  c Translate it by a vector of $\binom{-3}{2}$.

2 Give the equations of lines A and B.

3 Name these 3-D shapes.
  a　b　c

4 Simplify these ratios.
  a 4:2　b 3:9　c 2 m:10 cm　d 2:3½

119

Unit commentary

Aim of the unit
This unit aims to extend knowledge of transformations, and to link transformations to the concepts of congruence and similarity.

Introduction
Discuss the meaning of the word 'transformation'. Discuss the appearance in real life of:
▸ reflections
▸ rotations
▸ translations
▸ enlargements.

Framework references
This unit focuses on:
▸ Teaching objectives pages: 203–215
▸ Problem-solving objectives pages: 5, 15, 33

Differentiation

Core tier
Focuses on the relationship between transformations and congruence.

Check in activity

Use a 3, 4, 5 triangle on a grid to illustrate:
▸ reflection in a line
▸ rotation about a point
▸ translation up and along
▸ enlargement to a 6, 8, 10 triangle.
Discuss properties of the new shapes.

Mental starter

Using an OHP, discuss the movement of similar shapes on a grid.
Ask the class to describe these transformations.
Review the names of horizontal and vertical lines of reflection.

Useful resources

S3.1OHP shows the diagrams in the student book.
R8 is a blank coordinate grid from ⁻10 to 10.

Introductory activity

Discuss congruent shapes.

Demonstrate that two shapes can be identical in shape and size, but may have a different orientation or be in a different position.

Students should appreciate that two shapes are congruent if their sides and angles correspond. Illustrate with the two triangles in the student book.

All the diagrams in the student book are shown on **S3.1OHP**.

Show how congruent shapes can be mapped onto each other on a coordinate grid by a transformation.

Use the examples in the student book to recap these transformations:
▸ Translation
▸ Reflection
▸ Rotation

Revise the necessary information to fully describe a transformation:
▸ Describe a translation by a vector
▸ Describe a reflection by a mirror line
▸ Describe a rotation by its centre of rotation, angle and direction.

Use a blank coordinate grid as an OHT (you can use **R8**) to draw pairs of congruent shapes.
Ask the class to describe the transformations, using correct terminology.

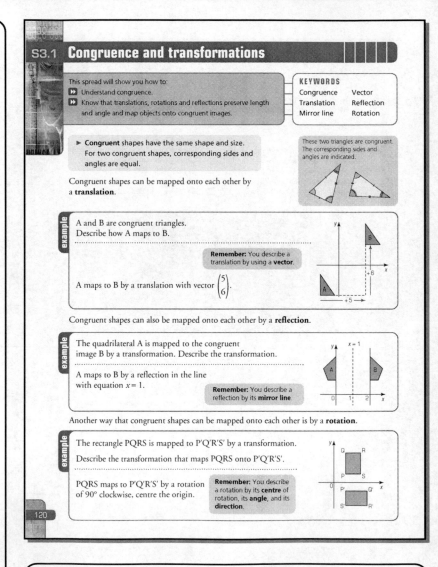

Plenary

Present a two-step transformation and discuss the transformations needed to return it to its starting point.
Can it be done in a single stage?

Further activities

In pairs, students can set each other problems like question 6. Ensure a mixture of reflection, rotation and translation, and use different shapes.

It may help to cut shapes out, to ensure that the images are congruent.

Differentiation

Extension questions:
▸ Questions 1, 2 and 4 consolidate understanding of transformations.
▸ Questions 3 and 5 requires students to identify congruent shapes on a grid.
▸ Question 6 reinforces the link between transformations and congruence.

Core tier: focuses on congruent shapes without a grid.

Exercise S3.1

1

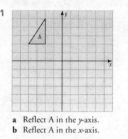

a Reflect A in the *y*-axis.
b Reflect A in the *x*-axis.

2

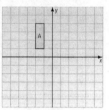

a Rotate A about (0, 0) through 90° in a clockwise direction.
b Rotate A about (0 ,0) through 90° in an anticlockwise direction.

3 Which shapes are congruent?

4 Translate the L-shape by these vectors:

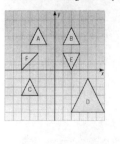

a $\begin{pmatrix} 2 \\ 3 \end{pmatrix}$ b $\begin{pmatrix} 4 \\ 5 \end{pmatrix}$

c $\begin{pmatrix} 1 \\ 0 \end{pmatrix}$ d $\begin{pmatrix} -2 \\ 0 \end{pmatrix}$

e $\begin{pmatrix} 0 \\ 5 \end{pmatrix}$

5 What transformations map A to each of the other shapes?

6 a Which shapes are congruent to A?
   b Describe the transformation that maps A to each of the **congruent** shapes.

Exercise commentary

The questions assess the objectives on Framework Pages 203 and 205.

Problem solving
Question 6 assesses the objectives on Framework Page 15.

Group work
Students should try the further activity in pairs.

Misconceptions
Students commonly draw images incorrectly as a result of a transformation, particularly rotation where the image is often incongruent.
Emphasise that with a reflection, corresponding points should be the same distance from the mirror line.
Encourage the use of tracing paper or cut-out shapes to help.

Links
This topic links to congruence and congruent triangles (Framework Page 191).

Homework

S3.1HW requires students to identify transformations of congruent shapes on a grid.

Answers

3 A, B and C
5 **B** Reflection in *y*-axis   **C** Reflection in *x*-axis   **D** Translation $\begin{pmatrix} -2 \\ 4 \end{pmatrix}$
   E Rotation through 180° about (⁻3, 4.5)
   F Rotation through 180° about (0, 0.5)
6 a B, C, E   b Reflection in *y*-axis, Translation $\begin{pmatrix} -1 \\ 6 \end{pmatrix}$, Rotation through 180° about (0, 2.5)

S3.2 Repeated transformations

Mental starter

Display a coordinate grid on the board or on an OHT (you could use **R8**).

Draw a simple polygon on the grid and ask students to describe the image after various transformations including:

▸ A translation with vector $\binom{3}{4}$

▸ A reflection in the line $x = y$

▸ A rotation of 90° clockwise, centre (1, 1)

Useful resources

R8 may be useful for the mental starter.
S3.2OHP shows how tessellations can be built up systematically.

Introductory activity

Revise tessellations.

Ask students to define the word using accurate terminology, and encourage volunteers to illustrate with examples. Emphasise that a tessellation is composed of **congruent** shapes.

Draw a simple shape on the board, such as a rectangle or square.

Draw a mirror line along one of its edges and ask students to describe:

▸ the image

▸ the resulting complete shape.

Draw another mirror line at right angles to the original one, and draw the new image. Build up the pattern.

Show that you can create a tessellation by repeated reflections.

S3.2OHP contains the diagrams in the student book. You can use this to discuss building up the square pattern.

Draw a scalene triangle on the board.

Ask students to describe the image when it is rotated by 180° about the midpoint of one of its sides.

Discuss how you can build up the pattern into a tessellation. Emphasise the importance of filling space – not just a single row of shapes.

Show that you can create a tessellation by repeated rotations.

S3.2OHP shows the result of rotating an equilateral triangle.

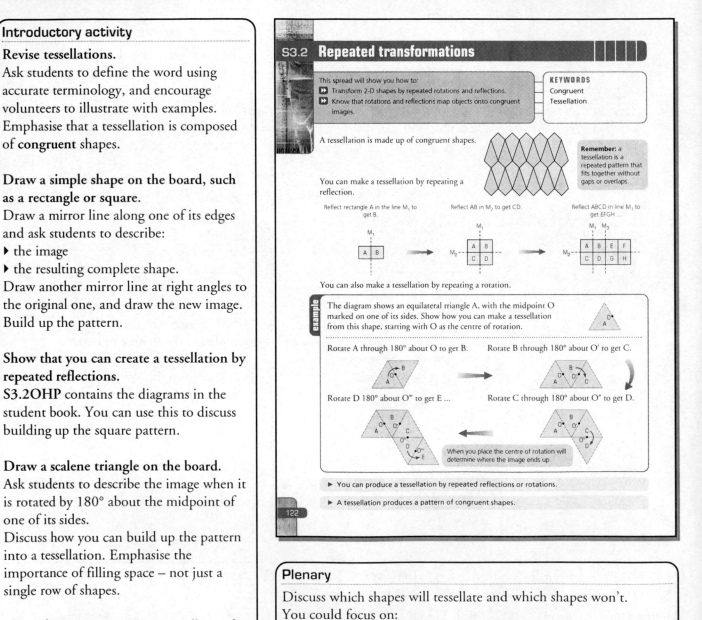

S3.2 Repeated transformations

This spread will show you how to:
▶ Transform 2-D shapes by repeated rotations and reflections.
▶ Know that rotations and reflections map objects onto congruent images.

KEYWORDS
Congruent
Tessellation

A tessellation is made up of congruent shapes.

You can make a tessellation by repeating a reflection.

Reflect rectangle A in the line M_1 to get B. Reflect AB in M_2 to get CD. Reflect ABCD in line M_3 to get EFGH ...

Remember: a tessellation is a repeated pattern that fits together without gaps or overlaps.

You can also make a tessellation by repeating a rotation.

example

The diagram shows an equilateral triangle A, with the midpoint O marked on one of its sides. Show how you can make a tessellation from this shape, starting with O as the centre of rotation.

Rotate A through 180° about O to get B. Rotate B through 180° about O′ to get C.

Rotate D 180° about O‴ to get E ... Rotate C through 180° about O″ to get D.

When you place the centre of rotation will determine where the image ends up.

▶ You can produce a tessellation by repeated reflections or rotations.

▶ A tessellation produces a pattern of congruent shapes.

Plenary

Discuss which shapes will tessellate and which shapes won't.
You could focus on:

▸ A tessellation of equilateral triangles (as in the example). Discuss the angles around a vertex.

▸ An attempt to tessellate regular pentagons. You will need to recap how to calculate interior angles of regular polygons from S1.

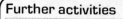

Further activities

Question 2 can be further extended using different shapes, such as a triangle, a pentagon or a hexagon.

Differentiation

Extension questions:
▸ Question 1 practises reflecting simple shapes on a grid.
▸ Question 2 and 3 are investigations based on repeated transformations.
▸ Question 4 explores tessellations.

Core tier: focuses on translations, reflections and rotations.

Exercise S3.2

1

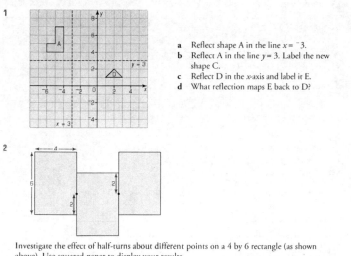

a   Reflect shape A in the line $x = {}^-3$.
b   Reflect A in the line $y = 3$. Label the new shape C.
c   Reflect D in the x-axis and label it E.
d   What reflection maps E back to D?

2

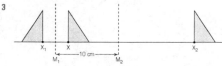

Investigate the effect of half-turns about different points on a 4 by 6 rectangle (as shown above). Use squared paper to display your results.

3

The triangle X is reflected in the mirror line M_1, and the image is X_1.
X_1 is then reflected in the mirror line M_2, and the image is X_2.

Mirror 1 (M_1) and mirror 2 (M_2) are 10 cm apart. $M_1X = 3$ cm. Copy the diagram and measure:
a M_1X_1
b XX_2
c X_1X_2
What do you notice? Investigate.

4   Make your own tessellation patterns using regular polygons.

Exercise commentary

The questions assess the objectives on Framework Page 203.

Problem solving

Questions 2 and 3 assess the objectives on Framework Page 15. Question 3 also assesses Page 33.

Group work

Questions 2 and 3 can be attempted in pairs, and extended to include different centres of rotation and different mirror lines.

Misconceptions

When building up a tessellation, students may find that their shapes are no longer congruent after a few iterations. Encourage students to use simple shapes, and aim to produce no more than ten shapes in their tessellation.
Tracing paper may help with reflections and rotations.

Links

This topic links to congruence (Framework Page 191).

Homework

S3.2HW is an investigation based on reflecting a shape in a series of parallel mirror lines.

Answers

1 d Reflection in x-axis
3 a 3 cm   **b** 20 cm   **c** 26 cm

Mental starter

Discuss what shape the combined object and image will form when:
▶ An equilateral triangle is reflected along one side
▶ A rectangle is rotated through 180° about a vertex
▶ An isosceles trapezium is reflected along its longer parallel side

Encourage students to describe the symmetry of the resulting shape.

Useful resources

R8 is a coordinate grid that can be used with the mental starter.
S3.3OHP shows three combined transformations for a whole-class activity.

Introductory activity

Refer to the mental starter.
Discuss strategies for identifying whether a transformation is a translation, a reflection or a rotation.
Focus on the **orientation** of the image.

Show an object and an image that can only be related to each other by a combined transformation.
S3.3OHP shows three such combined transformations for class discussion:
▶ A rotation followed by a translation
▶ A reflection followed by a translation
▶ A reflection followed by a rotation.
Challenge students to identify each of the transformations.
Encourage sketching to help aid visualisation.

Discuss the order of transformations.
Refer to the first example in the student book, and emphasise that the order of transformations generally does matter (draw an analogy with the order of arithmetical operations).

Demonstrate that a combined transformation can often be described using a single transformation.
This is simplest to demonstrate if you use two of the same type. For example two successive translations can be expressed using a single vector.

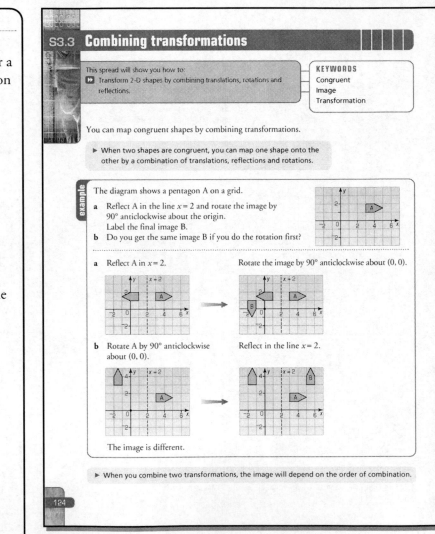

Plenary

Discuss the possibility of a reflection giving the same image as a rotation. When does this occur?
Discuss the possibility of two reflections giving a rotation. When does this occur?

Further activities

Students can investigate transformations applied to solid shapes, in particular:
▸ reflecting a solid in a plane
▸ rotating a solid about an axis.
This will lead into next lesson's work.

Differentiation

Extension questions:
▸ Question 1 practises combining two transformations of a simple shape.
▸ Questions 2–4 focus on the equivalence of combined and single transformations.

Core tier: focuses on simple combined transformations.

Exercise S3.3

Copy the grids in this exercise onto squared paper.

1 A is transformed by a reflection in the x-axis followed by a reflection in the y-axis.

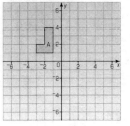

a What single transformation has the same effect?
b Do you get the same image if you do the reflections the other way round?

2 The diagram shows a shape A on a grid.

a Rotate A by 90° clockwise about the origin.
Label the image B.
b Reflect B in the x-axis.
Label the image C.
c What single transformation maps A to C?

3 The diagram shows a shape P on a grid.

a Reflect P in the y-axis.
Label the image Q.
b Translate Q by the vector $\begin{pmatrix} 0 \\ -3 \end{pmatrix}$.
Label the image R.
c Rotate R by 90° clockwise about the point $(1, 2)$.
Label the image S.

4 The diagram shows congruent shapes A to F on a grid.
Find a single transformation that will map:
a A to C	**b** C to D
c D to E	**d** E to F
e A to D	**f** D to F
g C to E.	

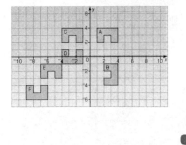

Exercise commentary

The questions assess the objectives on Framework Page 203.

Problem solving
The questions assess the objectives on Framework page 15.

Group work
Students can discuss question 4 in pairs. Can there be more that one possible answer?

Misconceptions
Students may position vertices incorrectly when applying transformations to shapes on a grid.
Also encourage a rough sketch or a mental image as to where a shape is likely to end up, particularly as a result of a combined transformation.
If students are unsure, it may help to label each vertex rather than just the whole shape.

Links
This topic links to congruence (Framework Page 191).

Homework

S3.3HW requires students to identify a variety of combined transformations.

Answers

1 a Rotation through 180° about $(0, 0)$ **b** Yes
2 c Reflection in $y = x$
4 a Reflection in y-axis   **b** Reflection in $y = 1.5$ **c** Rotation through 180° about $(^-4, ^-1)$ **d** Rotation through 180° about $(^-6.5, ^-3.5)$ **e** Rotation through 180° about $(0, 1.5)$ **f** Translation $\begin{pmatrix} ^-5 \\ ^-5 \end{pmatrix}$ **g** Translation $\begin{pmatrix} ^-3 \\ 5 \end{pmatrix}$

Mental starter

Ask students to identify quadrilaterals with 0, 1, 2, or 4 lines of symmetry.
Students can respond using mini-whiteboards, with a sketch of the shape and its name.

Ask: Can you have a quadrilateral with
▸ More than four lines of symmetry
▸ Three lines of symmetry?

Useful resources

Mini-whiteboards may be useful for the mental starter.

S3.4OHP illustrates the planes of symmetry of three solid shapes.

Introductory activity

Recap symmetry.
Briefly discuss symmetry in 2-D shapes, and focus on reflection symmetry.

Extend to 3-D shapes.
Discuss common solid objects that possess symmetry, and encourage students to describe what makes them symmetrical. You can use the example of semi-detached houses.

Define planes of symmetry.
Emphasise that a **plane** of symmetry divides a 3-D shape into two identical halves, and draw the analogy with a **line** of symmetry.

Illustrate the concept with examples of symmetrical solids, such as a cuboid. **S3.4OHP** shows the planes of symmetry of a cuboid, a square-based pyramid and a triangular prism.

Distinguish between the terms 'plane of symmetry' and 'cross-section'.
Use the second example in the student book to show that a triangular prism can have a cross-section that is **not** a plane of symmetry.

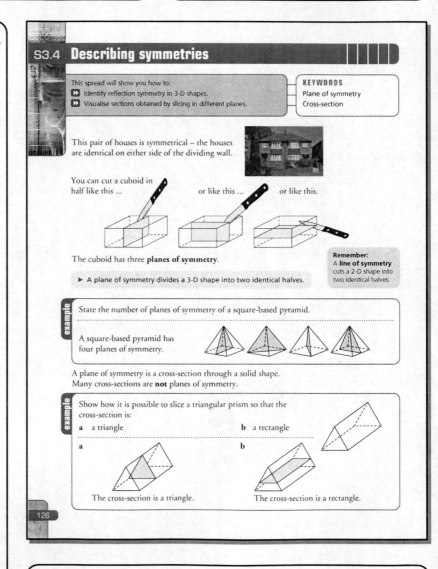

S3.4 **Describing symmetries**

This spread will show you how to:
▶ Identify reflection symmetry in 3-D shapes.
▶ Visualise sections obtained by slicing in different planes.

KEYWORDS
Plane of symmetry
Cross-section

This pair of houses is symmetrical – the houses are identical on either side of the dividing wall.

You can cut a cuboid in half like this ... or like this ... or like this.

The cuboid has three **planes of symmetry**.

▶ A plane of symmetry divides a 3-D shape into two identical halves.

Remember:
A **line of symmetry** cuts a 2-D shape into two identical halves.

example
State the number of planes of symmetry of a square-based pyramid.

A square-based pyramid has four planes of symmetry.

A plane of symmetry is a cross-section through a solid shape.
Many cross-sections are **not** planes of symmetry.

example
Show how it is possible to slice a triangular prism so that the cross-section is:
a a triangle   **b** a rectangle

a   **b**

The cross-section is a triangle.   The cross-section is a rectangle.

126

Plenary

Discuss question 8, in particular what it means to have an infinite number of planes of symmetry. Can students think of another shape that fits this description (sphere).

Discuss question 9, relating to the planes of symmetry of a cube. Encourage students to sketch their ideas.

Further activities

Early finishers can play this game in pairs.

Player 1 draws a symmetrical 3-D solid (keeping it fairly simple), and conceals it. Player 2 must guess the solid, by asking up to ten questions about it (number of planes of symmetry, number of faces ...) Players swap over.

Differentiation

Extension questions:

▶ Questions 1–8 require students to identify the planes of symmetry of various 3-D shapes.
▶ Question 9 extends to the planes of symmetry of a cube.
▶ Question 10 focuses on the different cross-sections of a square-based pyramid.

Core tier: focuses on the symmetries of a 2-D shape.

Exercise S3.4

Draw sketches of these solids showing all their planes of symmetry.

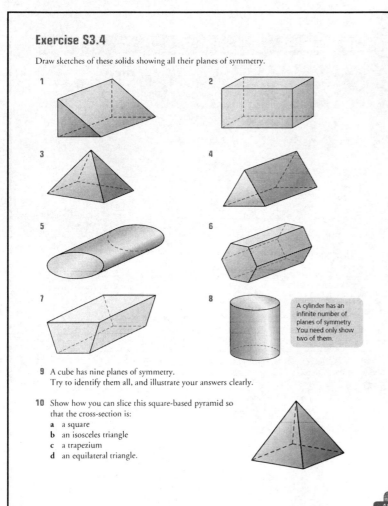

1
2
3
4
5
6
7
8

A cylinder has an infinite number of planes of symmetry. You need only show two of them.

9 A cube has nine planes of symmetry.
Try to identify them all, and illustrate your answers clearly.

10 Show how you can slice this square-based pyramid so that the cross-section is:
 a a square
 b an isosceles triangle
 c a trapezium
 d an equilateral triangle.

127

Exercise commentary

The questions assess the objectives on Framework Page 207.

Problem solving
Question 10 assesses the objectives on Framework Page 15.

Group work
Questions 9 and 10 can be attempted in pairs.

Misconceptions
Students will commonly assume that a diagonal plane through a rectangular cuboid is a plane of symmetry.

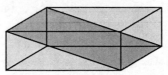

Show that this cannot be the case by taking the lower half of the shape together with the mirror line, and asking what the image will look like.

Links
This topic links to the properties of 3-D shapes (Framework pages 199 and 201).

Homework

S3.4HW is an investigation based on using four cubes to make different solid shapes.

Mental starter

Ask students to describe the dimensions of a given shape when it is enlarged to a shape 2×, 3×, 4× as big.
Use decimal and fraction lengths.

Useful resources

S3.5OHP illustrates the concept of negative enlargement and shows the example in the student book.

Introductory activity

Recap enlargement.
Describe enlargement as a **transformation** because you are making a change to a shape.
Unlike translation, reflection and rotation the image is not **congruent**. However, emphasise that the image is **similar**.

Identify the necessary information for an enlargement. You need a:
▸ centre of enlargement
▸ scale factor.

Highlight the key point with an example: the scale factor of an enlargement is the ratio of any two corresponding lengths.

Extend to negative enlargements.
S3.5OHP shows three different negative enlargements of a triangle.

Encourage students to describe the features of a negative enlargement: the image is reversed and on the opposite side of the centre of enlargement.

Discuss the example in the student book, which shows how to enlarge a shape using a negative scale factor.
The example is also shown on S3.5OHP.

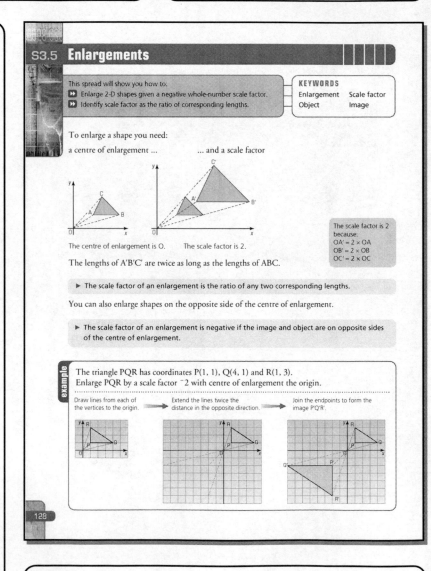

S3.5 Enlargements

This spread will show you how to:
▸▸ Enlarge 2-D shapes given a negative whole-number scale factor.
▸▸ Identify scale factor as the ratio of corresponding lengths.

KEYWORDS
Enlargement   Scale factor
Object   Image

To enlarge a shape you need:
a centre of enlargement ...   ... and a scale factor

The centre of enlargement is O.   The scale factor is 2.

The lengths of A'B'C' are twice as long as the lengths of ABC.

The scale factor is 2 because:
OA' = 2 × OA
OB' = 2 × OB
OC' = 2 × OC

▸ The scale factor of an enlargement is the ratio of any two corresponding lengths.

You can also enlarge shapes on the opposite side of the centre of enlargement.

▸ The scale factor of an enlargement is negative if the image and object are on opposite sides of the centre of enlargement.

example

The triangle PQR has coordinates P(1, 1), Q(4, 1) and R(1, 3).
Enlarge PQR by a scale factor ⁻2 with centre of enlargement the origin.

Draw lines from each of the vertices to the origin. → Extend the lines twice the distance in the opposite direction. → Join the endpoints to form the image P'Q'R'.

128

Plenary

Use question 4 to discuss the reverse of a negative enlargement, and extend to other examples such as the inverse of an enlargement scale factor ⁻2.

Emphasise the definition of scale factor as the ratio of corresponding lengths, and encourage students to introduce fractional scale factors.

Students can play this game in pairs.
Both players have an identical grid with an identical T-shape in the first quadrant.
Player 1 draws an enlarged T-shape and conceals it.
Player 2 must guess the scale factor and centre of enlargement. To enable this, Player 1 reads out the coordinates of the image one by one.
Player 2 scores points by guessing correctly with the minimum number of coordinates.

Differentiation

Extension questions:
▸ Question 1 recaps positive scale factor enlargements.
▸ Questions 2 and 3 focus on negative enlargements.
▸ Question 4 explores the inverse of a negative enlargement.

Core tier: focuses on positive scale factor enlargements.

Exercise S3.5

1 On a pair of axes scaled from ⁻8 to +8 plot ABCD where A is (0, 0), B is (2, 0), C is (2, 4) and D is (0, 3). Enlarge ABCD
   a centre (0, 0), scale factor 2
   b centre (3, 3), scale factor 2.

2 Copy these figures and, using centre O, enlarge them by the scale factor stated (label the image carefully).

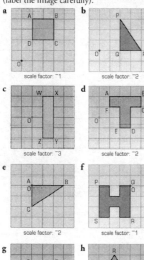

a scale factor: ⁻1
b scale factor: ⁻2
c scale factor: ⁻3
d scale factor: ⁻2
e scale factor: ⁻2
f scale factor: ⁻1
g scale factor: ⁻2
h scale factor: ⁻3

3 Copy these diagrams, mark the centres of enlargement and state the scale factors.

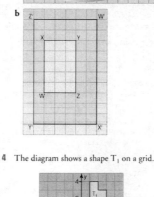

a
b

4 The diagram shows a shape T₁ on a grid.

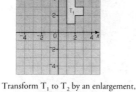

   a Transform T₁ to T₂ by an enlargement, centre (⁻1, 4), scale factor ⁻1.
   b What transformation maps T₂ to T₁?
   c What rotation maps T₁ to T₂?
   d What reflection maps T₁ to T₂?

Exercise commentary

The questions assess the objectives on Framework Page 213.

Problem solving
Question 4 assesses the objectives on Framework Page 15.

Group work
The further activity should be conducted in pairs.

Misconceptions
A common error with enlargements is to position the image incorrectly. Encourage students to check their image by seeing whether the shape is similar, either by checking corresponding angles or lengths.

Links
Enlargement links to ratio and proportion (Framework Page 81) and similarity (Page 193).

Homework

S3.5HW provides practice at both constructing negative enlargements and describing them.

Answers

3 a Scale factor ⁻2   b Scale factor ⁻2
4 b Enlargement centre (⁻1, 4) scale factor ⁻1
   c Rotation through 180° about (⁻1, 4)
   d None

S3.6 Enlargement and ratio

Mental starter

Practise simplifying ratios.
Write down some ratios for students to simplify, for example:
4 : 12, 28 : 16, 24 : 42, 27 : 96

Students can respond using digit cards or mini-whiteboards.

Useful resources

R1 digit cards or **mini-whiteboards** may be useful for the mental starter.

Introductory activity

Review enlargements from the previous lesson.

Discuss how enlargements result in a similar image, but are generally not congruent.

Discuss the special cases where the image is congruent, and encourage students to identify the scale factors 1 and ⁻1.

Emphasise that similarity means that enlargements preserve angle but not length.

Discuss the effect on the perimeter when you enlarge a shape.

Use the first example in the student book, and highlight the use of a **ratio** in describing the increase both in corresponding lengths and perimeter.

Use the example to show that when you enlarge a shape by scale factor k:

▸ All corresponding lengths are multiplied by k.
▸ The perimeter is multiplied by k.

Show how you can describe an enlargement by a percentage.

Use the second example in the student book to illustrate this.

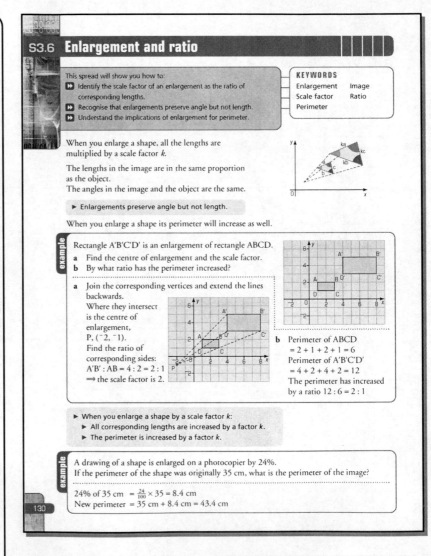

S3.6 Enlargement and ratio

This spread will show you how to:
- Identify the scale factor of an enlargement as the ratio of corresponding lengths.
- Recognise that enlargements preserve angle but not length.
- Understand the implications of enlargement for perimeter.

KEYWORDS
Enlargement   Image
Scale factor   Ratio
Perimeter

When you enlarge a shape, all the lengths are multiplied by a scale factor k.

The lengths in the image are in the same proportion as the object.
The angles in the image and the object are the same.

▸ Enlargements preserve angle but not length.

When you enlarge a shape its perimeter will increase as well.

Rectangle A'B'C'D' is an enlargement of rectangle ABCD.
a   Find the centre of enlargement and the scale factor.
b   By what ratio has the perimeter increased?

a   Join the corresponding vertices and extend the lines backwards.
Where they intersect is the centre of enlargement, P, (⁻2, ⁻1).
Find the ratio of corresponding sides:
A'B' : AB = 4 : 2 = 2 : 1
⟹ the scale factor is 2.

b   Perimeter of ABCD
= 2 + 1 + 2 + 1 = 6
Perimeter of A'B'C'D'
= 4 + 2 + 4 + 2 = 12
The perimeter has increased by a ratio 12 : 6 = 2 : 1

▸ When you enlarge a shape by a scale factor k:
  ▸ All corresponding lengths are increased by a factor k.
  ▸ The perimeter is increased by a factor k.

A drawing of a shape is enlarged on a photocopier by 24%.
If the perimeter of the shape was originally 35 cm, what is the perimeter of the image?

24% of 35 cm $= \frac{24}{100} \times 35 = 8.4$ cm
New perimeter $= 35$ cm $+ 8.4$ cm $= 43.4$ cm

130

Plenary

Discuss the proportions of metric paper sizes.
You could tack a piece of A5, A4 and A3 paper together on the board so that they are joined at the bottom right hand corner, to illustrate the relationship between them.
If students have a piece of A4 and A5 paper, they can confirm that the scale factor of enlargement is roughly 0.7.
This will be explored further in P1.

Further activities

Students can investigate the effect of enlargement on the volume of a cube.

S3.6ICT requires students to enlarge a triangle and quadrilateral using an interactive geometry package. Students explore the relationship between the lengths of corresponding sides.

Differentiation

Extension questions:

▶ Question 1 consolidates the relationship between ratio and scale factor.
▶ Question 2 focuses on the relationship between enlargement and perimeter.
▶ Questions 3 and 4 extend to the relationship between enlargement and area.

Core tier: focuses on enlarging shapes on a coordinate grid.

Exercise S3.6

1   Stamp B is an enlargement of stamp A by a scale factor 3.
   a Calculate the dimensions p and q of stamp B.
   b Calculate the perimeter of
      i stamp A
      ii stamp B
   c Compare the ratio of perimeter A to perimeter B.
      What do you notice?

2   Photo B is an enlargement by scale factor 2 of photo A.
   a What are the lengths x and y?
   b Find the area of photo A and the area of photo B.
      What is the ratio, area A : area B?
   c Compare this with the ratio, perimeter A : perimeter B.
      What do you notice?

3   Triangle X is a scale factor +3 enlargement of triangle Y.
   a Calculate the area of each triangle and hence find the ratio of their areas.
   b Compare the ratio of the perimeters to the ratio of areas.
      What do you notice?

4   The ratio of the perimeters of two shapes is 1 : 5.
   What is the ratio of their areas?

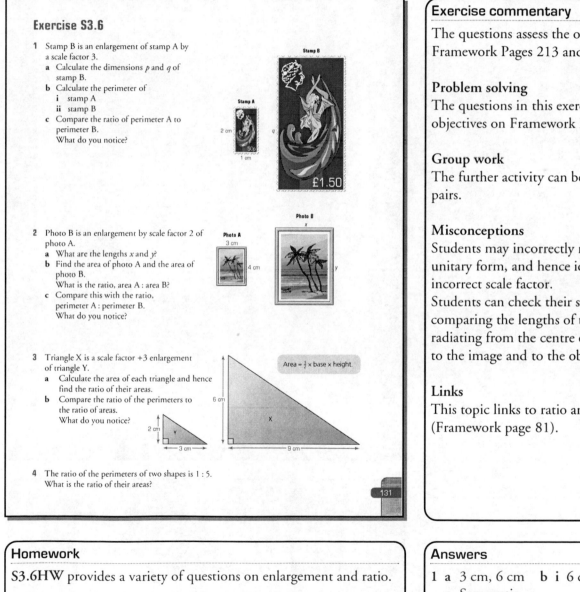

Area = ½ × base × height.

131

Exercise commentary

The questions assess the objectives on Framework Pages 213 and 215.

Problem solving
The questions in this exercise assess the objectives on Framework Pages 5 and 33.

Group work
The further activity can be conducted in pairs.

Misconceptions
Students may incorrectly reduce a ratio to unitary form, and hence identify an incorrect scale factor.
Students can check their scale factors by comparing the lengths of the lines radiating from the centre of enlargement to the image and to the object.

Links
This topic links to ratio and proportion (Framework page 81).

Homework

S3.6HW provides a variety of questions on enlargement and ratio.

Answers

1 **a** 3 cm, 6 cm   **b i** 6 cm   **ii** 18 cm
   c Same ratio
2 **a** 6 cm, 8 cm   **b** 48 cm, 12 cm, 4 : 1
   c Lengths are 1 : 2, areas are 1 : 4
3 **a** 3 cm, 27 cm, 1 : 9
4 1 : 25

Summary

The key objectives for this unit are:
▸ Know that translations, rotations and reflections preserve length and angle and map objects onto congruent images.
▸ Use proportional reasoning to solve a problem.

Plenary activity

Give out a grid board to each student, and a simple shape such as

Students should rotate, translate and reflect the shape according to your instructions.

Check out commentary

1   Check that students are familiar with the term congruent.
    In part **b**, there are two possible answers for the transformation
    A → B.
    However, in cases where the individual vertices are labelled there would be no ambiguity.

    In describing the rotation A → B, ensure that students identify all required information.
    Tracing paper may help in locating the centre of rotation.

2   Students may be unsure when to multiply and when to divide by the scale factor. Encourage students to look at their answer to see if it is sensible.

    Emphasise that the scale factor k relates to length, and the perimeter is also a length.

    Students should know that the ratio of lengths = k:1, and they should have discovered for themselves that the ratio of areas = k^2:1.

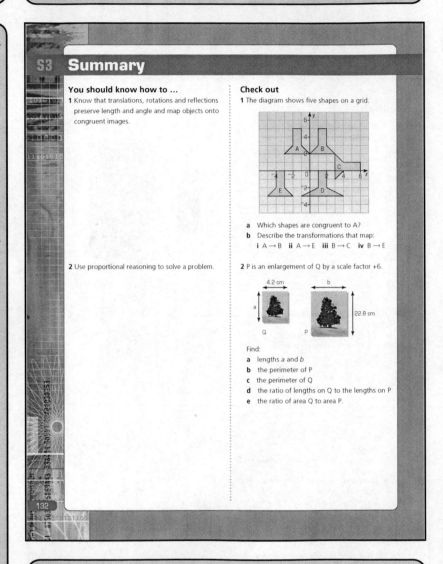

S3 Summary

You should know how to ...
1 Know that translations, rotations and reflections preserve length and angle and map objects onto congruent images.

2 Use proportional reasoning to solve a problem.

Check out
1 The diagram shows five shapes on a grid.

a   Which shapes are congruent to A?
b   Describe the transformations that map:
    i A → B   ii A → E   iii B → C   iv B → E

2 P is an enlargement of Q by a scale factor +6.

Find:
a   lengths a and b
b   the perimeter of P
c   the perimeter of Q
d   the ratio of lengths on Q to the lengths on P
e   the ratio of area Q to area P.

Development

Analysis of 3-D shapes is developed in S4, in the context of plans and elevations.
The link between ratio and enlargement is also developed in S4, in the context of maps and scale drawings.

Links

The concepts of transformation and symmetry can lead to interesting projects in art and design.

Objectives covered in this unit:

▸ Solve equations.
▸ Apply mental skills to solve simple problems.
▸ Know and use squares and square roots.
▸ Calculate using knowledge of multiplication and division facts.

Resources needed

* means class set needed

Essential:
A4.1OHP – Linear equations
A4.2OHP – Algebraic fractions
A4.3OHP – Solving problems
A4.4OHP – Inequalities
A4.5OHP – Formulae

Useful:
Mini-whiteboards
R6 – Number lines
R23 – Inequality cards
A4.6F – rearranging formulae
A4.1ICT – substitution

A4 Equations and formulae

This unit will show you how to:

▸▸ Distinguish the different roles played by letter symbols and equations, identities, formulae and functions.
▸▸ Add simple algebraic fractions.
▸▸ Construct and solve linear equations with integer coefficients, using an appropriate method.
▸▸ Use formulae from mathematics and other subjects.
▸▸ Substitute numbers into expressions and formulae.

▸▸ Derive a formula and, in simple cases, change its subject.
▸▸ Solve increasingly demanding problems and evaluate solutions.
▸▸ Explore connections in mathematics across a range of contexts.
▸▸ Represent problems and synthesise information in algebraic form.
▸▸ Solve substantial problems by breaking them into simpler tasks.

Computers allow you to substitute multiple values into a formula.

Before you start

You should know how to ...

1 Solve simple linear equations.

2 Substitute numbers into a formula.

3 Add and subtract numerical fractions.

Check in

1 Solve these equations:
 a $2x + 6 = 40$
 b $5y + 8 = 3y + 16$
 c $10 - 2z = 6z - 14$

2 a Which formula gives the biggest value of p when $r = 5$ and $t = 2$?
  $p = 3r + 2t$ $p = 10 - t^2$ $p = rt - 4$
 b If $C = 2w + 2l$, what is t when $C = 22$ and $w = 6$?

3 a Evaluate:
  i $\frac{1}{5} + \frac{1}{9}$ ii $\frac{2}{3} - \frac{1}{7}$ iii $3\frac{1}{6} - 2\frac{1}{2}$
 b Find the perimeter of this shape:

133

Unit commentary

Aim of the unit

This unit aims to develop skills in dealing with equations and formulae, and extends to algebraic fractions. Inequalities are introduced in this unit.

Introduction

Discuss the importance of algebra as a means of communicating mathematically and a powerful problem solving tool. Emphasise the importance of improving and extending students' techniques in algebra, to enable them to deal with more difficult problems.

Discuss the difference between an expression, an equation and a formula. Use examples to highlight the differences, such as: $5x + 4 = 6x + 1$, $A = 4w + 1$, $3x - 11$.

Students should identify which is which.

Framework references

This unit focuses on:
▸ Teaching objectives pages: 119, 123, 125, 131, 141 and 143
▸ Problem-solving objectives pages: 9, 27, 29, 35.

The whole unit assesses Page 27.

Check in activity

Let $x = 4$, and write down these two statements:
$3x + 1 = 20 - 2x$ (equation)
$A = 2x^2 + 6$ (formula)

▸ Does the value of x fit the equation?
▸ If not, what value of x would?
▸ What value of A results in the formula?
▸ What other value of x gives the same solution in the formula?

Differentiation

Core tier

Focuses on solving simple linear equations by removing brackets and collecting like terms. The unit extends to substituting integers into formulae, some involving powers.

Mental starter

Draw this spider diagram and ask students to copy it.

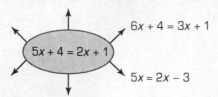

$6x + 4 = 3x + 1$

$5x + 4 = 2x + 1$

$5x = 2x - 3$

Invite students to add on variations of the main equation.

Useful resources

A4.1OHP contains the equations and worked solutions in the student book.
Mini-whiteboards are useful to reinforce inverse operations.

Introductory activity

Refer to the mental starter.
Ask students how they would solve the equation, and discuss strategies.

Extend to equations containing negatives.
The student book contains the equation $10 - 4x = 2$. Give students a minute, and then discuss strategies as a group.

Discuss the methods used by Jalina, Clare and Imran in the student book.
Focus on Clare's method and reinforce the first line of the solution with quick-fire practice, for example:
$4 - 2x = 8 \Longrightarrow 4 = 8 + 2x$
$15 = 3 - 2x \Longrightarrow 15 + 2x = 3$
Mini-whiteboards may be useful.
Complete the solution of $10 - 4x = 2$.
A4.1OHP contains the worked solutions to the equations in the student book.
Reinforce the use of inverse operations, and emphasise that you use addition to move a negative term to the other side.

Extend to fraction terms.
Use the example $\frac{5}{x} + 3 = 7$.
First ask for student suggestions, then show all three methods suggested in the book.
Focus on Jalina's method, which is outlined on **A4.1OHP**. Emphasise that you use multiplication to move a division term to the other side.

Finally, extend to cross-multiplication.
You can use the example in the student book.

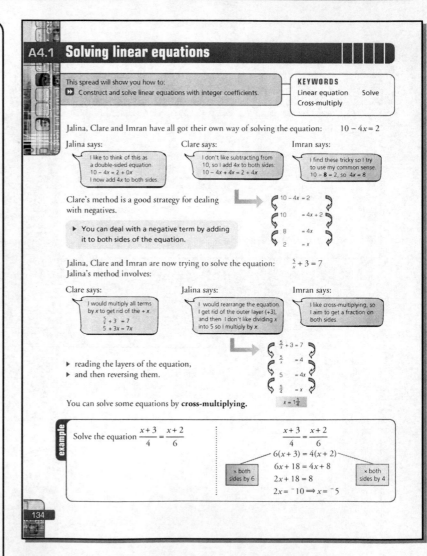

A4.1 Solving linear equations

This spread will show you how to:
▶ Construct and solve linear equations with integer coefficients.

KEYWORDS
Linear equation   Solve
Cross-multiply

Jalina, Clare and Imran have all got their own way of solving the equation: $10 - 4x = 2$

Jalina says:
I like to think of this as a double-sided equation.
$10 - 4x = 2 + 0x$
I now add $4x$ to both sides.

Clare says:
I don't like subtracting from 10, so I add $4x$ to both sides.
$10 - 4x + 4x = 2 + 4x$

Imran says:
I find these tricky so I try to use my common sense.
$10 - 8 = 2$, so $4x = 8$

Clare's method is a good strategy for dealing with negatives.

▶ You can deal with a negative term by adding it to both sides of the equation.

$10 - 4x = 2$
$10 = 4x + 2$
$8 = 4x$
$2 = x$

Jalina, Clare and Imran are now trying to solve the equation: $\frac{5}{x} + 3 = 7$
Jalina's method involves:

Clare says:
I would multiply all terms by x to get rid of the $\div x$.
$\frac{5}{x} + 3 = 7$
$5 + 3x = 7x$

Jalina says:
I would rearrange the equation. I get rid of the outer layer ($+3$), and then I don't like dividing x into 5 so I multiply by x.

Imran says:
I like cross-multiplying, so I aim to get a fraction on both sides.

▶ reading the layers of the equation,
▶ and then reversing them.

$\frac{5}{x} + 3 = 7$
$\frac{5}{x} = 4$
$5 = 4x$
$\frac{5}{4} = x$

You can solve some equations by **cross-multiplying**.

$x = 1\frac{1}{4}$

example

Solve the equation $\frac{x+3}{4} = \frac{x+2}{6}$

$\frac{x+3}{4} = \frac{x+2}{6}$
$6(x+3) = 4(x+2)$
$6x + 18 = 4x + 8$ ← × both sides by 6 / × both sides by 4
$2x + 18 = 8$
$2x = ^-10 \Longrightarrow x = ^-5$

134

Plenary

This is a challenging lesson that needs repetitive reinforcement. Invite students to show that all three of these equations have the same solution:
$10 - 3x = 4$
$\frac{12}{x} + 1 = 7$
$\dfrac{15}{x+3} = \dfrac{9}{x+1}$

Further activities

In pairs, students can invent word problems that practise the main skills of the lesson. They can present their word problems to the class.

A4.1ICT allows students to build a spreadsheet to substitute values into an expression and focusses on trial and improvement.

Differentiation

Extension questions:

▸ Questions 1 and 2 revisit previous skills in one- and two-sided equations.
▸ Questions 3, 4 and 5 use and consolidate the key skills involved in dealing with negative and fractional terms.
▸ Questions 6 and 7 apply the skills to solving problems.

Core tier: focuses on solving linear equations using function machines and extending to a standard technique.

Exercise A4.1

1 Solve these one-sided equations.

  a $3x - 4 = 29$ **b** $\frac{y}{4} + 8 = 14$ **c** $2(a - 4) = 12$ **d** $\frac{p^2 - 9}{10} = 4$

  e $3x + 2(x - 4) = 11$ **f** $\frac{2}{x}y - 2 = 12$

2 Solve these two-sided equations.

  a $5x - 4 = 2x + 17$ **b** $5x - 4 = 3 - x$ **c** $7 - 3x = 5 - 2x$
  d $2(3y - 1) = 3(y - 1)$ **e** $3(y + 2) + 5(y - 4) = 2(y + 5) - (y - 6)$

3 Solve these equations containing negative terms.

  a $10 - 3x = 7$ **b** $5 - 8x = 1$ **c** $6 - 9x = 24$ **d** $^-x - 4 = ^-3$
  e $^-3 - y = ^-5$ **f** $5a - 3(a - 1) = 39$

4 Solve these equations containing fractions.

  a $\frac{2x - 1}{3} = \frac{x}{2}$ **b** $\frac{3y + 2}{4} = \frac{y - 5}{3}$ **c** $\frac{z + 3}{2} = \frac{z - 4}{5}$

  d $\frac{5}{x} = 3$ **e** $^-4 = \frac{9}{y}$ **f** $2 = \frac{18}{x + 4}$

  g $\frac{4}{x} + 5 = 3$ **h** $\frac{9}{y} - 2 = 8$ **i** $5 - \frac{6}{z} = ^-1$

5 Find a pair of equations with matching solutions.

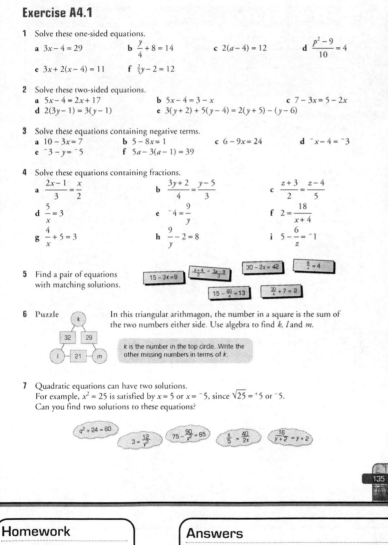

  $15 - 3x = 9$ $\frac{x + 4}{2} = \frac{3x + 9}{3}$ $30 - 2x = 42$ $\frac{8}{x} = 4$ $15 - \frac{90}{x} = 13$ $\frac{30}{x} + 7 = 2$

6 Puzzle   In this triangular arithmagon, the number in a square is the sum of the two numbers either side. Use algebra to find k, l and m.

  k is the number in the top circle. Write the other missing numbers in terms of k.

  (square: 32) (square: 29) (circle: l) (square: 21) (circle: m)

7 Quadratic equations can have two solutions.
  For example, $x^2 = 25$ is satisfied by $x = 5$ or $x = ^-5$, since $\sqrt{25} = ^+5$ or $^-5$.
  Can you find two solutions to these equations?

  $q^2 + 24 = 60$ $3 = \frac{12}{y^2}$ $75 = \frac{90}{x^2} = 65$ $\frac{x}{5} = \frac{40}{2x}$ $\frac{16}{y + 2} = y + 2$

Exercise commentary

The questions assess the objectives on Framework Pages 123 and 125.

Problem solving
Questions 5, 6 and 7 assess the objectives on Framework Page 9.

Group work
Questions 5–7 are good discussion problems and lend themselves to working in pairs.

Misconceptions
Students are often unsure which term to deal with first. A good rule is to subtract the smallest algebraic term from each side. However, this leads to a common mistake with negatives, for example
$10 - 2x = 6x + 5 \Rightarrow 10 = 4x + 5$.
Reinforce the idea that subtracting a negative is the same as adding.
When using cross-multiplication, students need two fractions first. This avoids
$\frac{5}{x} + 3 = 10 \Rightarrow 5 + 3 = 10x$

Links
This topic links to rearranging formulae.

Homework

A4.1HW provides further consolidation of the key techniques, including problems in context.

Answers

1 a $x = 11$ b $y = 24$ c $a = 10$ d $p = 7$ or $^-7$ e $x = 3.8$ f $y = 35$
2 a $x = 7$ b $x = 1\frac{1}{6}$ c $x = 2$ d $y = ^-\frac{1}{3}$ e $y = 4\frac{2}{7}$
3 a $x = 1$ b $x = \frac{1}{2}$ c $x = ^-2$ d $x = ^-1$ e $y = 2$ f $a = 18$
4 a $x = 2$ b $y = ^-5\frac{1}{5}$ c $z = ^-7\frac{2}{3}$ d $x = 1\frac{2}{3}$ e $y = ^-2\frac{1}{4}$ f $x = 5$ g $x = ^-2$
  h $y = \frac{9}{10}$ i $z = 1$
5 $30 - 2x = 42$ and $\frac{30}{x} + 7 = 2$, $15 - 3x = 9$ and $\frac{8}{x} = 4$, $\frac{x + 4}{2} = \frac{3x + 9}{3}$ and $15 - \frac{20}{x} = 13$
6 $K = 20$, $L = 12$, $M = 9$
7 $q = 6$ or $^-6$, $y = 2$ or $^-2$, $x = 3$ or $^-3$, $x = 10$ or $^-10$, $y = 2$ or $^-6$

Mental starter

Discuss this fraction sum: $\frac{1}{7} + \frac{1}{3} = \frac{2}{10}$. Is it true or false?
Encourage students to explain why it is false, and describe how it should be done.

Challenge students to find pairs of fractions that add to $\frac{5}{24}$.
Students can respond using mini-whiteboards.

Useful resources

A4.2OHP contains the examples from the student book.
Mini-whiteboards may be useful for the mental starter.

Introductory activity

Refer to the mental starter.
Emphasise that you need equal denominators before you can add or subtract fractions.

Extend to algebra.
Encourage students to treat algebra fractions in the same way as numerical fractions. Give an example:
$\frac{x}{4} + \frac{x}{3} = \frac{3x}{12} + \frac{4x}{12} = \frac{7x}{12}$
Reinforce the steps:
▸ Make the denominators equal
▸ Collect like terms in the numerator.

Progress to an example with no like terms. For example, $\frac{x}{4} + \frac{y}{3}$ or $\frac{x}{3} + \frac{2y}{7}$ which is in the student book. Emphasise that there is a limit to simplification.

Extend to examples where the algebra is contained in the denominator.
$\frac{10}{x} + \frac{11}{2x} = \frac{20}{2x} + \frac{11}{2x} = \frac{31}{2x}$
Then try the second example in the student book: $\frac{3}{x} - \frac{4}{y}$.

Extend to examples containing mixed numbers.
Start with $3\frac{1}{2}x - 2\frac{1}{3}y = \frac{7x}{2} - \frac{7y}{3}$
$= \frac{21x}{6} - \frac{14y}{6}$
$= \frac{21x - 14y}{6}$
Then try the last example in the student book. All of the examples are on **A4.2OHP**.

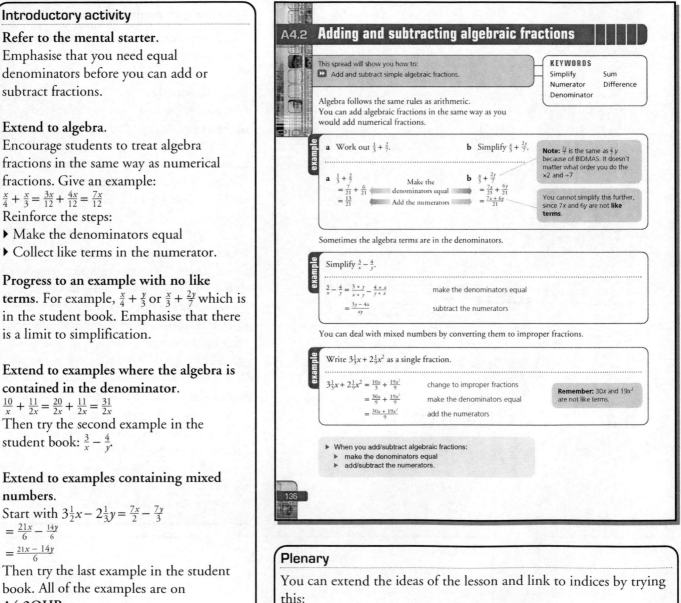

A4.2 Adding and subtracting algebraic fractions

This spread will show you how to:
▸▸ Add and subtract simple algebraic fractions.

KEYWORDS
Simplify   Sum
Numerator   Difference
Denominator

Algebra follows the same rules as arithmetic.
You can add algebraic fractions in the same way as you would add numerical fractions.

example
a Work out $\frac{1}{3} + \frac{2}{7}$.   **b** Simplify $\frac{x}{3} + \frac{2y}{7}$.

a $\frac{1}{3} + \frac{2}{7}$
$= \frac{7}{21} + \frac{6}{21}$ Make the denominators equal
$= \frac{13}{21}$ Add the numerators

b $\frac{x}{3} + \frac{2y}{7}$
$= \frac{7x}{21} + \frac{6y}{21}$
$= \frac{7x + 6y}{21}$

Note: $\frac{2y}{7}$ is the same as $\frac{2}{7}y$ because of BIDMAS. It doesn't matter what order you do the ×2 and ÷7

You cannot simplify this further, since $7x$ and $6y$ are not **like terms**.

Sometimes the algebra terms are in the denominators.

example
Simplify $\frac{3}{x} - \frac{4}{y}$.

$\frac{3}{x} - \frac{4}{y} = \frac{3 \times y}{x \times y} - \frac{4 \times x}{y \times x}$ make the denominators equal

$= \frac{3y - 4x}{xy}$ subtract the numerators

You can deal with mixed numbers by converting them to improper fractions.

example
Write $3\frac{1}{3}x + 2\frac{1}{9}x^2$ as a single fraction.

$3\frac{1}{3}x + 2\frac{1}{9}x^2 = \frac{10x}{3} + \frac{19x^2}{9}$ change to improper fractions
$= \frac{30x}{9} + \frac{19x^2}{9}$ make the denominators equal
$= \frac{30x + 19x^2}{9}$ add the numerators

Remember: $30x$ and $19x^2$ are not like terms

▸ When you add/subtract algebraic fractions:
  ▸ make the denominators equal
  ▸ add/subtract the numerators.

136

Plenary

You can extend the ideas of the lesson and link to indices by trying this:
$\frac{3}{x^2} + \frac{4}{x^7}$

Further activities

Students can practise multiplying and dividing numerical fractions, with a view to exploring how to multiply and divide algebraic fractions.

Differentiation

Extension questions:

▸ Question 1 practises adding and subtracting numerical fractions.
▸ Questions 2, 3 and 4 focus on understanding the processes behind adding and subtracting algebraic fractions.
▸ Questions 5 and 6 apply the skills learned in context.

Core tier: focuses on simplifying linear expressions.

Exercise A4.2

1 Calculate:

a $\frac{2}{5}+\frac{3}{4}$ b $\frac{2}{9}-\frac{1}{6}$ c $\frac{3}{4}\times\frac{1}{7}$ d $\frac{5}{6}\div\frac{1}{2}$ e $3\frac{1}{2}+2\frac{1}{5}$ f $7\frac{1}{3}-2\frac{1}{6}$

2 Simplify:

a $\frac{4}{x}+\frac{2}{x}$ b $\frac{x}{3}+\frac{2x}{3}$ c $\frac{x}{4}+\frac{x}{5}$ d $\frac{2x}{5}-\frac{x}{5}$ e $\frac{1}{5}x+\frac{3}{5}x$ f $\frac{4x}{5}+\frac{x}{4}$

g $\frac{x}{7}-\frac{x}{9}$ h $\frac{5}{x}+\frac{6}{y}$ i $\frac{7}{p}-\frac{2}{p}$ j $\frac{7}{p}-\frac{2}{q}$ k $5\frac{1}{2}x+2\frac{1}{2}x$ l $1\frac{1}{9}x-\frac{2}{7}x$

3 This is what a student wrote. Show that the student was wrong.

> For all numbers a and b:
> $\frac{1}{a}+\frac{1}{b}=\frac{2}{a+b}$

4 Find pairs of matching fraction cards.

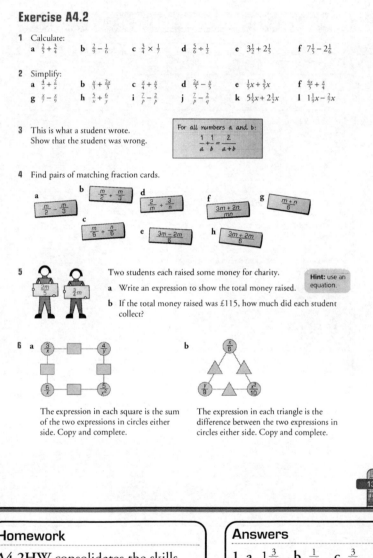

a $\frac{m}{2}-\frac{m}{3}$ b $\frac{m}{2}+\frac{m}{3}$ c $\frac{m}{6}+\frac{n}{6}$ d $\frac{2}{m}+\frac{3}{n}$ e $\frac{3m-2m}{6}$ f $\frac{3m+2n}{mn}$ g $\frac{m+n}{6}$ h $\frac{3m+2m}{6}$

5 Two students each raised some money for charity.

a Write an expression to show the total money raised.

b If the total money raised was £115, how much did each student collect?

[Left figure label: $\frac{2m}{5}$; right figure label: $\frac{3}{4}m$]

Hint: use an equation

6 a [square diagram with $\frac{3}{x}$, $\frac{4}{y}$, $\frac{6}{x}$, $\frac{5}{x}$]   b [triangle diagram with $\frac{x}{6}$, $\frac{x}{9}$, $\frac{x}{10}$]

The expression in each square is the sum of the two expressions in circles either side. Copy and complete.

The expression in each triangle is the difference between the two expressions in circles either side. Copy and complete.

137

Exercise commentary

The questions assess the objectives on Framework Page 119.

Problem solving

Question 5 assesses the objectives on Framework Page 27.
Question 6 assesses Page 9.

Group work

Questions 3, 4, 5 and 6 can be attempted in small groups.

Misconceptions

Students may try to simplify an expression too far, for example $\frac{x+y}{6}=\frac{xy}{6}$. Reinforce like and unlike terms.
Students often perform this loop without realising it:
$\frac{x}{2}+\frac{y}{3}=\frac{3x}{6}+\frac{2y}{6}=\frac{x}{2}+\frac{y}{3}$
Encourage students to check over their working at the end.

Links

These techniques can be used to solve equations containing algebraic fractions.

Homework

A4.2HW consolidates the skills learned and provides a link to equations.

Answers

1 a $1\frac{3}{20}$ b $\frac{1}{18}$ c $\frac{3}{28}$ d $1\frac{2}{3}$ e $5\frac{7}{10}$ f $5\frac{1}{30}$

2 a $\frac{6}{x}$ b x c $\frac{9x}{20}$ d $\frac{x}{5}$ e $\frac{4x}{5}$ f $\frac{21x}{20}$ g $\frac{2x}{63}$ h $\frac{5y+6x}{xy}$ i $\frac{5}{p}$ j $\frac{7q-2p}{pq}$

k $8x$ l $\frac{52x}{63}$

3 Should be $\frac{(b+a)}{ab}$

4 $\frac{m}{2}-\frac{m}{3}=\frac{3m-2m}{6}$, $\frac{m}{2}+\frac{m}{3}=\frac{3m+2m}{6}$, $\frac{m}{6}+\frac{n}{6}=\frac{m+n}{6}$, $\frac{2}{m}+\frac{3}{n}=\frac{3m+2n}{mn}$

5 a $\frac{23m}{20}$ b £40, £75

6 a $\frac{3y+4x}{xy}$, $\frac{9}{x}$, $\frac{4x^2+5y}{x^2y}$, $\frac{6x+5}{x^2}$ b $\frac{3x-2y}{18}$, $\frac{5x-3x^2}{30}$, $\frac{10y-9x^2}{90}$ [answers may be negative of these]

Mental starter

Organise students into groups of five or six, each with an A3 sheet and marker pens. Instruct groups to write as many equations as possible, each with solution $x = \frac{1}{2}$.

Share some of these as a class, and select equations to recap solving single- and double-sided equations.

Useful resources

A4.3OHP contains the examples in the student book.

Introductory activity

Introduce the scenario:

Charlie makes three equal jumps, then runs 2 metres, making a total of $9\frac{1}{2}$ metres. How big are Charlie's jumps?

Allow students time to guess the answer, then show how you can **construct and solve an equation**.

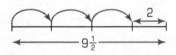

Let one jump be x metres.

Then $3x + 2 = 9\frac{1}{2}$, and proceed to solve the equation.

Extend to problems involving double-sided equations.

Emily does two jumps and runs 5 m. Charlie does 3 jumps and runs 1 m. They both cover the same distance and their jumps are the same size. How big are their jumps?

▸ Construct an equation: $2x + 5 = 3x + 1$
▸ Solve it: $x = 4$

You can adapt the 'jumping' scenario to increasing complexity.

Show how algebra can be used to solve problems in shape.

A4.3OHP contains the examples in the student book. Emphasise that you may need to use facts that are not given explicitly, such as properties of angles and shapes.

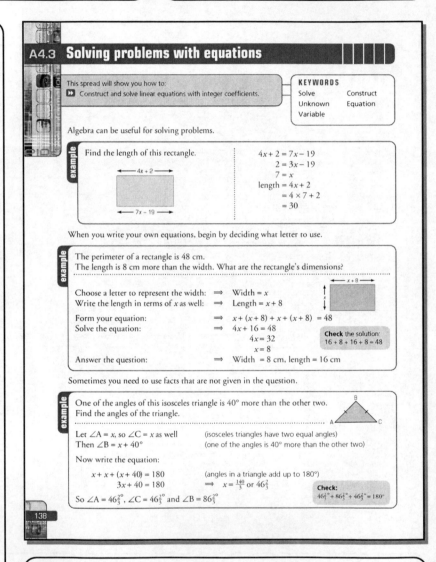

A4.3 Solving problems with equations

This spread will show you how to:
▶▶ Construct and solve linear equations with integer coefficients.

KEYWORDS
Solve   Construct
Unknown   Equation
Variable

Algebra can be useful for solving problems.

example
Find the length of this rectangle.

$4x + 2 = 7x - 19$
$2 = 3x - 19$
$7 = x$
length $= 4x + 2$
$= 4 \times 7 + 2$
$= 30$

When you write your own equations, begin by deciding what letter to use.

example
The perimeter of a rectangle is 48 cm.
The length is 8 cm more than the width. What are the rectangle's dimensions?

Choose a letter to represent the width: ⇒ Width $= x$
Write the length in terms of x as well: ⇒ Length $= x + 8$
Form your equation: ⇒ $x + (x + 8) + x + (x + 8) = 48$
Solve the equation: ⇒ $4x + 16 = 48$
$4x = 32$
$x = 8$
Answer the question: ⇒ Width $= 8$ cm, length $= 16$ cm

Check the solution:
$16 + 8 + 16 + 8 = 48$

Sometimes you need to use facts that are not given in the question.

example
One of the angles of this isosceles triangle is 40° more than the other two.
Find the angles of the triangle.

Let ∠A $= x$, so ∠C $= x$ as well   (isosceles triangles have two equal angles)
Then ∠B $= x + 40°$ (one of the angles is 40° more than the other two)

Now write the equation:
$x + x + (x + 40) = 180$ (angles in a triangle add up to 180°)
$3x + 40 = 180$ ⇒ $x = \frac{140}{3}$ or $46\frac{2}{3}$

Check:
$46\frac{2}{3}° + 86\frac{2}{3}° + 46\frac{2}{3}° = 180°$

So ∠A $= 46\frac{2}{3}°$, ∠C $= 46\frac{2}{3}°$ and ∠B $= 86\frac{2}{3}°$

13B

Plenary

Pose this problem:
Each bag contains some 10p and some 50p coins.

Bag A contains the same number of 50p's as Bag B has 10p's. Bag A has 30p in 10p coins, and Bag B has three 50p coins.
If Bag A contains 40p more than Bag B, how much is in each bag?

Further activities

Students could invent some more 'jumping' scenarios for a partner to solve. For example:

Emily makes 6 jumps whereas Charlie makes 3 jumps and runs 2 m, but is still 7 m behind Emily $(6x - 7 = 3x + 2)$.

Differentiation

Extension questions:

▸ Question 1 uses a simple context to allow students to construct and solve an equation.
▸ Questions 2–7 focus on constructing and solving equations, and include the recall of knowledge of other mathematical concepts.
▸ Questions 8–10 are set in more challenging contexts.

Core tier: focuses on balancing linear equations.

Exercise A4.3

1 In a Hot Cross, the horizontal and vertical lines have the same total. Find the value of the unknown in these Hot Crosses.

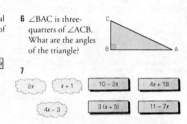

a
	5	
-11	x	2x
	6	

b
	7y	
-4	3y	18
	-6	

c
	7	
10z	-5z	2z
	3	

d
	2	
3	10	-6x
	5	

2 Try these 'Think of a number' problems by forming an equation.
 a I think of a number. When I treble it and add 6 I get the same answer as when I multiply it by 7 and subtract 14.
 b I think of a number. When I double it and subtract it *from* 15 I get the same answer as when I treble it and add 4.
 c I think of a number. I get 30 when I subtract four lots of it from 50.
 d I think of a number. When I divide 30 by this number I get 45.
 e I think if a number. When I add 5 to its reciprocal I get 11.

3 Ben and Vicky have £250 between them. Ben gives Vicky £50. Vicky now has four times as much money as Ben. Use an equation to find how much money each person had to start with.

4 The area of the L-shape is equal to the area of the square that has been removed. What length square has been removed?

5

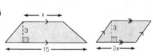

The areas of the shapes are equal. What is this area?

6 ∠BAC is three-quarters of ∠ACB. What are the angles of the triangle?

7

2x   x + 1   10 – 2x   4x + 18

4x – 3   3 (x + 5)   11 – 7x

The mean of the expressions in the clouds is the same as the mean of the expressions in the boxes. What numbers are in each cloud and box?

8 Sarah is 11 years younger than Jan who, in turn, is eight years younger than Pauline. Between them, they have been alive 197 years. How old is Jan?

9 The angle at the centre of a regular polygon is dependent on the number of sides.
 a Write an expression for the central angle in an *n*-sided polygon.
 b If the central angle is 15°, how many sides does the polygon have?

10 a Six times the reciprocal of a number is 18. What is the number?
 b When 7 is added to 10 times the reciprocal of a number, you get 12. What is the number?
 c When 8 times the reciprocal of a number is subtracted from 13, you get 3. What is the number?

Exercise commentary

The questions assess the objectives on Framework Pages 123 and 125.

Problem solving

All of the questions in this exercise provide the opportunity to represent problems in algebraic form, assessing the objectives on Framework Page 27. Questions 3, 4, 5, 6 and 7 also assess Page 29.

Group work

Students should be encouraged to work in pairs for the whole exercise.

Misconceptions

Beware of too many variables in constructing equations. In question 8 students may write: $x + y + z = 197$. Encourage students to use a single variable. When writing x is 4 greater than y, students may write $x + 4 = y$. Encourage students to try out statements with actual numbers, for example $11 = 7 + 4$ since 11 is 4 greater than 7.

Links

These techniques link to all problem-solving work where algebra is the best resort.

Homework

A4.3HW consolidates all equation skills in 'think of a number' contexts, and provides further problems to solve using algebraic techniques.

Answers

1 a $x = 5\frac{1}{2}$ **b** $y = 2\frac{6}{7}$ **c** $z = \frac{5}{6}$ **d** $x = \frac{-2}{3}$
2 a 5 **b** 2.2 **c** 5 **d** $\frac{2}{3}$ **e** $\frac{1}{6}$
3 Ben £50, Vicky £200   **4** 6   **5** 30
6 38.6°, 51.4°, 90°   **7** 10, 6, 17; 0, 38, 30, ⁻24
8 66 years   **9 a** $\frac{360°}{n}$ **10 a** $\frac{1}{3}$ **b** 2 **c** 0.8

Mental starter

Recap the four inequality signs, ≤, ≥, < and >.
Write down these pairs of mental calculations:
10% of 360, and 20% of 230 $\qquad \frac{1}{9} + \frac{1}{7}$, and $\frac{2}{3} - \frac{2}{5}$
13^2 and $12 \times 14 \qquad\qquad 0.3^2$ and 0.2×0.5
$n^2 + 2$ and $2n^2 - 10$, where $n = 4$
Students respond with inequality cards (you can use **R23**).

Useful resources

A4.4OHP shows the number lines in the student book.
R6 number lines are useful for demonstrating inequalities.
R23 inequality cards may be useful for the mental starter.

Introductory activity

Invite students to think of real-life situations involving inequalities. Offer suggestions if students are unsure, such as 'A lift can only take up to 12 people'. Guide students in abbreviating these statements, for example $p \le 12$, where p is the number of people in the lift.

Extend to statements involving two inequalities, for example:
Entrants to a beauty contest must be between 18 and 25 years of age ($18 < a < 25$).

Illustrate how to show inequalities on a number line.
You can use the example of $18 < a < 25$.

```
  18  19  20  21  22  23  24  25
  ○───┼───┼───┼───┼───┼───┼───○
```

Discuss why a number line is useful, and why you can't just list the numbers 19, 20, 21, 22, 23 and 24.
A4.4OHP shows the number lines in the student book. Reinforce the conventions using number lines on an OHT (you can use **R6** number lines).

Extend to solving linear inequalities.
The student book gives the example of $3x + 2 > 47$.
Reinforce the similarity with solving an equation, and emphasise how the answer should be interpreted (x is any number greater than 15).

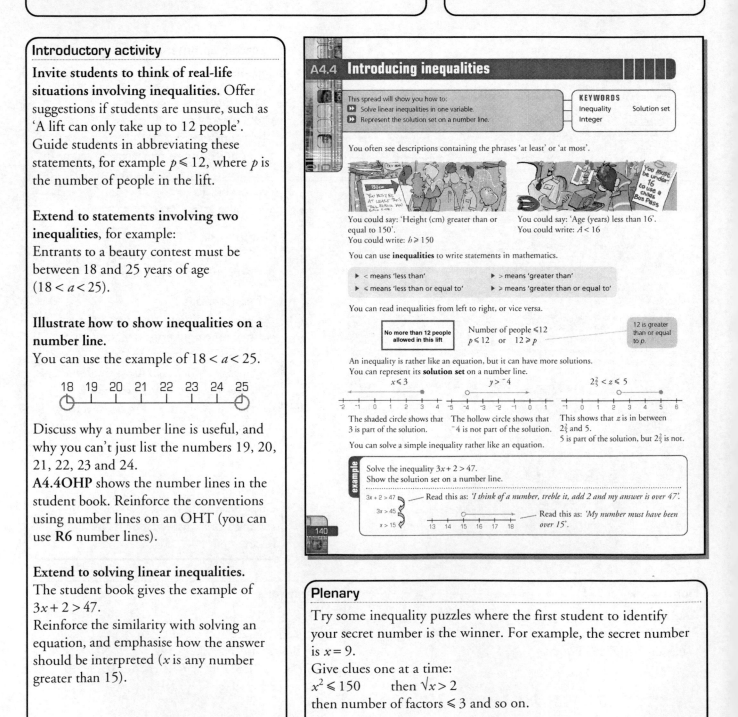

A4.4 **Introducing inequalities**

This spread will show you how to:
▶ Solve linear inequalities in one variable.
▶ Represent the solution set on a number line.

KEYWORDS
Inequality   Solution set
Integer

You often see descriptions containing the phrases 'at least' or 'at most'.

You could say: 'Height (cm) greater than or equal to 150'.
You could write: $h \ge 150$

You could say: 'Age (years) less than 16'.
You could write: $A < 16$

You can use **inequalities** to write statements in mathematics.

▶ < means 'less than'
▶ ≤ means 'less than or equal to'
▶ > means 'greater than'
▶ ≥ means 'greater than or equal to'

You can read inequalities from left to right, or vice versa.

No more than 12 people allowed in this lift

Number of people ≤12
$p \le 12$ or $12 \ge p$

12 is greater than or equal to p.

An inequality is rather like an equation, but it can have more solutions.
You can represent its **solution set** on a number line.

$x \le 3$
The shaded circle shows that 3 is part of the solution.

$y > ^-4$
The hollow circle shows that $^-4$ is not part of the solution.

$2\frac{2}{5} < z \le 5$
This shows that z is in between $2\frac{2}{5}$ and 5.
5 is part of the solution, but $2\frac{2}{5}$ is not.

You can solve a simple inequality rather like an equation.

example

Solve the inequality $3x + 2 > 47$.
Show the solution set on a number line.

$3x + 2 > 47$ — Read this as: *'I think of a number, treble it, add 2 and my answer is over 47'.*
$3x > 45$
$x > 15$ — Read this as: *'My number must have been over 15'.*

140

Plenary

Try some inequality puzzles where the first student to identify your secret number is the winner. For example, the secret number is $x = 9$.
Give clues one at a time:
$x^2 \le 150 \qquad$ then $\sqrt{x} > 2$
then number of factors ≤ 3 and so on.

Students can invent their own inequality problems to test on others. For example:
$100 \leqslant$ square number $\leqslant 1000$
What could the square number be?

Differentiation

Extension questions:
▸ Questions 1 and 2 require students to apply inequalities to real contexts.
▸ Questions 3–5 require a mathematical understanding of inequality.
▸ Questions 6 and 7 focus on solving inequalities.

Core tier: focuses on expanding brackets in linear expressions.

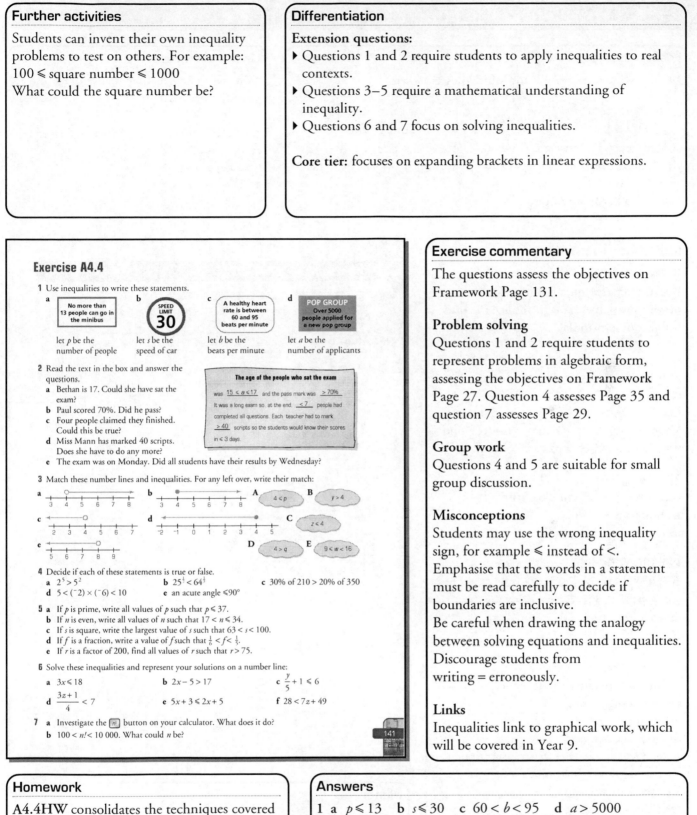

Exercise A4.4

1 Use inequalities to write these statements.

a
No more than 13 people can go in the minibus
let p be the number of people

b
SPEED LIMIT **30**
let s be the speed of car

c
A healthy heart rate is between 60 and 95 beats per minute
let b be the beats per minute

d
POP GROUP
Over 5000 people applied for a new pop group
let a be the number of applicants

2 Read the text in the box and answer the questions.
a Bethan is 17. Could she have sat the exam?
b Paul scored 70%. Did he pass?
c Four people claimed they finished. Could this be true?
d Miss Mann has marked 40 scripts. Does she have to do any more?
e The exam was on Monday. Did all students have their results by Wednesday?

The age of the people who sat the exam
was $15 < a \leqslant 17$ and the pass mark was $> 70\%$
It was a long exam so, at the end, < 7 people had completed all questions. Each teacher had to mark > 40 scripts so the students would know their scores in < 3 days.

3 Match these number lines and inequalities. For any left over, write their match:

a [number line 3 to 8, open circle]
b [number line 3 to 8, filled circle]
c [number line 2 to 7, open circle]
d [number line −2 to 5, filled circle]
e [number line 5 to 9, open circle]

A $4 < p$
B $y > 4$
C $z < 4$
D $4 > q$
E $9 \leqslant w < 16$

4 Decide if each of these statements is true or false.
a $2^5 > 5^2$
b $25^{\frac{1}{2}} < 64^{\frac{1}{3}}$
c 30% of 210 > 20% of 350
d $5 < (^-2) \times (^-6) < 10$
e an acute angle $\leqslant 90°$

5 a If p is prime, write all values of p such that $p \leqslant 37$.
b If n is even, write all values of n such that $17 < n \leqslant 34$.
c If s is square, write the largest value of s such that $63 < s < 100$.
d If f is a fraction, write a value of f such that $\frac{1}{6} < f < \frac{1}{5}$.
e If r is a factor of 200, find all values of r such that $r > 75$.

6 Solve these inequalities and represent your solutions on a number line.
a $3x \leqslant 18$
b $2x - 5 > 17$
c $\frac{y}{5} + 1 \leqslant 6$
d $\frac{3z + 1}{4} < 7$
e $5x + 3 \leqslant 2x + 5$
f $28 < 7z + 49$

7 a Investigate the $\boxed{n!}$ button on your calculator. What does it do?
b $100 < n! < 10\,000$. What could n be?

141

Exercise commentary

The questions assess the objectives on Framework Page 131.

Problem solving

Questions 1 and 2 require students to represent problems in algebraic form, assessing the objectives on Framework Page 27. Question 4 assesses Page 35 and question 7 assesses Page 29.

Group work

Questions 4 and 5 are suitable for small group discussion.

Misconceptions

Students may use the wrong inequality sign, for example $\leqslant$ instead of $<$. Emphasise that the words in a statement must be read carefully to decide if boundaries are inclusive.
Be careful when drawing the analogy between solving equations and inequalities. Discourage students from writing $=$ erroneously.

Links

Inequalities link to graphical work, which will be covered in Year 9.

Homework

A4.4HW consolidates the techniques covered in the lesson, and provides further opportunity for solving basic inequalities.

Answers

1 a $p \leqslant 13$ b $s \leqslant 30$ c $60 < b < 95$ d $a > 5000$
2 a Yes b No c Yes d Yes e Not necessarily
3 a B b A c E d C e $x < 8$ (any letter)
4 a True b False c False d False e False
5 a 2, 3, 5, 7, 11, 13, 17, 19, 23, 29, 31, 37
    b 18, 20, 22, 24, 26, 28, 30, 32, 34 c 81 e 100, 200
6 a $x \leqslant 6$ b $x > 11$ c $y \leqslant 25$ d $z < 9$ e $x \leqslant \frac{2}{3}$ f $x > ^-3$
7 b 5, 6, 7

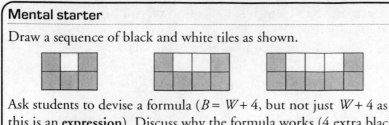

Mental starter

Draw a sequence of black and white tiles as shown.

Ask students to devise a formula ($B = W + 4$, but not just $W + 4$ as this is an **expression**). Discuss why the formula works (4 extra black tiles on the end).
Extend to harder examples.

Useful resources

Mini-whiteboards may be useful for the introductory activity.

A4.5OHP shows the spider diagram in the student book.

Introductory activity

Revise formulae.
Recap that a formula is a relationship between variables, and introduce the term 'subject of a formula'.

Encourage students to suggest formulae that they already know (for example, $S = \frac{D}{T}$).
Let $D = 100\,m$ and $T = 40\,s$, and ask for S. Vary the numbers – students can respond using mini-whiteboards.

Now let $S = 4\,m/s$ and $D = 64\,m$. Ask for the value of T. Discuss the difficulty of working out T in this form, and the need to **rearrange the formula**.

Refer back to the mental starter.
$B = W + 4$. To make W the subject, W has had 4 added to it so this must now be taken away from both sides:
$B - 4 = W$.

Illustrate how you can rearrange a formula by reading the layers of algebra and reversing them.
For example, $tw + r = s$
Make w the subject.
▸ Read the layers: 'w has been multiplied by t, then r is added on'.
▸ Reverse them:
  ▸ subtract r: $tw = s - r$
  ▸ then divide by t: $w = \dfrac{s - r}{t}$

Draw the analogy with getting dressed and undressed: shirt on, jumper on then jumper off, shirt off.
The first layer on will be the last layer to come off.

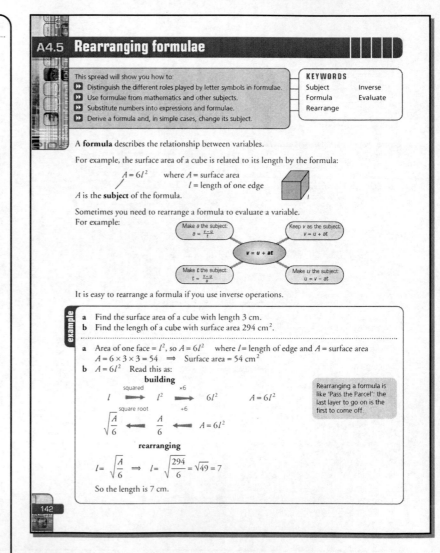

Plenary

Discuss question 6, and in particular part d where students are required to justify the formula in the context of the original diagram.

Further activities

Ask students to rearrange harder formulae involving skills acquired in other lessons, such as:

▸ Negative terms

$w - x = p$ (x)

▸ Factorising

$p = at^2 + ka$ (a)

▸ Fractions

$p = \frac{1}{2}x - k$ (x)

Differentiation

Extension questions:

▸ Question 1 recaps using a formula.

▸ Questions 2–5 focus on deriving and rearranging formulae, including in context.

▸ Question 6 is a longer task that is broken down into smaller parts.

Core tier: focuses on substituting integers into a simple formula.

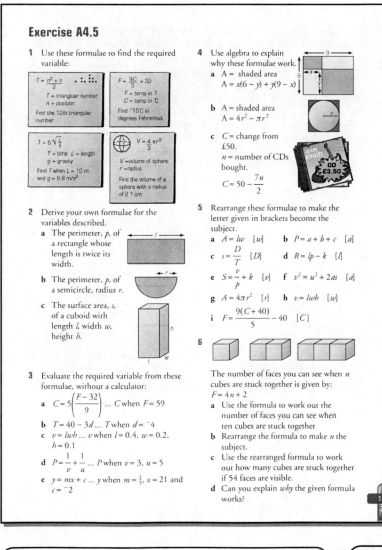

Exercise A4.5

1 Use these formulae to find the required variable:

$T = \frac{n^2 + n}{2}$

T = triangular number
n = position
Find the 12th triangular number

$F = \frac{9C}{5} + 32$

F = temp in F
C = temp in C
Find ⁻15 C in degrees Fahrenheit

$T = 6\sqrt{\frac{L}{g}}$

T = time L = length
g = gravity
Find T when L = 10 m and g = 9.8 m/s²

$V = \frac{4}{3}\pi r^3$

V = volume of sphere
r = radius
Find the volume of a sphere with a radius of 2.1 cm

2 Derive your own formulae for the variables described.

a The perimeter, p, of a rectangle whose length is twice its width.

b The perimeter, p, of a semicircle, radius r.

c The surface area, s, of a cuboid with length l, width w, height h.

3 Evaluate the required variable from these formulae, without a calculator:

a $C = 5\left(\frac{F - 32}{9}\right)$ ... C when F = 59

b $T = 40 - 3d$ ... T when d = ⁻4

c $v = lwh$ ... v when l = 0.4, w = 0.2, h = 0.1

d $P = \frac{1}{v} + \frac{1}{u}$ ... P when v = 3, u = 5

e $y = mx + c$ ... y when m = $\frac{1}{3}$, x = 21 and c = ⁻2

4 Use algebra to explain why these formulae work.

a A = shaded area
A = $x(6 - y) + y(9 - x)$

b A = shaded area
A = $4r^2 - \pi r^2$

c C = change from £50.
n = number of CDs bought.

$C = 50 - \frac{7n}{2}$

5 Rearrange these formulae to make the letter given in brackets become the subject.

a $A = lw$ $[w]$ **b** $P = a + b + c$ $[a]$

c $s = \frac{D}{T}$ $[D]$ **d** $R = lp - k$ $[l]$

e $S = \frac{v}{p} + k$ $[v]$ **f** $v^2 = u^2 + 2as$ $[a]$

g $A = 4\pi r^2$ $[r]$ **h** $v = lwh$ $[w]$

i $F = \frac{9(C + 40)}{5} - 40$ $[C]$

6

The number of faces you can see when n cubes are stuck together is given by:
$F = 4n + 2$

a Use the formula to work out the number of faces you can see when ten cubes are stuck together

b Rearrange the formula to make n the subject.

c Use the rearranged formula to work out how many cubes are stuck together if 54 faces are visible.

d Can you explain *why* the given formula works?

Exercise commentary

The questions assess the objectives on Framework Page 141

Problem solving

Question 6 requires students to represent a problem algebraically and to move from one form to another, assessing the objectives on Framework Page 27. Question 2 assesses Page 31.

Group work

Encourage students to compare their formulae from question 2 in pairs, and to discuss in small groups why the formulae in questions 4 and 6 work.

Misconceptions

A common mistake is to undo the 'first on' layer first, for example:
$pw + k = r \Rightarrow w + k = r/p$
Reinforce the need to read the algebra layers and peel them off **in reverse**.

Links

The skills required to rearrange a formula link to straight-line graphs and later work in simultaneous equations.

Homework

A4.5HW provides practice in rearranging a formula, and also includes constructing and justifying a formula.

Answers

1 78, 5 °F, 6.1 s, 38.8 cm³

2 **a** $p = 3t$ **b** $p = 2r + \pi r$ **c** $s = 2lw + 2lh + 2wh$

3 **a** 15   **b** 52   **c** 0.008   **d** $\frac{8}{15}$ **e** 5

5 **a** $w = \frac{A}{l}$ **b** $a = P - b - c$ **c** $D = sT$ **d** $l = \frac{R + k}{p}$

   e $v = p(S - k)$ **f** $a = \frac{v^2 - u^2}{2s}$ **g** $r = \sqrt{\frac{A}{4\pi}}$ **h** $w = \frac{v}{lh}$

   i $c = \frac{5(F + 40)}{a} - 40$

6 **a** 42   **b** $n = \frac{F - 2}{4}$ **c** 13

Mental starter

Write down these workings and ask students to spot the errors:

$px + r = w$ $kx^2 - t = q$ $\sqrt{x} + t = z$

$x + r = \frac{w}{p}$ $kx^2 = q + t$ $\sqrt{x} = z - t$

$x = \frac{w}{p} - r$ $kx = \sqrt{q + t}$ $x = z - t^2$

$$x = \frac{\sqrt{q + t}}{k}$$

Student can show the correct workings on mini-whiteboards.

Useful resources

Mini-whiteboards may be useful for the mental starter.

Introductory activity

Recap the main theme of the previous lesson: many formulae can be rearranged by reading the layers of algebra and undoing them in reverse order.

Choose a formula with a negative term.
For example, rearrange $k = w - px$ to make x the subject.

Read the layers : x has been multiplied by p and then **subtracted from** w.

Deal with the 'subtract from' operation: add px to both sides, so that $k + px = w$
This removes the negative term, and now you can proceed to read the **new** layers, and then reverse them.

Choose a formula with a fractional term.
For example, rearrange $k = w + \frac{p}{x}$ to make x the subject.

Read the layers: x has been **divided into** p and then w is added.

Deal with the 'divide into' operation:
Multiply both sides by x, so that
$x(k - w) = p$
This removes the fractional term, and now you can proceed to read the **new** layers, and then reverse them.

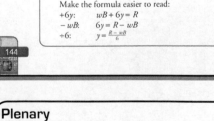

A4.6 Rearranging harder formulae

This spread will show you how to:
- Use formulae from mathematics and other subjects.
- Derive a formula and change its subject.

KEYWORDS
Rearrange   Formula
Subject   Inverse

You can rearrange formulae by using inverse operations.
Remember that you perform any operations on **both** sides of the formula.

example

Rearrange:

a $p = 2(l + w)$ to make w the subject.   **b** $R = 8l^3$ to make l the subject.

a $+l$ then $\times 2$ **b**   cube l, then $\times 8$ ► **Read** the
$\div 2$: $\frac{p}{2} = l + w$ $\div 8$: $\frac{R}{8} = l^3$ formula
$-l$: $\frac{p}{2} - l = w$ cube root $\sqrt[3]{\frac{R}{8}} = l$ ► **Reverse** the
$\frac{p}{2} - l$ operations

There are two types of inverse operations that you need to be careful with.

example

Rearrange the formulae to make x the subject.

a $V = p - qx$ **b** $k = \frac{p}{x}$

a   ► $V = p - qx$ $\times q$ then **subtract from** p **b** $k = \frac{p}{x}$ **divide into** p
  ► Make the formula easier to read: ► Make the formula easier to read:
  $V + qx = p$ $\times q$ then $+V$ $kx = p$ $\times k$
  ► Now reverse: ► Now reverse:
  $-V$: $qx = p - V$ $\div k$: $x = \frac{p}{k}$
  $\div q$: $x = \frac{p - V}{q}$

The operations 'subtract from' and 'divide into' may be part of a larger formula.

example

Rearrange each formula to make the letter in brackets become the subject.

a $B = \frac{R - 6y}{w}$ (y) **b** $R = \frac{S}{T} - U$ (T)

a   Read the formula: **b**   Read the formula:
  $\times 6$, then subtract from R, then divide by w Divide into S, then $-U$
  $\times w$: $wB = R - 6y$ $+U$: $R + U = \frac{S}{T}$
  Make the formula easier to read: Make the formula easier to read:
  $+6y$: $wB + 6y = R$ $\times T$: $T(R + U) = S$
  $-wB$: $6y = R - wB$ $\div (R + U)$: $T = \frac{S}{R + U}$
  $\div 6$: $y = \frac{R - wB}{6}$

144

Plenary

To consolidate all the skills in this area, ask students to complete this spider diagram:

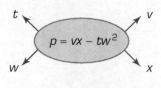

Further activities

Give students further practice at rearranging formulae where the required subject appears twice, for example:

$s = ut + 5t$ (make t the subject).

A4.6F contains a variety of formulae to rearrange.

Differentiation

Extension questions:

▶ Questions 1 and 2 provide an opportunity to reinforce the techniques learned in the previous lesson on formulae.
▶ Questions 3–6 focus on negative and fractional terms.
▶ Questions 7 and 8 provide more challenging formulae to rearrange.

Core tier: focuses on deriving and rearranging simple formulae.

Exercise A4.6

1 Make the letter in brackets the subject of each formula.

a $p = s + r$ (s) **b** $z = qp + k$ (q) **c** $w = \frac{p}{x} - m$ (p)

d $z = \sqrt{x}$ (x) **e** $m = p^2 + 3$ (p) **f** $r = x^3 + z$ (x)

g $k = px^2 - q$ (x) **h** $m = p^4$ (p) **i** $p(x - r) = w$ (x)

2 These formulae involve negative terms. Rearrange them to make x the subject by first adding the inverse of the negative term to both sides.

a $p = w - x$ **b** $z = r - px$ **c** $q = z - x^2$ **d** $p - wx = r$

e $m = l - \sqrt{x}$ **f** $m = z - x^3$ **g** $w = r - mx^2$ **h** $m = p - \frac{x}{q}$

3 These formulae involve fraction terms. Rearrange them to make y the subject by first multiplying both sides by the denominator.

a $p = \frac{w}{y}$ **b** $z = \frac{m}{y}$ **c** $q = \frac{z}{3y}$ **d** $r = \frac{p}{y^2}$

e $k = \frac{1}{\sqrt{y}}$ **f** $w = \frac{m^2}{y^3}$ **g** $m = \frac{p}{my}$ **h** $t = \frac{s}{wy^2}$

4 Read the layers carefully to rearrange these formulae to make m the subject.

a $pm - q = x$ **b** $q - m = px$ **c** $\frac{z}{m} = q$ **d** $\frac{m^2}{p} - q = k$

e $q - m^2 = r$ **f** $\frac{q}{m^2} + 2 = p$ **g** $\frac{p - m}{q} = z$ **h** $k(q - m^2) = t$

5 Rearrange these formulae to show all given subjects:

6 Challenge
Rearrange these formulae to make x the subject.

a $k = \frac{z(px + q)}{w} + c$ **b** $m = \frac{[\frac{z}{p} - q]}{w} + k$ **c** $w = \frac{px^2 - k + c}{m}$

7 $m = p - \frac{w}{x}$ ... this formula involves a negative term and a fraction. How would you make x the subject?

8 $x = \frac{\sqrt{b^2 - 4ac} - b}{2a}$ This is a well-known formula in mathematics.
a Can you make c the subject?
b Find out what the formula is for and see if you can use it.

145

Exercise commentary

The questions assess the objectives on Framework Page 141.

Problem solving
Question 8 requires students to explore connections in mathematics, assessing the objectives on Framework Page 7.

Group work
Questions 7 and 8 can be discussed in pairs as they are challenging.

Misconceptions
Students may continue to deal with layers in the wrong order. This may either be due to forgetting to reverse the order when rearranging, or it may be more fundamental: students may read the layers incorrectly in the first instance.
You may need to recap the correct order of operations, and algebraic notation.

Links
Rearranging formulae links to solving equations (Framework Pages 123 and 125).

Homework

A4.6HW provides practice at rearranging challenging formulae.

Answers

1 a $p - r$ **b** $\frac{z - k}{p}$ **c** $x(w + m)$ **d** z^2 **e** $\sqrt{m - 3}$ **f** $\sqrt[3]{r - z}$ **g** $\sqrt{\frac{k + q}{p}}$ **h** $\sqrt[4]{m}$ **i** $\frac{w}{p} + r$

2 a $w - p$ **b** $\frac{r - z}{p}$ **c** $\sqrt{z - q}$ **d** $\frac{p - r}{w}$ **e** $(1 - m)^2$ **f** $\sqrt[3]{z - m}$ **g** $\sqrt{\frac{r - w}{m}}$ **h** $q(p - m)$

3 a $\frac{w}{p}$ **b** $\frac{m}{z}$ **c** $\frac{z}{3q}$ **d** $\sqrt{\frac{p}{r}}$ **e** $\frac{1}{k^2}$ **f** $\sqrt[3]{\frac{m^2}{w}}$ **g** $\frac{p}{m^2}$ **h** $\sqrt{\frac{s}{tw}}$

4 a $\frac{x + q}{p}$ **b** $q - px$ **c** $\frac{z}{q}$ **d** $\sqrt{p(k + q)}$ **e** $\sqrt{q - r}$ **f** $\sqrt{q(p - 2)}$ **g** $p - qz$ **h** $\sqrt{q - \frac{t}{k}}$

5 $p = \frac{v + x}{q}$, $q = \frac{v + x}{p}$, $x = pq - v$, $x = \sqrt{\frac{m + t}{w}}$, $w = \frac{m + t}{x^2}$, $t = wx^2 - m$; $m = x(w + k)$, $x = \frac{m}{w + k}$, $k = \frac{m}{x} - w$

6 a $\frac{w(k + c)}{zp} - \frac{q}{p}$ **b** $p(mw + k + q)$ **c** $\sqrt{\frac{z(wm - c) + k}{p}}$ **7** $\frac{w}{p - m}$ **8 a** $c = \frac{b^2 - (2ax + b)^2}{4a}$

145

Summary

The key objectives for this unit are:
▶ Construct and solve linear equations with integer coefficients using an appropriate method.
▶ Present a concise reasoned argument.

Plenary activity

Ask students to carry out these mixed examples, focusing on the distinction between equation, expression formulae and inequality:
▶ Solve the **equation**: $10 - 5x = 3$
▶ Simplify the **expression**: $\frac{2}{x} + \frac{3}{y}$
▶ Rearrange the **formula** to make z the subject: $p - wz^2 = q$
▶ Find positive integer solutions of the **inequality**: $4x + 7 \leqslant 25$

Check out commentary

1 **a** Students should cope with **i** and **ii** easily but may need to be reminded how to deal with negative and fractional terms in **iii–v**.
Reading algebraic layers must also be encouraged to avoid errors, especially in the final stage, for example
$$10x = 3$$
$$\therefore x = \frac{3}{10} \text{ not } \frac{10}{3}.$$

b Students may need to recap on adding and subtracting fractions in **ii** and **iii**.
In **iii**, encourage students to add length and width to get 10 cm.

c Encourage students to use a number line to display the obtained solutions.

2 Students must remember to write formulae, not expressions.
They should be encouraged to think of BIDMAS in **a**.
In part **b**, advise students to take care in dealing with negative terms.
You may need to reinforce these techniques with simple examples.

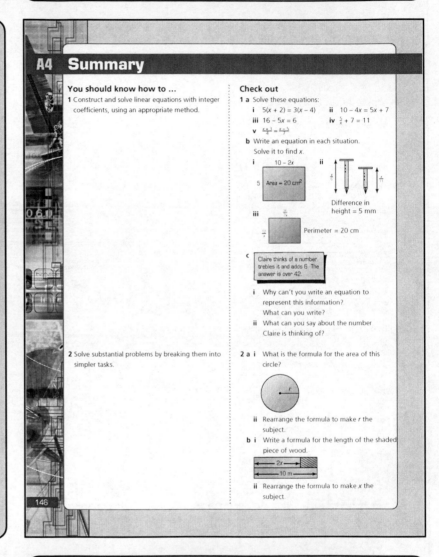

Development

All techniques are revisited and extended in Year 9.
Equations and formulae are used in a problem-solving context in P1.

Links

Formulae are used in science, for example in $V = IR$.

Mental starters

Objectives covered in this unit:
▶ Find fractions and percentages of quantities.
▶ Order, add, subtract, multiply and divide integers.
▶ Visualise, describe and sketch 2-D shapes.

Resources needed

* means class set needed
Essential:
Compasses*
Protractors*
D2.2OHP – stem-and-leaf diagrams
D2.3OHP – stem and leaf diagram
D2.4OHP – scatter graph
D2.5OHP – tables
D2.6OHP – line graph and multiple
bar chart
Useful:
Mini-whiteboards
Radio and TV listings
R11* – tally chart
D2.2ICT* – statistical graphs

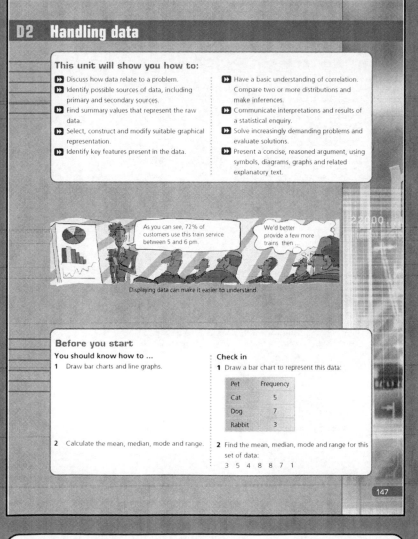

D2 Handling data

This unit will show you how to:
▶▶ Discuss how data relate to a problem.
▶▶ Identify possible sources of data, including primary and secondary sources.
▶▶ Find summary values that represent the raw data.
▶▶ Select, construct and modify suitable graphical representation.
▶▶ Identify key features present in the data.

▶▶ Have a basic understanding of correlation. Compare two or more distributions and make inferences.
▶▶ Communicate interpretations and results of a statistical enquiry.
▶▶ Solve increasingly demanding problems and evaluate solutions.
▶▶ Present a concise, reasoned argument, using symbols, diagrams, graphs and related explanatory text.

As you can see, 72% of customers use this train service between 5 and 6 pm.

We'd better provide a few more trains then ...

Displaying data can make it easier to understand.

Before you start

You should know how to ...
1 Draw bar charts and line graphs.

Check in
1 Draw a bar chart to represent this data:

Pet	Frequency
Cat	5
Dog	7
Rabbit	3

2 Calculate the mean, median, mode and range.

2 Find the mean, median, mode and range for this set of data:
3 5 4 8 8 7 1

147

Unit commentary

Aim of the unit
This unit focuses on methods used in statistical enquiries, from identifying areas for research and sources of data and methods of data collection, through drawing suitable diagrams and calculating appropriate statistics to summarising the data and producing statistical reports.

Introduction
Ask students to describe the three types of average.
Discuss situations where it would be appropriate to use each of them.

Framework references
This unit focuses on:
▶ Teaching objectives Pages: 249, 251, 257, 263, 267–273
▶ Problem-solving objectives Pages: 25, 27, 31 and 33.

Differentiation

Core tier
Focuses on displaying and interpreting data, and includes stem-and-leaf diagrams. There is less emphasis on scatter graphs than in the extension tier.

Check in activity

Write down these nine numbers:
2, 8, 7, 2, 10, 9, 11, 3, 2.
Ask students to find the mode, median, mean and range of this data.
Which average gives an untypical value for the data?

Mental starter

Brainstorm sources of data and write them on the board. Include books, magazines, newspapers, Internet, computer databases, questionnaires, surveys of samples of people, published surveys (for example Census), etc.

Ask students to sort these into primary and secondary data sources.

Useful resources

Mini-whiteboards may be useful for the mental starter.
R11 – tally charts
Radio and TV listings may be useful for Further Activities.

Introductory activity

Recap the terms 'primary data' and 'secondary data' introduced in Year 7.

Discuss the scenario in the student book.

Ask students to think about the street they live in:

▸ How many houses are there? What do the house numbers go up to?
▸ Do any numbers appear twice, for example 11 and 11A?
▸ Is their street typical of nearby streets, of the local area, of the county, nationally?

Discuss possible sources of data for the area, the whole town or village they live in, the county, the whole country. For example, the electoral roll, street directories or primary research in the form of surveys.

Discuss ways of recording the data collected, including tally charts (you could use **R11**).

Emphasise the importance of the points in the bulleted list at the bottom of the student's page.

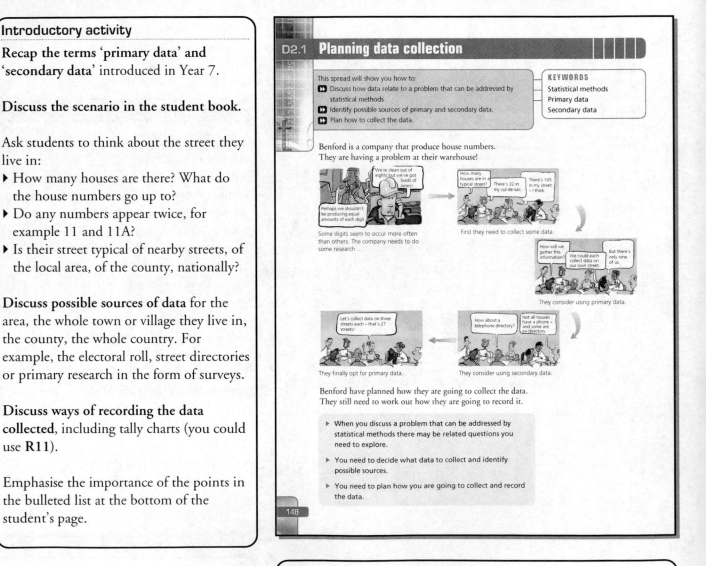

Plenary

Discuss ideas generated from question 3 and suggested further activity about the timing of sports slots on radio and television.

Ensure that the discussion includes the three points listed at the bottom of the student's page.

Further activities

Students could extend question 3 by thinking about the best timing for sports slots on radio and television programmes. Ask students to identify related questions to explore.

Differentiation

Extension questions:

▸ Question 1 provides an opportunity to use real data for discussion.
▸ Question 2 explores the context of question 1 in greater depth.
▸ Question 3 sets up a problem that can be addressed by statistical methods.

Core tier: focuses on planning data collection using the context of football.

Exercise D2.1

1 A survey of house numbers in five streets produced the following data.

Street	A	B	C	D	E
Even numbers	2 to 34	2 to 60	2 to 38	2 to 52	2 to 44
Odd numbers	1 to 33	1 to 45	1 to 55	1 to 51	1 to 59

Note: Sometimes streets are curved so that there are more houses on one side than the other

a Do you think this data is typical?
  Give reasons for your answer.
b Using the data for street C complete a tally chart to show the distribution of all the digits that will be needed to number each house.
c Using the data for street C complete a tally chart to show the distribution of the **first** digit needed for each house.
d Compare the two tally charts and comment on your results.
e Do you think that the other streets will show similar results?
  Give a reason for your answer.

2 a Choose one of streets A, B, D or E from the table in question 1 and draw two tally charts to show the distribution of: **i** all digits **ii** the first digits of house numbers.
  b Compare the two tally charts and comment on your results.
  c Compare your observations on house number digits for this street with your observations for street C. What do you notice?

3 A hospital radio station can only broadcast 15 minutes of sport each Saturday.
  They can do this either in:
  i three 5-minute slots
  ii a 10-minute slot and a 5-minute slot, or
  iii one 15-minute slot.
  The sporting news they transmit is from the game played by the local football team.
  a If they want to be on air when a goal is scored, which of the three options should they choose?
  b Identify what data you would need to collect and where you might collect this data.

149

Exercise commentary

The questions assess the objectives on Framework Pages 249 and 251.

Problem solving

Question 3 provides an opportunity to assess Framework Page 25, as long as little or no guidance is given.
The whole exercise assesses Page 33.

Group work

Question 2 can be attempted by groups of four. Each member of the group should choose a different street.

Misconceptions

With the house numbers students may initially think that all the digits will be used the same number of times.
Planning is a skill that may come easier to some students than others. Reinforce planning as an integral part of the data handling process, not an optional extra.

Links

Problem solving: Framework Page 25.

Homework

D2.1HW requires students to consider data collection methods in the context of a jigsaw puzzle.

Answers

1 b Frequencies for 0–9: 3, 16, 14, 16, 9, 9, 4, 5, 4, 5
  c Frequencies for 0–9: 0, 11, 11, 11, 6, 4, 1, 1, 1, 1

Mental starter

Write these fractions on the board:

$$\frac{5}{36} \qquad \frac{6}{30} \qquad \frac{15}{90} \qquad \frac{11}{60} \qquad \frac{7}{24}$$

Ask students to calculate these fractions of 360.

Useful resources

Mini-whiteboards may be useful for the mental starter.

Protractors and compasses for drawing pie charts.

D2.2OHP – a stem-and-leaf diagram drawn with data in the order it appears, and then with data in numerical order. It also contains a back-to-back diagram.

Introductory activity

Recap pie charts from Year 7.
Use the example in the student book to remind students how to calculate the angles for each sector.

Introduce stem-and-leaf diagrams.

Emphasise that with stem-and-leaf diagrams none of the actual data values are lost. The shape of the diagram shows how the data values are distributed.

Discuss how to choose the stem values. They can be the first digit (or digits for 100s, 1000s, etc). The 'leaves' are the rest of the number and they decorate the 'stem'.

Emphasise that the simplest way to draw stem-and-leaf diagrams is to add leaves to the stem in the order the data appears. Then use this to draw a second stem-and-leaf diagram with the leaves in numerical order. See **D2.2OHP**.

Remind students to write a key.
Discuss back-to-back stem-and-leaf diagrams. There is an example of one on **D2.2OHP**.

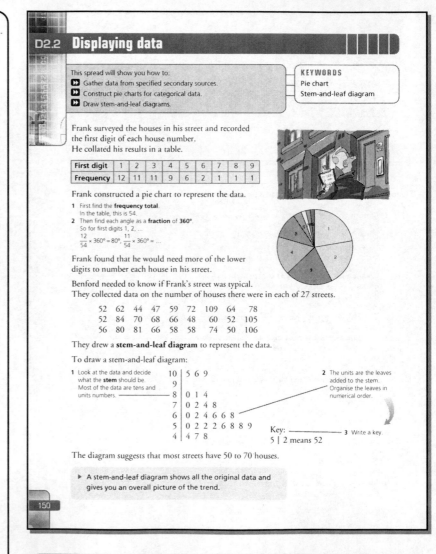

D2.2 Displaying data

This spread will show you how to:
- Gather data from specified secondary sources.
- Construct pie charts for categorical data.
- Draw stem-and-leaf diagrams.

KEYWORDS
Pie chart
Stem-and-leaf diagram

Frank surveyed the houses in his street and recorded the first digit of each house number.
He collated his results in a table.

First digit	1	2	3	4	5	6	7	8	9
Frequency	12	11	11	9	6	2	1	1	1

Frank constructed a pie chart to represent the data.

1 First find the **frequency total**.
In the table, this is 54.
2 Then find each angle as a **fraction** of 360°.
So for first digits 1, 2, …
$\frac{12}{54} \times 360° = 80°$, $\frac{11}{54} \times 360° = $ …

Frank found that he would need more of the lower digits to number each house in his street.

Benford needed to know if Frank's street was typical.
They collected data on the number of houses there were in each of 27 streets.

```
52   62   44   47   59   72   109   64    78
52   84   70   68   66   48    60   52   105
56   80   81   66   58   58    74   50   106
```

They drew a **stem-and-leaf diagram** to represent the data.

To draw a stem-and-leaf diagram:

1 Look at the data and decide what the **stem** should be. Most of the data are tens and units numbers.

```
10 | 5 6 9
 9 |
 8 | 0 1 4
 7 | 0 2 4 8
 6 | 0 2 4 6 6 8
 5 | 0 2 2 2 6 8 8 9
 4 | 4 7 8
```

2 The units are the leaves added to the stem. Organise the leaves in numerical order.

Key:
5 | 2 means 52

3 Write a key.

The diagram suggests that most streets have 50 to 70 houses.

▶ A stem-and-leaf diagram shows all the original data and gives you an overall picture of the trend.

Plenary

Discuss the advantages and disadvantages of using pie charts and stem-and-leaf diagrams to display data.
▶ Could you draw stem-and-leaf diagrams for questions 1 and 2?
▶ Could you draw pie charts for questions 3–5?

Further activities

Using the stem-and-leaf diagrams drawn for questions 4 and 5, students could comment on the similarities in the number of points scored in the three divisions.

D2.2ICT involves using a spreadsheet program to construct statistical graphs.

Differentiation

Extension questions:
▶ Questions 1 and 2 involve constructing pie charts.
▶ Questions 3 and 4 involve constructing stem-and-leaf diagrams.
▶ Question 5 extends to constructing a back-to-back stem-and-leaf diagram.

Core tier: focuses on pie charts and scatter graphs.

Exercise D2.2

1   Here is part of the data from Exercise D2.1 on page 149, relating to house numbers on a street.

Street	Even numbers	Odd numbers
C	2 to 38	1 to 55

Use the data for street C to draw pie charts to show:
a the distribution of all the digits
b the distribution of the first digit of each house number.

2   The doors of the houses in street C are painted in the following colours:

Colour	Blue	Green	Red	White	Yellow
Frequency	12	8	6	18	3

Draw a pie chart to represent these data.

3   Gemma surveyed 27 streets in her area, counting the number of houses in each street. They were

47 54 68 39 32 57 54 44 60
56 53 48 62 38 58 38 46 58
36 48 62 54 64 66 42 44 50

Draw a stem-and-leaf diagram to represent these data.

4   The data shows the total points scored by the teams in the premier league division for the 2001/2002 season.

87 80 77 71 66 64 53 50 50 46 45 45 44 44 43 40 40 36 30 28

Draw a stem-and-leaf diagram to represent these data.

5   Draw a back to back stem-and-leaf diagram to show the overall points scored by Division One and Division Two in the Nationwide league for the 2001/2002 season.

Division One
99 89 86 77 76 75 75 72 67 66 66 64
60 59 55 54 53 51 50 50 49 49 48 26

Division Two
90 84 83 83 80 78 73 71 70 64 64 63
59 58 57 56 55 52 50 49 44 44 43 34

A **back-to-back** stem-and-leaf diagram looks like this:

```
        Boys         Girls
      9 5 2 | 8 | 0 5 9
  9 9 3 3 1 | 7 | 1 2 4 6 6 7
9 5 4 4 4 2 | 6 | 2 3 3 8 9
  8 3 2 2 1 | 5 | 0 4 5 5 6
      7 5 0 | 4 | 2 3 7
      8 4 1 | 3 | 0 2
```

You can use it to **compare** two data sets.

151

Exercise commentary

The questions assess the objectives on Framework Page 263.

Problem solving
Commenting on the stem-and-leaf diagrams for questions 4 and 5 (see Further activities) provides an opportunity to assess Framework Page 25.
The whole exercise assesses Page 27.

Group work
None of the questions in this exercise are specifically suitable for group work.

Misconceptions
In a back-to-back stem-and-leaf diagram, students do not always put the lowest values closest to the stem, particularly for the data arranged on the left of the stem. Remind students to include a key. Also encourage a neat layout, so that the columns line up. It is then easier to see the shape of the distrubution.

Links
Problem solving: Framework Page 25.

Homework

D2.2HW requires students to draw a back-to-back stem-and-leaf diagram.

Answers

1 **a** 13°, 68°, 59°, 68°, 38°, 38°, 17°, 21°, 17°, 21°
   b 0°, 84°, 84°, 84°, 46°, 31°, 8°, 8°, 8°, 8°

2 92°, 61°, 46°, 138°, 23°

3 30: 2, 6, 8, 8, 9; 40: 2, 4, 4, 6, 7, 8, 8; 50: 0, 3, 4, 4, 4, 6, 7, 8, 8; 60: 0, 2, 2, 4, 6, 8

4 20: 8; 30: 0, 6; 40: 0, 0, 3, 4, 4, 5, 5, 6; 50: 0, 0, 3; 60: 4, 6; 70: 1, 7; 80: 0, 7

5 **One** 20: 6; 40: 8, 9, 9; 50: 0, 0, 1, 3, 4, 5, 9; 60: 0, 4, 6, 6, 7; 70: 2, 5, 5, 6, 7; 80: 6, 9; 90: 9
   Two 30: 4; 40: 3, 4, 4, 9; 50: 0, 2, 5, 6, 7, 8, 9; 60: 3, 4, 4; 70: 0, 1, 3, 8; 80: 0, 3, 3, 4; 90: 0

Mental starter

Write two sets of numbers on the board and ask students to find the mean, median, mode and range of each set.
For example

      10, 8, 14, 8, 13, 6, 11

      24, 16, 33, 24, 35, 33, 27

Useful resources

Mini-whiteboards may be useful for the mental starter.

D2.3OHP – stem-and-leaf diagram from the student book.

Introductory activity

Recap range and averages from Year 7 (using the mental starter).

Discuss how to find the median, mode, mean and range when data is presented in a stem-and-leaf diagram, using D2.3OHP.

Emphasise that to find the median from a stem-and-leaf diagram, the entries must be in numerical order.

Discuss the interpretation of the statistics in the students' book.

Emphasise that statistics should be interpreted in terms of what they represent (house numbers in this case).

Emphasise that the interpretation should answer the initial question (not prove its truth).
It should aim to use all the available results.
Interpretation should also include discussion about the shape of the diagram.

D2.3 Interpreting data

This spread will show you how to:
- Interpret tables, graphs and diagrams.
- Draw inferences to support or cast doubt on initial conjectures.

KEYWORDS
Interpret   Infer
Mean   Median
Mode   Range
Stem-and-leaf diagram

Benford want to analyse the results of their house number survey.
They can calculate statistics from their stem-and-leaf diagram.

```
10 | 5 6 9
 9 |
 8 | 0 1 4
 7 | 0 2 4 8
 6 | 0 2 4 6 6 8
 5 | 0 2 2 2 6 8 8 9
 4 | 4 7 8
```

Median: the 27 pieces of data are arranged in order so just count to find the middle (14th) value:
The median is 64 houses.

Mode: Just scan the rows to find the most frequent number:
The mode is 52 houses.

Key:
5 | 2 means 52

Mean: Add up all the values (1821) and divide by the number of values (27): $1821 \div 27 = 67.4$
The mean is 67.4 houses.

Range: Subtract the smallest number (44) from the largest (109):
The range is 65.

> The mean is 67.4 houses
> What's 0.4 of a house – a shed?

Benford need to interpret the statistics that they have calculated.

The part of the stem with most leaves is 50, and the mode supports this.

    5 | 0 2 2 2 6 8 8 9

> The mean is not necessarily a data value.

However, the median and mean are both in the sixties, suggesting a slightly higher average value for house numbers.

The shape of the diagram shows that most of the data is in the fifties and sixties. All three averages are within these values.

```
10 | 5 6 9
 9 |
 8 | 0 1 4
 7 | 0 2 4 8
 6 | 0 2 4 6 6 8
 5 | 0 2 2 2 6 8 8 9
 4 | 4 7 8
```

The range is 65, suggesting that there is a lot of variation in the number of houses in each street.

Three streets contained more than 100 houses.
Benford did not have enough data to decide whether streets with over 100 houses were **extreme** values, or were likely to be **typical**.

152

Plenary

Reinforce how an ordered stem-and-leaf graph simplifies the work to find mode, median and range and to describe skew.

Further activities

Students could explore ways of finding the mean from a stem-and-leaf diagram using an estimate of the mean.

Differentiation

Extension questions:
▶ Question 1 requires students to draw and interpret a stem-and-leaf diagram.
▶ Question 2 extends to data given in units of time.
▶ Question 3 requires students to draw and interpret a back-to-back stem-and-leaf diagram.

Core tier: focuses on interpreting pie charts, population pyramids and line graphs.

Exercise D2.3

1 The masses, in milligrams, of a sample of pebbles are:

169 178 164 182 194 204 186 192 201
182 164 175 179 173 182 172 180 171
168 172 169 185 182 200 166 198 170

a Draw a stem-and-leaf diagram to represent these data.
b Find the range, mean, median and mode of these data.
c Comment on the shape of the diagram and your results in **b**.

2 The winning times, in hours, minutes and seconds, of the New York marathon from 1976 to 1992 for men are given in this set of data.

2 : 10 : 10   2 : 11 : 29   2 : 12 : 12   2 : 11 : 42   2 : 09 : 41
2 : 08 : 13   2 : 09 : 29   2 : 08 : 59   2 : 14 : 53   2 : 11 : 34
2 : 11 : 06   2 : 11 : 01   2 : 08 : 20   2 : 08 : 01   2 : 12 : 39
2 : 09 : 24   2 : 09 : 29

a Draw a stem-and-leaf diagram to represent these data.
b Find the range, mean, median and mode of these data.
c Comment on the shape of the diagram and your results in **b**.

3 The data shows the total points scored by the football teams in Division 1 and Division 2 at the end of the 1980/1981 season.

Division One
60 56 53 52 51 50 50 48 44 43 42
39 38 37 36 35 35 35 35 33 32 19

Division Two
66 53 50 50 48 45 45 43 43 42 42
40 40 39 39 38 38 38 36 36 30 23

a Draw a back-to-back stem-and-leaf diagram to represent these data.
b Find the mean, median and range of these data sets.
c Comment on the shape of the diagram and your results in **b**.

153

Exercise commentary

The questions assess the objectives on Framework Pages 269, 271 and 273.

Problem solving
The exercise provide an opportunity to assess Framework Pages 25, 27 and 33.

Group work
The suggested further activities are suitable for working in pairs or alone.

Misconceptions
In question 2 students may have problems finding the mean as they are dealing with times.
Encourage students to think sensibly about how they will organise their diagram. An obvious choice would be hours and minutes in the stem, and seconds in the leaves.

Links
This topic links to practical investigations in science.

Homework

D2.3HW requires students to calculate statistics from a stem-and-leaf diagram and to interpret their results.

Answers

1 b 40; 179.9; 179; 182;   c Skewed to lower values.

2 b 6 : 52; 2 : 10 : 30; 2 : 10 : 10; 2 : 9 : 29;
   c One outlier.

3 b Div.1 41.95, 40.5, 41; Div.2 42, 41, 43
   c Similar summary statistics, but graph shape different.

Mental starter

To practise plotting points on graphs, write the coordinates of the corners of recognisable shapes on the board.
For example, trapezium (2, 2) (4, 5) (8, 5) (10, 2); parallelogram (3, 1) (5, 4) (7, 4) (5, 1); pentagon (2, 5) (4, 7) (6, 5) (5, 3) (3, 3). Ask the students to plot the points on a sketch graph and name the shapes.

Useful resources

Mini-whiteboards may be useful for the mental starter.
D2.4OHP – scatter graphs from the student book.

Introductory activity

Discuss the data given in the student book, relating to wheel-clamps and tows. Show how the data can be represented on a scatter graph.
Discuss what the data shows, and use the terms trend and correlation. Discuss reasons as to why the two variables are linked, and encourage students to suggest a third (hidden) variable: an increased number of parking attendents perhaps, or an increased number of vehicles.

Discuss the different types of correlation – D2.4OHP shows the scatter graphs in the student book.

Discuss what a scatter graph shows you about the two variables.

Emphasise that although the trend on the scatter graph between two variables may show positive or negative correlation, there may not be a connection between the two variables.

For example, a scatter graph of the number of storks nesting on rooftops and the number of babies born in those houses may show positive correlation, but there is no real connection between the variables.

D2.4 Scatter graphs

This spread will show you how to:
▶ Construct scatter graphs.
▶ Identify scatter graphs.
▶ Develop basic understanding of correlation.

KEYWORDS
Scatter graph
Variable
Correlation
Trend

The table gives data about the number of vehicles wheel-clamped and towed away by London police.

All figures are given in thousands to the nearest thousand.

Year	1984	1985	1986	1987	1988	1989	1990	1991	1992	1993	1994
Wheel-clamps	44	35	34	114	130	153	163	128	100	55	21
Tows	48	45	47	46	102	117	136	138	123	81	31

Each column in the table shows a pair of variables, linked by year, for example (34, 47) in 1986.

1986
34
47

▶ You can show data containing two linked variables on a **scatter graph**.

The graph shows that generally as the number of wheel-clamps increased, so did the number of vehicles towed away.
The **trend** is generally increasing. There is a relationship, or **correlation**, between wheel-clamps and tows.

Plot the data as (x, y) points. For example, (153, 117).

Show one variable on each of the axes.

▶ You use a scatter graph to show if there is any **correlation** between two variables.

When the trend is increasing, there is **positive correlation** between the variables.

When the trend is decreasing, there is **negative correlation** between the variables.

When there is no apparent trend, there is **no correlation** between the variables.

154

Plenary

Discuss how a description of a trend (a comment on correlation) is not the same as an interpretation (a comment on the relationship between the variables), as in the example of the storks.

Discuss why it is not always appropriate to extrapolate – refer to question 3.

Further activities

Students could extend question 4 by collecting data from the class about the time taken for people to spell their first names both forwards and backwards.

Differentiation

Extension questions:
▸ Questions 1 and 2 require students to draw and comment on a scatter graph.
▸ Question 3 includes discussion of extrapolation.
▸ Question 4 includes discussion of an outlier.

Core tier: focuses on stem-and-leaf diagrams.

Exercise D2.4

1 The data shows the goals scored for and the goals scored against each team in the premier division in 2001/2002.

For	79	67	87	74	53	66	48	46	49	55	46	35	36	38	45	44	29	41	33	30
Against	36	30	45	52	37	38	57	47	53	51	54	47	44	49	57	62	51	64	63	64

a Draw a scatter graph to display this data.
b Comment on any trend shown by your graph.

2 The results of two tests X and Y, taken by 12 students are given in the table.

Test X	16	12	18	15	6	14	18	4	16	8	13	16
Test Y	14	13	15	15	8	12	19	6	17	5	11	19

a Draw a scatter diagram to show these results.
b What can you say about the performance of the students in each of the tests X and Y?

3 Josh was training hard to improve his 100 m sprint times.
He was timed, to the nearest tenth of a second, each Sunday morning over a period of ten weeks.

Week	1	2	3	4	5	6	7	8	9	10
Time (s)	34.6	33.9	33.0	32.2	31.5	31.8	31.0	30.6	30.2	29.6

a Draw a scatter diagram to show these results.
b Comment on the correlation shown by your graph.
c Use your graph to comment on what Josh's 100 m time might be after 50 weeks training. Is your answer sensible?

4 The table shows how long, in seconds, it took eight people to spell their name backwards.

Name	Hugh	Helen	Harry	Hannah	Hamish	Horatio	Heather	Henrietta
Number of letters in name	4	5	5	6	6	7	7	9
Time in (seconds)	3.3	5.4	4.9	3.6	6.2	7.8	7.6	9.6

a Plot a scatter diagram of these data.
b One of the points plotted does not seem to fit with the rest of the data.
Circle this point and suggest a reason why it may not follow the trend.

Exercise commentary

The questions assess the objectives on Framework Pages 267 and 271.

Problem solving
Questions 3 and 4 provide an opportunity to assess Framework Page 25, as long as little or no guidance is given. The whole exercise assesses Pages 27 and 33.

Group work
The final parts of questions 3 and 4 are suitable for discussion in pairs.
The suggested further activity is most suitable for large group or whole class work.

Misconceptions
In questions 3 and 4 students may be tempted to put time on the *x*-axis, but the independent variable is the week number in question 3 and the number of letters in question 4.

Links
This topic links to practical investigations in science.

Homework

D2.4HW requires students to draw and interpret a scatter graph.

Answers

1 b Weak negative correlation
2 b The results in both tests are similar, i.e. good in one indicates good in the other.
3 b Negative correlation c 10 seconds; not sensible
4 b Hannah; name is a palindrome

Mental starter

Write these data sets on the board.

Set A: 6 4 2 5 13
Set B: 12 20 16 13 16 18 16
Set C: 16 7 12 11 7 11 9 15

Ask students which is the most appropriate average to use for each set. Discuss reasons why.

Useful resources

D2.5OHP contains the tables from the student book.

Introductory activity

Recap on how to choose the most appropriate average to use for a data set (using the mental starter). Remind students that the best value to use is the one that is most 'typical' for that data set.

Emphasise that you use an average and the range to summarise a set of data.
If there are some rogue values, or if some values are missing, then these summary values will be distorted.

Discuss how you can use summary values to compare sets of data.

Emphasise that you need a large set of data to analyse real situations and so you need to be able to summarise it effectively.

Discuss the statistics represented in the student book. **D2.5OHP** shows the tables, relating to wheel-clamps and tows.
Use the discussion to emphasise that a statistical comparison should take into account variation as well as average.

D2.5 Comparing data

This spread will show you how to:
- Select statistics most appropriate to the problem.
- Compare two or more distributions using appropriate statistics.
- Analyse the effect that minimal changes in data can have on graphs and statistical measures.

KEYWORDS
Mean   Range
Median   Variation

The table gives data on vehicles wheel-clamped and towed away by London police, as shown on page 154. All figures are given in thousands, to the nearest thousand.

Year	1984	1985	1986	1987	1988	1989	1990	1991	1992	1993	1994
Wheel-clamps	44	35	34	114	130	153	163	128	100	55	21
Tows	48	45	47	46	102	117	136	138	123	81	31

You can use statistics to compare the data.

	Mean	Median	Range
Wheelclamps	88.8	100	142
Tows	83.1	81	107

The two means and medians suggest that there are more vehicles wheel-clamped than towed away.
The two ranges suggest that there is greater variation in numbers of vehicles clamped than towed away.

When people deal with large amounts of data, sometimes data can get lost or distorted.

▶ Minimal changes in data can affect graphs and statistical measures.

If data had been lost for 1987 to 1990, then the calculated statistics would be:

	Mean	Median	Range
Wheelclamps	59.6	44	107
Tows	73.3	48	107

The median number of wheel-clamps is more typical of the data.
The mean is distorted by the large values in 1991 and 1992.

These statistics suggest that on average more vehicles are towed away than clamped, and that there is no difference in variation.

▶ The average you choose for a data set is the one that should best represent all the data.

Statistics summarise the original data, or **raw data**.
When you calculate statistics, you lose some of the information in the raw data.

▶ Statistics can help compare data, but the original values are also important.

156

Plenary

Discuss the different answers that students found for questions 4, 5 and 6. Use these to emphasise why original data values can be important, while the summary statistics give an overall picture.

Further activities

Students could find further examples of data sets for questions 4 and 5.
They could find a rule for the data sets in question 5.

Differentiation

Extension questions:

▸ Question 1 requires students to select and justify appropriate statistics.
▸ Question 2–5 focus on comparing sets of discrete data.
▸ Question 6 extends to continuous data.

Core tier: focuses on calculating statistics and comparing distributions.

Exercise D2.5

1 For each of the following sets of data write down, with reasons:
   i which average, mean, median or mode, is the most appropriate to find, and
   ii find this average.
   a 14.6   19.3   12.0   15.7   31.7
   b 2003   2005   2008   2011   2006   2003   2008   2003
   c 101   102   102   102   108   108   108   109
   d 49   44   44   47   38   36   44   44   49   44   43

2 **a** Find the mean, median, mode and range of this set of data.
      7   7   24   5   2
   b One number is added to the data set. What could that number be if
      i   the mode remains the same
      ii   the median remains the same
      iii the mean remains the same
      iv   the mean is increased by 2?
   c One number is removed from the data set. What could that number be if
      i   the range and median remain unchanged
      ii   the mean is increased by 1?

3 **a** Calculate the mean, median, mode and range of each of these two data sets.
      Set A: 0, 99, 99, 100, 100, 100, 100, 100, 101, 101, 200
      Set B: 0, 0, 99, 99, 100, 100, 100, 101, 101, 200, 200
   b Compare your results. What do the statistics suggest about the two data sets?

4 Find two sets of numbers that have the same mean, median, mode and range.
   (Do not use the data in question 3.)

5 Find data sets with the following properties
   a Set A: Whatever number you remove, the mode remains the same.
   b Set B: When you remove one number, the mean stays the same.
   c Set C: When you remove one number, the mean is half the original value.
   d Set D: When you remove a number, the mean and range remain the same.

6 The table shows grouped data of heights of a sample of Year 8 pupils.

Height, h cm	$135 \leqslant h < 145$	$145 \leqslant h < 150$	$150 \leqslant h < 155$	$155 \leqslant h < 170$
Frequency	2	4	6	3

   Find two different sets of raw data that:
   a could be grouped in this frequency table
   b have the same mean, median and range.

Exercise commentary

The questions assess the objectives on Framework Pages 257 and 273.

Problem solving

Questions 2–6 provide an opportunity to assess Framework Page 25, as long as little or no guidance is given. Question 3 also assesses Page 31.

Group work

Students can work in pairs to compare and check their answers to questions 4, 5 and 6.

Misconceptions

Data needs to be ordered to find the median.
Range is a single number, calculated by
      largest value – smallest value
Students may find identifying an appropriate average difficult. Encourage students to ask themselves: 'Is this value typical of the data?'

Links

This topic links to practical investigations in science.

Homework

D2.5HW requires students to compare two distributions.

Answers

1 **a i** Median **ii** 15.7   **b i** Mean or median
      ii 2005.875 or 2005.5   **c i** Mean or median
      ii 105   **d i** mode **ii** 44
2 **a** 9, 7, 7, 22   **b i** Not 2, 5 or 24 **ii** 7 or more **iii** 9
      iv 21   **c i** 5 **ii** 5.
3 **a** 100, 100, 100, 200; 100, 100, 100, 200
      b Suggests they are the same, but they are not.

D2.6 Statistical reports

Mental starter

Write sets of data on the board. For example:
- Discrete data, shoe sizes: 5, $4\frac{1}{2}$, 3, 5, 4, 5, $6\frac{1}{2}$
- Continuous data, temperatures (°C): 18, 20, 24, 19, 18, 20
- Bivariate data, rainfall (mm) in two towns

A	4.2	3.6	2.1	2.7
B	3.2	3.5	2.6	4.2

Ask students to list suitable graphs in each case.

Useful resources

Mini-whiteboards may be useful for the mental starter.

D2.6OHP – line graph and multiple bar chart from the student book.

Introductory activity

Discuss the function of a statistical report – to explain to others what you have discovered through a statistical enquiry.

Emphasise that you should draw the most appropriate graph for the data.
A graph should highlight a key point that you want to show.
It is pointless to draw several different types of graph that all highlight the same point. Different graphs should be drawn to highlight different key points.

Remind students that not all types of graph are appropriate for both discrete and continuous data.

Emphasise that when interpreting graphs showing two or more variables (scatter diagram, multiple bar chart) you should make comparisons between the variables.

Discuss the scenario in the student book, which relates to the changing patterns in air travel over a 10-year period.
Show how you can use graphs to extract key points.
D2.6OHP shows the graphs in the student book.

D2.6 Statistical reports

This spread will show you how to:
- Communicate interpretations and results of a statistical enquiry.
- Select tables, graphs and diagrams to support findings.

KEYWORDS
Line graph
Bar chart
Statistical enquiry

Sharon conducted a survey into how air traffic had changed from 1984 to 1994.

The table shows the numbers of air terminal passengers in millions.

	Heathrow Airport	Gatwick Airport	Luton Airport	Stanstead Airport	Scheduled flight	Charter flight
1984	29.2	14.0	1.8	0.5	35.0	10.5
1985	31.3	14.9	1.6	0.5	38.0	10.4
1986	31.3	16.3	2.0	0.5	38.5	11.7
1987	34.7	19.4	2.6	0.7	43.8	13.7
1988	37.5	20.7	2.8	1.0	48.1	14.2
1989	39.6	21.2	2.8	1.3	51.9	13.3
1990	42.6	21.0	2.7	1.2	56.4	11.5
1991	40.2	18.7	2.0	1.7	52.2	10.6
1992	45.0	19.8	2.0	2.3	58.0	11.4
1993	47.6	20.1	1.8	2.7	61.0	11.5
1994	51.4	21.0	1.8	3.3	65.5	12.4

Sharon used statistics from the Department of Transport, which is secondary data. You should always state where the data has come from.

Sharon drew a line graph to show changes in passenger numbers at the airports.

Sharon drew a **multiple bar chart** to show passenger numbers at Luton and Stanstead.

- Generally numbers were increasing.
- There was a drop in numbers in 1991 (an effect of the Gulf War).
- Stanstead had a rapid increase in passenger numbers after 1991 (a new terminal opened).

- The bar chart shows clearly how passenger numbers at Stanstead had overtaken passenger numbers at Luton.

In a statistical report you should
- Explain how the data was collected
- Analyse the data
- Use graphs to highlight key points
- Look to see if there are other related questions to explore.

Plenary

Discuss students' different interpretations of the data sets in the questions.

Discuss what other related questions may be important or interesting to explore.

Further activities

Students could use the Internet to search for data on road accidents and casualties in the local area.

They could compare the findings with the data given in questions 2 and 3.

Differentiation

Extension questions:
▸ Question 1 requires students to draw and interpret a line graph.
▸ Question 2 requires students to draw a multiple bar graph, and write a short report.
▸ Question 3 allows students to choose a suitable diagram to display the data.

Core tier: focuses on writing up a statistical report, using simpler contexts and analysis.

Exercise D2.6

1 a Use the air traffic data on the opposite page to draw a line graph showing how scheduled flights and charter flights changed over the 11-year period.
   b Comment on the changes in scheduled and charter flights from 1984–1994.

2 The numbers of people injured in road traffic accidents in a London borough over a three-year period sorted into male and female drivers are given in the table.

	Pedestrians	Car drivers	Car passengers	Pedal cyclists	Motorcyclists	Bus passengers	Goods vehicles	Other
Male Drivers	400	1200	550	250	550	200	100	100
Female Drivers	200	1000	325	50	50	10	20	20

   a Draw a multiple bar graph to analyse these data.
   b Write a short report on the differences between the vehicle types that are involved in accidents by men and women.

3 The table shows the number of road casualties in a London borough by month of the year and by age.

Month	Under 15	16–39 years	40–59 years	Over 60
January	35	130	120	50
February	55	180	150	35
March	65	135	135	60
April	45	115	125	45
May	60	150	130	50
June	45	135	115	45
July	60	130	120	40
August	60	140	120	50
September	70	160	140	45
October	60	160	155	60
November	50	180	170	70
December	30	180	130	65

   a Draw a diagram to represent these data.
   b Why do you think that the number of casualties aged under 15 gets lower through the Autumn while for other age groups it gets higher?
   c Write a short report about the monthly variations in the number of road casualties for the different age groups.
      Suggest reasons for the monthly variations in the number of road casualties.

159

Exercise commentary

The questions assess the objectives on Framework Pages 263, 265, 269, 271 and 273.

Problem solving
Questions 1–3 provide an opportunity to assess Framework Page 31.

Group work
For question 3, students may like to discuss their thoughts on the data provided in small groups.

Misconceptions
Students' reports on bivariate data often only comment on one of the variables, rather than making comparisons between the variables.
Encourage students to use correct terminology where appropriate, such as 'trend' and 'correlation'.

Links
This topic links to practical investigations in science, PHSE and geography.

Homework

D2.6HW provides data on the effects of congestion charging in London, and requires students to write a short report.

Answers

1 b Scheduled flights nearly doubled; limited growth in charter flights.
2 b If goods vehicle, bus passenger, motorcyclist or pedal cyclist involved driver much more likely to be male.
3 b Taken to school by parents in winter months.

The key objectives for this unit are:
▶ Communicate interpretations and results of a statistical enquiry using selected tables, graphs and diagrams in support.
▶ Present a concise, reasoned argument, using symbols, diagram, graphs and related explanatory text.

Plenary activity

Describe different experiments in which data is collected and ask students to describe the most appropriate graph to use to display the data.

Check out commentary

1. Students should take care in drawing their stem-and-leaf diagram, paying particular attention to:
   ▶ Choosing sensible values for the stem, based on the range of data
   ▶ Ordering the digits on the leaves
   ▶ Arranging the digits on the leaves in orderly columns
   ▶ Including a key.
   The student book suggests expanding the stem into the digits 0–1 and 5–9.

   In part **b**, emphasise that the median can only be found from ordered data.

   In part **c**, encourage students to refer to both the shape of the diagram and the statistics that they calculated in part **b**.

2. Encourage students to choose their own statistics and diagrams in this question. The mean or the median are valid averages to use, but remind students to include the range in their comparison.

   A multiple bar chart may be the simplest graph to produce for this data.

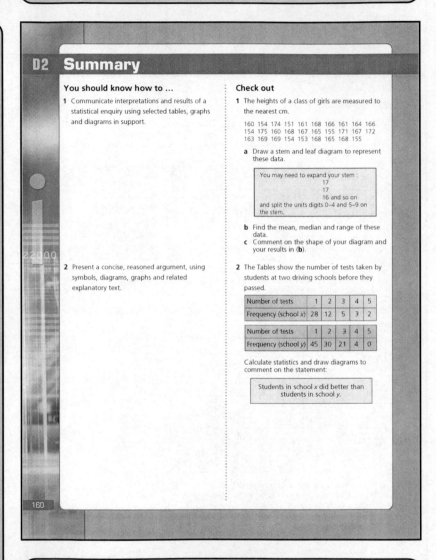

Development

The themes of this unit are developed in units D3 and D4.

Links

Handling data links to work in other subjects such as science.

Mental starters

Objectives covered in this unit:
▸ Convert between fractions, decimals and percentages.
▸ Find fractions and percentages of quantities.
▸ Use jottings to support addition and subtraction of whole numbers and decimals.
▸ Multiply and divide a two-digit number by a one-digit number.
▸ Calculate using knowledge of multiplication and division facts and place value.
▸ Convert between m, cm and mm, km and m, kg and g, litres and ml, cm^2 and mm^2.

Resources needed

* means class set needed
Essential:
Counters*
Dice*
N4.3OHP – BIDMAS chart
N4.4OHP – board for game in N4.4 question 6
N4.6OHP – metric and imperial/metric relationships
Useful:
OHP calculator
Mini-whiteboards
Measuring tape or metre rule
N4.4ICT – long multiplication

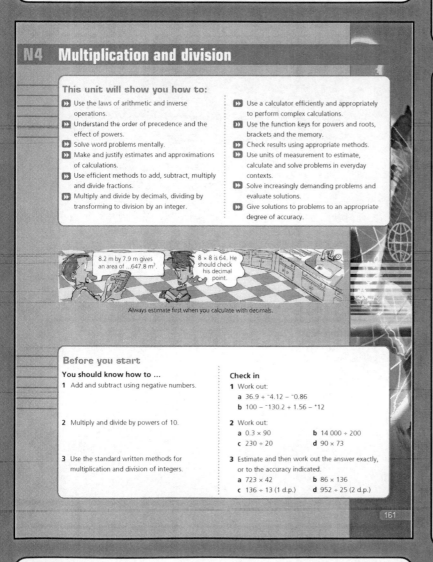

N4 Multiplication and division

This unit will show you how to:

- Use the laws of arithmetic and inverse operations.
- Understand the order of precedence and the effect of powers.
- Solve word problems mentally.
- Make and justify estimates and approximations of calculations.
- Use efficient methods to add, subtract, multiply and divide fractions.
- Multiply and divide by decimals, dividing by transforming to division by an integer.

- Use a calculator efficiently and appropriately to perform complex calculations.
- Use the function keys for powers and roots, brackets and the memory.
- Check results using appropriate methods.
- Use units of measurement to estimate, calculate and solve problems in everyday contexts.
- Solve increasingly demanding problems and evaluate solutions.
- Give solutions to problems to an appropriate degree of accuracy.

8.2 m by 7.9 m gives an area of ...647.8 m².

8 × 8 is 64. He should check his decimal point.

Always estimate first when you calculate with decimals.

Before you start

You should know how to ...
1 Add and subtract using negative numbers.

2 Multiply and divide by powers of 10.

3 Use the standard written methods for multiplication and division of integers.

Check in
1 Work out:
  a 36.9 + ⁻4.12 – ⁻0.86
  b 100 – ⁻130.2 + 1.56 – ⁺12
2 Work out:
  a 0.3 × 90   b 14 000 ÷ 200
  c 230 ÷ 20   d 90 × 73
3 Estimate and then work out the answer exactly, or to the accuracy indicated.
  a 723 × 42   b 86 × 136
  c 136 ÷ 13 (1 d.p.)   d 952 ÷ 25 (2 d.p.)

161

Unit commentary

Aim of the unit

This unit extends mental and written methods of addition, subtraction, multiplication and division to positive and negative fractions and decimals. The commutative nature of addition and multiplication, the order of operations, inverse operations and reciprocals are all considered. The final lesson uses the multiplication and division skills developed in metric and metric/imperial conversions.

Introduction

Discuss the variety of strategies that students have learned to enable them to solve calculations mentally. Focus on strategies that are particularly efficient.

Framework references

This unit focuses on:
▸ Teaching objectives pages: 85–87, 97–101, 105–107, 228–231
▸ Problem-solving objectives pages: 7, 19–21, 29, 35.

Check in activity

Write a variety of calculations on the board.
For example: 22 × 18, 3.5 – 2.8, 4.2 ÷ 7, 9.1 + 0.91
In pairs, students divide up the calculations between themselves and demonstrate to each other how to solve them.

Differentiation

Core tier

Focuses on multiplication and division, using simpler numbers.

Mental starter

Write these sums on the board:

$5.6 - 2.3$	$6.7 + 3.9$	$5.6 - 0.3$	$3.9 + 6.7$
$2.3 - 5.6$	$0.03 - 5.6$	$0.004 - 0.3$	$0.004 + 0.3$
$0.3 + 0.004$	$0.3 - 0.004$		

Students should pair up the sums that give the same answer.
What do they notice about the sums that are left?

Useful resources

Mini-whiteboards may be useful for the mental starter.
OHP calculator

Introductory activity

Discuss the sums that were left in the starter. Encourage students to describe them in their own words, for example back to front, reversible. Introduce the term 'commutative'.

Demonstrate how the commutative law applies to problems involving negatives, decimals and fractions, using an OHP calculator.

$$\tfrac{7}{8} - 1.91 \neq 1.91 - \tfrac{7}{8}$$
but
$$\tfrac{7}{8} + 1.91 = 1.91 + \tfrac{7}{8}$$
$$^-\tfrac{7}{8} - {}^-1.91 \neq {}^-1.91 - {}^-\tfrac{7}{8}$$
$$^-\tfrac{7}{8} + {}^-1.91 = {}^-1.91 + {}^-\tfrac{7}{8}$$

Recap the strategies previously learned for making calculations more manageable.

Emphasise that you should convert both numbers either to fractions or to decimals, in problems like the ones given above.

Recap formal written methods for addition and subtraction.
Demonstrate the use of 'redundant zeros' as place holders.

$$4.5 + 5.006 \rightarrow \quad \begin{array}{r} 4.500 \\ \underline{5.006} \end{array}$$

Encourage students to try a mental approach first, and to estimate the answer before they do the calculation, so they can check their exact answer.

Demonstrate the use of inverse operations to check calculations.

$$4.5 + 3.7 = 8.2$$
So it follows that
$$8.2 - 3.7 = 4.5$$

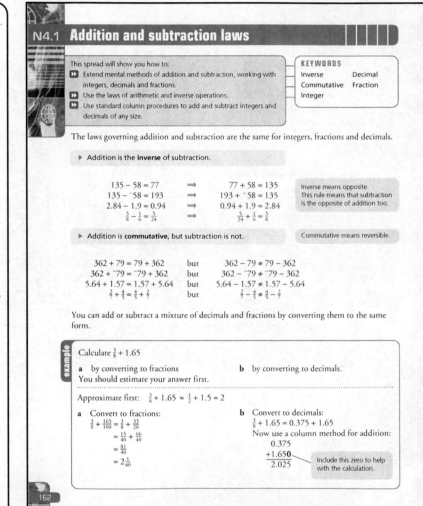

N4.1 Addition and subtraction laws

This spread will show you how to:
- Extend mental methods of addition and subtraction, working with integers, decimals and fractions.
- Use the laws of arithmetic and inverse operations.
- Use standard column procedures to add and subtract integers and decimals of any size.

KEYWORDS
Inverse   Decimal
Commutative Fraction
Integer

The laws governing addition and subtraction are the same for integers, fractions and decimals.

▶ Addition is the **inverse** of subtraction.

$135 - 58 = 77$	$\Rightarrow$	$77 + 58 = 135$
$135 - {}^-58 = 193$	$\Rightarrow$	$193 + {}^-58 = 135$
$2.84 - 1.9 = 0.94$	$\Rightarrow$	$0.94 + 1.9 = 2.84$
$\tfrac{3}{8} - \tfrac{1}{6} = \tfrac{5}{24}$	$\Rightarrow$	$\tfrac{5}{24} + \tfrac{1}{6} = \tfrac{3}{8}$

Inverse means opposite.
This rule means that subtraction is the opposite of addition too.

▶ Addition is **commutative**, but subtraction is not.

Commutative means reversible.

$362 + 79 = 79 + 362$	but	$362 - 79 \neq 79 - 362$
$362 + {}^-79 = {}^-79 + 362$	but	$362 - {}^-79 \neq {}^-79 - 362$
$5.64 + 1.57 = 1.57 + 5.64$	but	$5.64 - 1.57 \neq 1.57 - 5.64$
$\tfrac{2}{5} + \tfrac{4}{7} = \tfrac{4}{7} + \tfrac{2}{5}$	but	$\tfrac{2}{5} - \tfrac{4}{7} \neq \tfrac{4}{7} - \tfrac{2}{5}$

You can add or subtract a mixture of decimals and fractions by converting them to the same form.

example

Calculate $\tfrac{3}{8} + 1.65$

a by converting to fractions   **b** by converting to decimals.
You should estimate your answer first.

Approximate first: $\tfrac{3}{8} + 1.65 \approx \tfrac{1}{2} + 1.5 = 2$

a Convert to fractions:
$$\tfrac{3}{8} + \tfrac{165}{100} = \tfrac{3}{8} + \tfrac{33}{20}$$
$$= \tfrac{15}{40} + \tfrac{66}{40}$$
$$= \tfrac{81}{40}$$
$$= 2\tfrac{1}{40}$$

b Convert to decimals:
$$\tfrac{3}{8} + 1.65 = 0.375 + 1.65$$
Now use a column method for addition:
$$\begin{array}{r} 0.375 \\ +1.650 \\ \hline 2.025 \end{array}$$
Include this zero to help with the calculation.

162

Plenary

Write on the board:
$$\tfrac{6}{7} + 0.23 - {}^-0.44$$
Ask students to suggest strategies to make this calculation easier.

Discuss the use of mental methods, simple jottings, written methods, calculators.

Further activities

Students can develop question 6 by setting similar puzzles for a partner to solve.

Alternatively, students can invent addition pyramids for each other as in questions 4 and 7.

Differentiation

Extension questions:
- Questions 1 and 2 introduce addition and subtraction of mixed decimals and fractions.
- Questions 3–6 focus on addition and subtraction, using problem-solving activities.
- Questions 7 is a harder version of question 4, extending to four-tier pyramids.

Core tier: consolidates addition and subtraction strategies, using simpler numbers.

Exercise N4.1

1. Work out each of these giving your answer as a decimal.
   a $\frac{2}{5} + 1.69$ b $0.8 - \frac{1}{8}$
   c $\frac{4}{5} - 2.6$ d $5\frac{1}{2} - 7.03$
   e $1.3 - 4\frac{7}{8}$ f $16.2 + \frac{1}{8} + 1\frac{1}{5}$

2. Give each of your answers to question 1a–f as a fraction in its lowest terms.

3. Look at the shape. Work out
   a an estimate for its perimeter
   b its exact perimeter as a decimal.

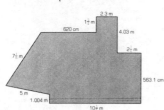

4. Puzzle
   In each pyramid the brick that sits directly above two bricks is the sum of these two bricks.
   Copy and complete these pyramids.

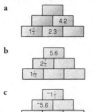
   a
   b
   c

5. The difference between two numbers is 0.12 and the sum of the two numbers is 1.32.
   Write the two numbers as decimals and as fractions.

6. Puzzle
   Use the numbers in the rectangle to complete the number sentences below.

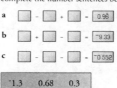
   a
   b
   c

⁻1.3	0.68	0.3
⁻8.4	⁻¼	⅞
⅔	⁻0.023	⁻1

7. Puzzle
   In these pyramids the brick which sits directly above two bricks is the sum of those two bricks, for example.

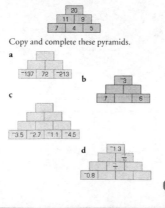

   Copy and complete these pyramids.
   a
   b
   c
   d

Exercise commentary

The questions assess the objectives on Framework Pages 85 and 105.

Problem solving
Questions 4–7 assess the objectives on Framework Page 9.

Group work
The further activity can be done in pairs.

Misconceptions
Students who have problems understanding the term 'commutative' should think of it as meaning 'reversible'. Encourage student to try simpler examples to reinforce the concept.

Links
Converting decimals and fractions: Framework Page 65
Question 3 links to metric units and perimeter.

Homework

N4.1HW requires students to add and subtract a mixture of decimals and fractions.

Answers

1 a 2.09 b 0.675 c ⁻1.8 d ⁻1.53 e ⁻3.575 f 17.525
2 a $2\frac{9}{100}$ b $\frac{27}{40}$ c $⁻1\frac{4}{5}$ d $⁻1\frac{53}{100}$ e $⁻3\frac{23}{40}$ f $17\frac{21}{40}$ 3 b 46.115 m
4 8.25, 4.05, 1.9; 2.725, $1\frac{13}{16}$, 0.9125; 4.1, ⁻7.4, 1.8 (or fraction/decimal equivalents)
5 0.72, 0.6; $\frac{18}{25}, \frac{3}{5}$ 6 a ⁻1 − ⁻1.3 + $\frac{2}{3}$ b ⁻8.4 + ⁻¼ − 0.68 c 0.3 − ⅞ − ⁻0.023
7 a ⁻206; ⁻65, ⁻141 b ⁻1, ⁻2; ⁻8 c ⁻19.4; ⁻10, ⁻9.4; ⁻6.2, ⁻3.8, ⁻5.6
   d ⁻0.6; ⁻0.4, ⁻0.5; 0.4, ⁻0.6, 0.1

Mental starter

Write on the board:
$A \times B = C$

Students should calculate the values of C in the table. Discuss the mental strategies for multiplication that they used.

A	B	C
0.4	$-\frac{1}{5}$	
$^-0.12$	$^-7$	
0.4	$^-10$	
$\frac{3}{8}$	0.2	
$-\frac{6}{7}$	60	

Useful resources

Mini-whiteboards may be useful for the mental starter and the plenary activity.

Introductory activity

Discuss the answers to the calculations in the starter activity.
Encourage students to describe what they notice, e.g.
'multiplying by a number less than 1 (but greater than 0) makes the number smaller'
'A, B and C cannot all be negative'

Demonstrate how multiplication is commutative but division is not, using an OHP calculator.
$$3 \times 4 = 4 \times 3$$
but $\qquad 3 \div 4 \neq 4 \div 3$

Emphasise that any division can be written as a multiplication if it makes the calculation easier.
$$0.7 \div \tfrac{1}{5} = 0.7 \times 5$$
because multiplication is the inverse of division, and 5 is the inverse (reciprocal) of $\frac{1}{5}$.

Give students practice in finding reciprocals, for example of $\frac{2}{5}$, $^-8$, 0.25

Encourage students to look for strategies to simplify calculations, such as converting between decimals and fractions.

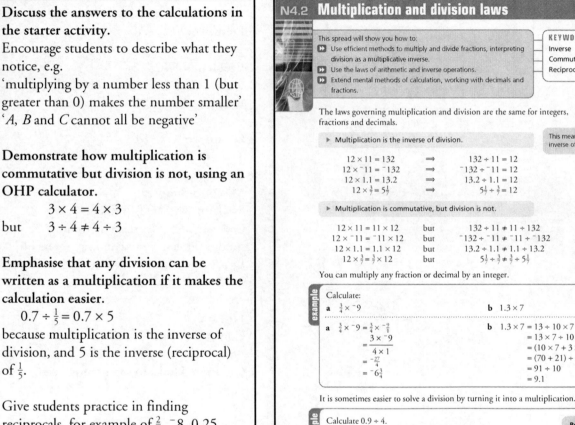

N4.2 Multiplication and division laws

This spread will show you how to:
- Use efficient methods to multiply and divide fractions, interpreting division as a multiplicative inverse.
- Use the laws of arithmetic and inverse operations.
- Extend mental methods of calculation, working with decimals and fractions.

KEYWORDS
Inverse
Commutative
Reciprocal

The laws governing multiplication and division are the same for integers, fractions and decimals.

▸ Multiplication is the inverse of division.

$12 \times 11 = 132$	$\Rightarrow$	$132 \div 11 = 12$
$12 \times {}^-11 = {}^-132$	$\Rightarrow$	$^-132 \div {}^-11 = 12$
$12 \times 1.1 = 13.2$	$\Rightarrow$	$13.2 \div 1.1 = 12$
$12 \times \frac{3}{7} = 5\frac{1}{7}$	$\Rightarrow$	$5\frac{1}{7} \div \frac{3}{7} = 12$

This means that division is the inverse of multiplication as well.

▸ Multiplication is commutative, but division is not.

$12 \times 11 = 11 \times 12$	but	$132 \div 11 \neq 11 \div 132$
$12 \times {}^-11 = {}^-11 \times 12$	but	$^-132 \div {}^-11 \neq {}^-11 \div {}^-132$
$12 \times 1.1 = 1.1 \times 12$	but	$13.2 \div 1.1 \neq 1.1 \div 13.2$
$12 \times \frac{3}{7} = \frac{3}{7} \times 12$	but	$5\frac{1}{7} \div \frac{3}{7} \neq \frac{3}{7} \div 5\frac{1}{7}$

You can multiply any fraction or decimal by an integer.

example

Calculate:

a $\frac{3}{4} \times {}^-9$ $\qquad\qquad$ **b** 1.3×7

a $\frac{3}{4} \times {}^-9 = \frac{3}{4} \times \frac{-9}{1}$
$\qquad = \frac{3 \times {}^-9}{4 \times 1}$
$\qquad = \frac{-27}{4}$
$\qquad = {}^-6\frac{3}{4}$

b $1.3 \times 7 = 13 \div 10 \times 7$
$\qquad = 13 \times 7 \div 10$
$\qquad = (10 \times 7 + 3 \times 7) \div 10$
$\qquad = (70 + 21) \div 10$
$\qquad = 91 \div 10$
$\qquad = 9.1$

It is sometimes easier to solve a division by turning it into a multiplication.

example

Calculate $0.9 \div 4$.

Remember: 4 is the reciprocal of $\frac{1}{4}$.

$0.9 \div 4 = 0.9 \times \frac{1}{4}$ $\qquad = 9 \times 25 \div 1000$
$\qquad = 0.9 \times 0.25$ $\qquad = 225 \div 1000$
$\qquad = 9 \div 10 \times 25 \div 100$ $\qquad = 0.225$

Plenary

Write on the board:
$$C \div B = A$$
Give the students questions to test their mental division strategies:
$A = 0.7$, find a value for B so that $^-1 < C < 0$
$B = {}^-\frac{3}{4}$, $A = 10$, find C as a decimal
$C = {}^-\frac{1}{100}$, $B = 40$, find A

Differentiation

Extension questions:

▸ Question 1 practices basic multiplication and division of decimals and fractions.

▸ Questions 2 to 7 focus on multiplication and division, including problem-solving activities.

▸ Question 8 is an investigation requiring students to generalise and justify their findings.

Core tier: focuses on strategies for multiplication and division, including the use of place value and factors to divide decimals.

Exercise N4.2

1 Calculate these:
  a $^-16 \times 0.7$ **b** $^-35 \div 0.5$
  c $\frac{5}{8} \times ^-14$ **d** $^-6 \div 0.03$
  e $\frac{2}{5} \times 18$ **f** $^-1.5 \div 0.3$
  g 25×1.8 **h** 4.8×0.3

2 Complete these multiplication pyramids. The number in each cell is the product of the two below it:

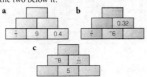

a   **b**

c

3 **a** Work out the value of each of the expressions **i** to **iv** if

  $a = 2$ $b = 0.6$ $c = ^-0.8$ $d = ^-9$

  Write your answers as fractions where appropriate.

  i $\frac{a}{c}$ **ii** $3b + 2c$ **iii** $3b - 2c$ **iv** d^2

  b Convert each of your fractions in part **a** to decimals.

4 The result

  $^-0.6 \times 0.37 = ^-0.222$

  allows you to write the solution to

  $^-0.222 \div ^-0.6 = 0.37$

  a Write down any **one** other problem and its solution that you can derive from knowing that $^-0.6 \times 0.37 = ^-0.222$

  b Write two problems and their solutions that you can derive from

  $0.003 \div ^-0.2 = ^-0.015$

5 Explain in your own words what the term 'commutative' means and write down two examples that show that multiplication is commutative but division is not.

6 Puzzle
  If $m = 0.5$, $n = \frac{1}{9}$ and $p = \frac{2}{3}$, write expressions using any combination of m, n and p that produce the following answers.
  You may only use the letters **once** in any one expression, but you may not have to use them all.

  a $^-\frac{1}{18}$ **b** $1\frac{1}{3}$
  c $\frac{2}{27}$ **d** $^-\frac{2}{9}$

7 Puzzle
  a Using the digits 1 to 10 once only complete the boxes:

  b Write the final answer as a decimal to 3 decimal places.
  c Is there more than one correct solution to this problem? Explain your answer.

8 Investigation
  Investigate this statement:

  > Multiplying makes things bigger and dividing makes things smaller.

  a Using only positive numbers, do some calculations to test this statement. You may use a calculator to help you.
  b Write a paragraph to explain whether you agree or disagree with this statement.

165

Exercise commentary

The questions assess the objectives on Framework Pages 105–107.

Problem solving

Question 7 provides an opportunity to assess Framework Page 7. Question 8 provides an opportunity to assess Framework Page 29. Question 5 assesses Page 35.

Group work

The further activity can be attemped in pairs.

Misconceptions

Students may need the commutativity of multiplication reinforced by using simpler examples, such as $3 \times 5 = 5 \times 3$. Similarly, use $4 \div 2 \neq 2 \div 4$ to highlight the non-commutative nature of division.

Links

Estimating calculations: Framework Page 103.
Use formulae from mathematics and other subjects: Framework Page 139.

Answers

1 **a** $^-11.2$ **b** $^-70$ **c** $^-5\frac{1}{4}$ **d** $^-200$ **e** $7\frac{1}{5}$ **f** $^-5$ **g** 45 **h** 1.44
2 **a** $10.8, 3, 3.6$; **b** $^-1.536, ^-4\frac{4}{5}, ^-\frac{4}{75}$ **c** $^-\frac{2}{5}, ^-1\frac{3}{5}, \frac{1}{100}$ (or fraction/decimal equivalents)
3 **a i** $^-2\frac{1}{2}$ **ii** $\frac{1}{5}$ **iii** $3\frac{2}{5}$ **iv** 81 **b i** $^-2.5$ **ii** 0.2 **iii** 3.4
7 **a** $\frac{7}{400}$ **b** 0.018 **c** Yes, numerators/denominators can be rearranged

Mental starter

Write on the board:
$$3 - 3 \times 5$$
$$\frac{26 - 1.5^2}{43 + 23}$$

Discuss how to solve each problem.
In which order do you carry out the operations? Does the order matter? Why?

Useful resources

Mini-whiteboards may be useful for the starter activity.
OHP calculator
N4.3OHP – BIDMAS chart

Introductory activity

Discuss the importance of a convention for the order of operations.

Recap BIDMAS (Brackets, Indices, Division and Multiplication, Addition and Subtraction).

Discuss the idea of 'hidden brackets' that surround the numerator and denominator, such as in the second example in the starter.

Demonstrate how to enter complex calculations involving powers and brackets into a calculator, using an OHP calculator.

Encourage pupils to work out the denominator part first, and store it in the memory, then the numerator and divide this by the memory.
If students prefer they can calculate the denominator first and write down the answer, then calculate the numerator and divide by the denominator.
Discuss the merits of these two methods.

Encourage students to do as much of a problem mentally as they can, before they use a calculator or formal written methods.

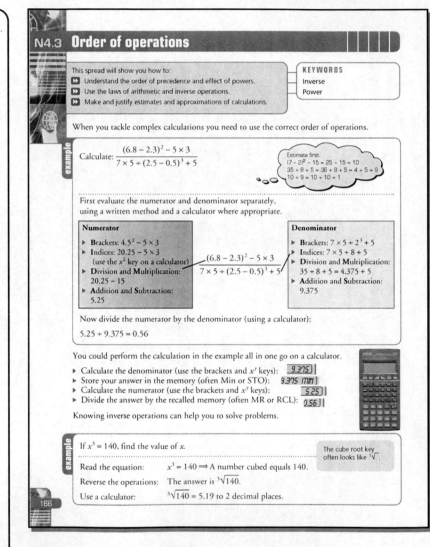

N4.3 Order of operations

This spread will show you how to:
- Understand the order of precedence and effect of powers.
- Use the laws of arithmetic and inverse operations.
- Make and justify estimates and approximations of calculations.

KEYWORDS
Inverse
Power

When you tackle complex calculations you need to use the correct order of operations.

example

Calculate: $\dfrac{(6.8 - 2.3)^2 - 5 \times 3}{7 \times 5 \div (2.5 - 0.5)^3 + 5}$

Estimate first:
$(7 - 2)^2 - 15 = 25 - 15 = 10$
$35 \div 8 + 5 = 36 \div 9 + 5 = 4 + 5 = 9$
$10 \div 9 = 10 \div 10 = 1$

First evaluate the numerator and denominator separately, using a written method and a calculator where appropriate.

Numerator
- Brackets: $4.5^2 - 5 \times 3$
- Indices: $20.25 - 5 \times 3$ (use the x^2 key on a calculator)
- Division and Multiplication: $20.25 - 15$
- Addition and Subtraction: 5.25

$\dfrac{(6.8 - 2.3)^2 - 5 \times 3}{7 \times 5 \div (2.5 - 0.5)^3 + 5}$

Denominator
- Brackets: $7 \times 5 \div 2^3 + 5$
- Indices: $7 \times 5 \div 8 + 5$
- Division and Multiplication: $35 \div 8 + 5 = 4.375 + 5$
- Addition and Subtraction: 9.375

Now divide the numerator by the denominator (using a calculator):

$5.25 \div 9.375 = 0.56$

You could perform the calculation in the example all in one go on a calculator.
- Calculate the denominator (use the brackets and x^y keys): `9.375`
- Store your answer in the memory (often Min or STO): `9.375 min`
- Calculate the numerator (use the brackets and x^y keys): `5.25`
- Divide the answer by the recalled memory (often MR or RCL): `0.56`

Knowing inverse operations can help you to solve problems.

example

If $x^3 = 140$, find the value of x.

The cube root key often looks like $\sqrt[3]{}$.

Read the equation: $\quad x^3 = 140 \Rightarrow$ A number cubed equals 140.
Reverse the operations: The answer is $\sqrt[3]{140}$.
Use a calculator: $\quad \sqrt[3]{140} = 5.19$ to 2 decimal places.

166

Plenary

Write calculations involving a variety of operations on the board:

e.g. $\dfrac{3(2^2 + 3) - 4 \times 2}{6 \div 3(1.5 - 0.75)}$

Encourage students to demonstrate how they would tackle the calculation, mentally and using a calculator.

Further activities

Students can repeat question 4, using different values of *m*.

They can develop question 9 by devising similar puzzles for a partner.

Differentiation

Extension questions:
- Question 1 provides simple practice at using the correct order of operations.
- Questions 2–9 focus on order of operations, including algebra.
- Question 10 requires students to identify a special case.

Core tier: focuses on the order of operations, including using a calculator for brackets, powers and roots.

Exercise N4.3

1 Work out:
a $3 \times 4 + 6 \times 2$ **b** $(4 + 2)^2 - 3$
c $6 \times 2 \div (5 - 3)^2$ **d** $3.5^2 - 4 \div 2$

2 Add one pair of brackets to each of the following to make the equations correct.
a $6 + 7^2 - 9 - 4 = 50$
b $7 \times 6 \div {}^-4 - 10 = {}^-3$
c $8 \times 7 + 8 - 7 = 64$

3 Insert the operations that produce the answer shown in each calculation below.
a $(8 ? 7) ? 6 ? 5 = 18$
b $1 ? 2 ? 3 ? 4 = {}^-1$
c $(10 ? 9 ? 8) ? 7 = {}^-49$
d $(5 ? 10 ? 15) ? 2 = 310$

4 Work out each of these when $m = 6.5$, giving your answer correct to 2 decimal places.
a $\dfrac{4m^2 - 6(m - 5.7)}{2m}$
b $\sqrt{(m^2 - m)} - 3(4 - \sqrt{m})$
c $\dfrac{({}^-m)^3 - 5(m - 5.4)}{7 - m}$

5 The expression $6(7 - 8) = 6({}^-1) = {}^-6$ can be thought of as
$6(7 - 8) = 6 \times 7 - 6 \times 8 = {}^-6$
Complete the expressions below:
a $10(7 - 3) = 10 \times ? - 10 \times ? = 40$
b $8(? + ?) = 8 \times 7 + ? \times 4 = 88$
c $7(? - ?) = ? \times 10 - ? \times ? = 28$
d $6(a + b) = 6a + ?$
e $5(f + {}^-5) =$

6 a Write this expression in full (as a sentence).
$\dfrac{(6 + {}^-7)^2 - 6 \times {}^-9}{7^2 - 8}$
b Estimate the answer to this expression.
c Work out the exact answer to the expression, giving your answer to 2 d.p.

7 Calculator investigation
The key sequence for calculating $\sqrt[3]{1728}$ is

$\boxed{\sqrt[3]{}}\ \boxed{1}\ \boxed{7}\ \boxed{2}\ \boxed{8}\ \boxed{=}$

However, there is another possible key sequence using the y^x key. Can you discover it? Write it out using the notation shown above.

8 Puzzle
All the people who live in Elzzup Avenue love to play mathematical tricks on their postman.
They regularly change house numbers and then give the postman puzzles he must solve to work out their new house numbers.
Help the postman by working these out for him – remember all door numbers are integers:
a My number lies between 70 and 80, and is made up by using the following ${}^-3, 9, {}^-5, (\quad)$, ^{2}
b My number lies between 1 and 10 and uses the number 5 four times, and the operations + and ÷ once each.

Hint: one of the 5s is used as a power.

9 Puzzle
Use the digits 1 to 5, each of the operations ×, ÷, +, − once only, and two pairs of brackets to produce the answer $\frac{1}{15}$.

10 a Explain in words the difference between the expressions:
$({}^-h)^2$ and ${}^-h^2$
b Show using values for *h* that these two expressions produce different results.
c Are there any values for which $({}^-h)^2$ and ${}^-h^2$ are equal? Explain your answer.

Exercise commentary

The questions assess the objectives on Framework Pages 85–87.

Problem solving
Questions 8 and 9 provide an opportunity to assess the objective on Framework Page 7. Question 10 assesses Page 35.

Group work
Students can work in pairs on the Further Activities.

Misconceptions
Student may forget that there are 'hidden brackets' around the numerator and denominator.

Links
Order of algebraic operations: Framework Page 115

Use formulae from mathematics and other subjects: Pages 139–141
Multiplying a bracket: Page 117.

Homework

N4.3HW requires students to use a calculator efficiently to carry out a complex calculation.

Answers

1 a 24   **b** 33   **c** 3   **d** 10.25
2 a $6 + 7^2 - (9 - 4) = 50$ **b** $7 \times 6 \div ({}^-4 - 10) = {}^-3$
   c $8 \times (7 + 8 - 7) = 64$
3 a +, ×, ÷   **b** −, ×, +   **c** −, −, ×   **d** +, ×, ×   **4 a** ${}^-18.63$ **c** 24.75
5 a $10 \times 7 - 10 \times 3$ **b** $8(7 + 4) = 8 \times 7 + 8 \times 4$ **c** $7(10 - 6) = 7 \times 10 - 7 \times 6 = 28$
   d $6b$ **e** $5f - 25$
6 c 1.34   **8 a** $({}^-5 + {}^-3)^2 + 9 = 73$ **b** $(5 \div 5)^5 + 5 = 6$
9 $4 - (2 + 1) \div (5 \times 3)$ **10 c** 0

Mental starter

Draw a spider diagram on the board, with eight legs.
Write an integer value in the body, for example $^-5$.
Write fractions or decimals along the legs.

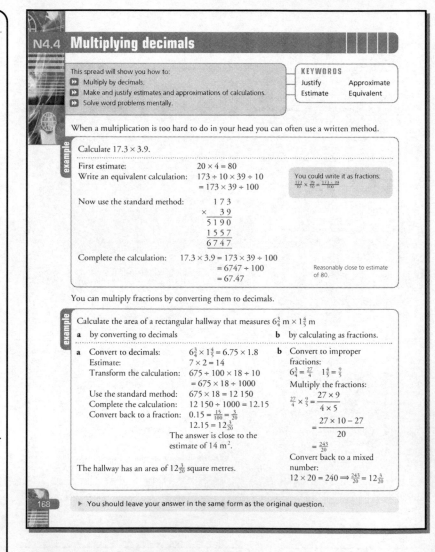

Encourage students to multiply the body-value by the fraction for each leg.

Useful resources

Mini-whiteboards may be useful for the mental starter.
Dice
Counters
N4.4OHP – board for game in question 6
Measuring tape or metre rule for further activities

Introductory activity

Recap strategies for multiplying an integer by a fraction or decimal.

Emphasise that students should always look for mental strategies for solving a problem first.
However, some calculations are too difficult to do mentally, so we need a written method.

Go over the first example in the students' book.
Recap the written method for multiplying a 3-digit number by a 2-digit number, and dividing by a multiple of 10.

Go over the second example in the student book.
Ensure students are clear on the two methods for multiplying a fraction by a fraction.
For further practice, work through an example with a negative mixed number, for example $6\frac{1}{4} \times {}^-1\frac{3}{5}$.

Emphasise the importance of estimating the answer before you do the calculation, and checking your answer against your estimate.

N4.4 Multiplying decimals

This spread will show you how to:
▶ Multiply by decimals.
▶ Make and justify estimates and approximations of calculations.
▶ Solve word problems mentally.

KEYWORDS
Justify   Approximate
Estimate   Equivalent

When a multiplication is too hard to do in your head you can often use a written method.

example

Calculate 17.3×3.9.

First estimate: $20 \times 4 = 80$
Write an equivalent calculation: $173 \div 10 \times 39 \div 10$
                                   $= 173 \times 39 \div 100$

You could write it as fractions:
$\frac{173}{10} \times \frac{39}{10} = \frac{173 \times 39}{100}$

Now use the standard method:
$$\begin{array}{r} 1\,7\,3 \\ \times \quad 3\,9 \\ \hline 5\,1\,9\,0 \\ 1\,5\,5\,7 \\ \hline 6\,7\,4\,7 \end{array}$$

Complete the calculation: $17.3 \times 3.9 = 173 \times 39 \div 100$
                            $= 6747 \div 100$
                            $= 67.47$

Reasonably close to estimate of 80.

You can multiply fractions by converting them to decimals.

example

Calculate the area of a rectangular hallway that measures $6\frac{3}{4}$ m $\times 1\frac{4}{5}$ m

a by converting to decimals   **b** by calculating as fractions.

a Convert to decimals: $6\frac{3}{4} \times 1\frac{4}{5} = 6.75 \times 1.8$
   Estimate: $7 \times 2 = 14$
   Transform the calculation: $675 \div 100 \times 18 \div 10$
                               $= 675 \times 18 \div 1000$
   Use the standard method: $675 \times 18 = 12\,150$
   Complete the calculation: $12\,150 \div 1000 = 12.15$
   Convert back to a fraction: $0.15 = \frac{15}{100} = \frac{3}{20}$
                               $12.15 = 12\frac{3}{20}$
   The answer is close to the estimate of 14 m^2.

The hallway has an area of $12\frac{3}{20}$ square metres.

b Convert to improper fractions:
   $6\frac{3}{4} = \frac{27}{4}$ $1\frac{4}{5} = \frac{9}{5}$
   Multiply the fractions:
   $\frac{27}{4} \times \frac{9}{5} = \frac{27 \times 9}{4 \times 5}$
   $= \frac{27 \times 10 - 27}{20}$
   $= \frac{243}{20}$
   Convert back to a mixed number:
   $12 \times 20 = 240 \Rightarrow \frac{243}{20} = 12\frac{3}{20}$

168

▶ You should leave your answer in the same form as the original question.

Plenary

Discuss any difficulties students encountered in the exercise, and how they solved them.
Write some fractions, decimals and mixed numbers on the board:

$\frac{1}{2}$ $^-\frac{3}{4}$ 0.4 $\frac{5}{8}$ $3\frac{1}{8}$

$^-4\frac{1}{4}$ 0.3 $6\frac{2}{3}$ $\frac{1}{5}$

Encourage students to choose two values from the board to multiply together and demonstrate their method.

Further activities

N4.4ICT uses a spreadsheet to generate the standard column method for long multiplication.

Differentiation

Extension questions:
▸ Question 1 allows students to practise straight-forward multiplication of decimals.
▸ Question 2 to 5 focus on multiplying decimals, using various contexts.
▸ Question 6 is a game for a small group activity.

Core tier: focuses on using the standard written method to multiply decimals, using simpler numbers.

Exercise N4.4

1 Use a written method to work out the exact answer to each of the following. You must write an approximate answer before you begin each calculation.

a 13.2×6.3	**b** 766×0.34
c 3.45×34	**d** 0.34×561
e 5.21×92	**f** 72×154
g 452×0.56	**h** 3.5×6.91
i 30.2×9.2	**j** 84.6×0.014
k 0.45×0.0341	**l** 0.87×851

2 Work out the areas of these rectangles.

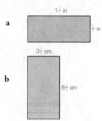

3 Investigation
Two plants, X and Y, are 15 cm tall and 20 cm tall respective. The rate of growth for the two plants is:

Plant X $\frac{1}{20}$ of its height each year
Plant Y $\frac{1}{22}$ of its height each year

How many years will it take until Plant X is taller than Plant Y?

4 Investigation
$(1\frac{1}{2})^2 = 1\frac{1}{2} \times 1\frac{1}{2} = \frac{3}{2} \times \frac{3}{2} = \frac{9}{4} = 2\frac{1}{4}$
$(1\frac{1}{2})^3 = 1\frac{1}{2} \times 1\frac{1}{2} \times 1\frac{1}{2} = \frac{3}{2} \times \frac{3}{2} \times \frac{3}{2} = \frac{27}{8} = 3\frac{3}{8}$
Find the lowest value of n such that $(1\frac{1}{2})^n$ is:
a greater than 10
b greater than 20
c equal to 1.

5 The table shows the salaries of five workers. The figure in brackets shows the fraction by which each employee's wage increases each year.

Complete the table:

Employee	A($\frac{1}{17}$)	B($\frac{1}{12}$)	C($\frac{1}{10}$)	D($\frac{1}{25}$)	E($\frac{1}{23}$)
Current salary	£21 000	£32 000	£12 000	£17 000	£40 000
After 1 year					
After 2 years					

6 **Game** (for two or three players)
Copy this board.

Place counters on the Start square.
Roll a dice to establish a 'start number'.
Each player takes it in turns to roll the dice.
Players move the number of squares rolled and multiply their 'current total' by what it says in the square. For example if Player 1 throws 3 as his start number and his next two throws are 4 and 1 his current total would be $3 \times \bar{6} \times \frac{2}{3} = \bar{1}2$.

The game ends when one player reaches or passes the Finish square.
The player with the highest total is the winner.

Exercise commentary

The questions assess the objectives on Framework Pages 97–101.

Problem solving
Questions 3 and 4 assess the objectives on Framework Page 7.

Group work
Question 6 is a game for two or three players.
Use N4.4OHP as a worksheet for students to use as the game board.

Misconceptions
Students may not realise that many fraction problems can be solved by first converting to decimals.
Emphasise however, that the final answer should be given in the form of the numbers in the question.

Links
Measures and mensuration: Framework Pages 228–31
Finding fractions of quantities: Framework Pages 66–7

Homework

N4.4HW requires students to multiply decimals and fractions.

Answers

1 **a** 83.16 **b** 260.44 **c** 117.3 **d** 190.74 **e** 479.32 **f** 11 088
g 253.12 **h** 24.185 **i** 277.84 **j** 1.1844 **k** 0.015345 **l** 740.37
2 **a** $\frac{15}{16}$ in^2 **b** 20 cm^2
3 68 years
4 **a** 6 **b** 8 **c** 0
5 £23 470.59, £36 266.67, £13 200, £17 680, £41 702.13; £26 231.84, £41 102.22, £14 520, £18 387.20, £43 476.69

Mental starter

This starter checks that students are confident in multiplying fractions by fractions (from N4.4).

Write on the board:

1 $\frac{4}{5}$ 2 $3\frac{6}{7}$ 3 $\frac{4}{13}$

4 $1\frac{4}{8}$ 5 $^{-}5\frac{1}{9}$ 6 $\frac{2}{11}$

Roll a dice twice. The numbers the dice lands on select the two values to be multiplied together.

Useful resources

Mini-whiteboards may be useful for the mental starter.
Dice

Introductory activity

Write on the board:

$$? \times \frac{4}{5} = 3\frac{3}{35}$$

Encourage students to suggest how they could work on the missing number in this problem:

$$? = 3\frac{3}{35} \div \frac{4}{5}$$

Discuss how you could carry out this calculation.

Recap inverse operations and reciprocals, for example $\div\frac{1}{2}$ is the same as $\times 2$

Discuss the first example in the student book.
Ask students how they would convert calculations such as $3.1 \div 6.7$ to equivalent calculations ($31 \div 67 \div 100$).

Recap remainders and rounding to a given number of decimal places.

Emphasise the importance of estimating the answer before you do the calculation, and checking your answer against your estimate.

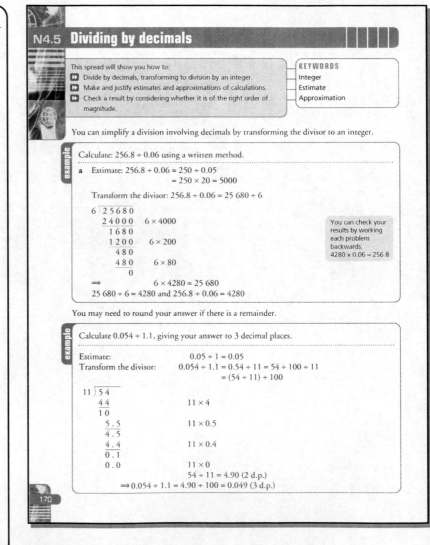

N4.5 Dividing by decimals

This spread will show you how to:
- Divide by decimals, transforming to division by an integer.
- Make and justify estimates and approximations of calculations.
- Check a result by considering whether it is of the right order of magnitude.

KEYWORDS
Integer
Estimate
Approximation

You can simplify a division involving decimals by transforming the divisor to an integer.

example

Calculate: $256.8 \div 0.06$ using a written method.

a   Estimate: $256.8 \div 0.06 \approx 250 \div 0.05$
$= 250 \times 20 = 5000$

Transform the divisor: $256.8 \div 0.06 = 25\,680 \div 6$

```
6 ) 25680
    24000     6 × 4000
     1680
     1200     6 × 200
      480
      480     6 × 80
        0
⇒           6 × 4280 = 25 680
25 680 ÷ 6 = 4280 and 256.8 ÷ 0.06 = 4280
```

You can check your results by working each problem backwards.
$4280 \times 0.06 = 256.8$

You may need to round your answer if there is a remainder.

example

Calculate $0.054 \div 1.1$, giving your answer to 3 decimal places.

Estimate: $0.05 \div 1 = 0.05$
Transform the divisor: $0.054 \div 1.1 = 0.54 \div 11 = 54 \div 100 \div 11$
$= (54 \div 11) \div 100$

```
11 ) 54
     44          11 × 4
     10
      5 . 5       11 × 0.5
      4 . 5
      4 . 4       11 × 0.4
      0 . 1
      0 . 0       11 × 0
                  54 ÷ 11 = 4.90 (2 d.p.)
⇒ 0.054 ÷ 1.1 = 4.90 ÷ 100 = 0.049 (3 d.p.)
```

170

Plenary

Write on the board:
$$? \times 0.8 = 3.24$$
So
$$? = 3.24 \div 0.8$$

Encourage students to demonstrate different methods for finding the value of ?

Further activities

Students could use a spreadsheet or graphical calculator to explore the effects of repeated division.
Encourage students to justify what happens (answer tends to zero if the divisor is greater then 1).

Differentiation

Extension questions:
▸ Question 1 practises multiplying and dividing decimals.
▸ Questions 2–5 focus on dividing decimals, and include fractions and mixed units.
▸ Question 6 links to shape and algebra.

Core tier: focuses on using the standard method to divide decimals, involving simpler numbers.

Exercise N4.5

1 The numbers in the table follow this rule:

$a \times b = c$

a Copy and complete the table.
b Check your answers by mentally working out an approximate answer.

a	b	c
2.41	2.3	
6		21.9
	0.045	0.2025
0.004	0.45	
	0.57	1.5732
0.0078		6.552

2 Calculate these quantities, giving your answer to 2 decimal places where appropriate:
a 70.3 m ÷ 3.4   b 0.456 mm ÷ 71
c 5.03 cm ÷ 0.42   d 0.21 kg ÷ 0.56
e 4536 m ÷ 0.8   f 5.003 g ÷ 0.43

3 The table shows how a problem that involves dividing by a fraction can be written as a multiplication problem.

Copy and complete the table.

Division problem	Equivalent multiplication problem
$3\frac{1}{2} \div \frac{6}{7}$	$3\frac{1}{2} \times \frac{7}{6}$
$4\frac{1}{8} \div \frac{6}{5}$	
$\frac{5}{8} \div 3\frac{1}{5}$	
$6 \div \frac{-1}{20}$	
$5\frac{7}{10} \div 4\frac{7}{9}$	

4 One of the world's highest buildings is in Colorado, USA. It is 1053 ft high. Using 3.25 ft = 1 m work out its height in metres.

5 Work out the missing numbers.
Following the left and right paths should both lead to 0.2.

6 The diagram shows a trapezium with parallel sides labelled a and b.
Find the length of side b when:
Area = 56.7 m²
$a = 3.4$ m
$h = 7$ m

Use the formula: Area = $\frac{(a+b)h}{2}$ and show all your working out.

171

Exercise commentary

The questions assess the objectives on Framework Pages 97–101.

Problem solving
Question 5 assesses the objectives on Framework Page 5.

Group work
The further activity can be tackled in pairs.

Misconceptions
Students may have problems working to given levels of accuracy and dealing with remainders.
Refer to the examples in the introductory activity for guidance.
Some students may not have grasped the inverse nature of multiplication and division.
Reinforce this concept with simple integers, then extend to decimals as in question 1.

Links
Measures and mensuration: Framework Pages 228–31

Homework

N4.5HW requires students to divide by decimals and fractions.

Answers

1 a $c = 5.543$, $b = 3.65$, $a = 4.5$, $c = 0.0018$, $a = 2.76$, $b = 840$
2 a 20.68 m b 0.01 mm c 11.98 cm d 0.38 kg e 5670 m
   f 11.63 g
3 $4\frac{1}{8} \times \frac{7}{6}$, $\frac{5}{8} \times \frac{5}{16}$, $6 \times {}^-20$, $5\frac{7}{10} \times \frac{-19}{79}$
4 324 m
5 0.4, 0.625; 3125, 0.00008
6 12.8 m

Mental starter

Make sets of cards with these measurements on them:

1000 cm   5 kg   0.5 kg   10 m   1 m   5000 g   500 g
10 000 mm   0.001 km   5 000 000 mg   100 cm

Give students one card each and instruct them to get into groups with equivalent values.

Can they think of any other values that could join their group?

Useful resources

N4.6OHP – metric and imperial/metric relationships

OHP calculator

Introductory activity

Recap metric measurements and the relationships between them.

Discuss quantities that are measured in metric units.

Encourage students to learn metric relationships by heart.
Test them using **N4.6OHP** by covering up different values.

Discuss quantities that are measured in imperial units and when you may need to convert between the two.
Using the rough imperial/metric conversions on **N4.6OHP**, discuss how you could convert for example metres to inches, litres to gallons.

Demonstrate how to use a calculator for calculations of time, using an OHP calculator.
Encourage students to suggest how to enter times such as 25 min, 30 min, $\frac{3}{4}$ h, 10 min in a calculator.
Discuss the second example in the student book. Reinforce the meaning of the words 'hectare' and 'capacity'.

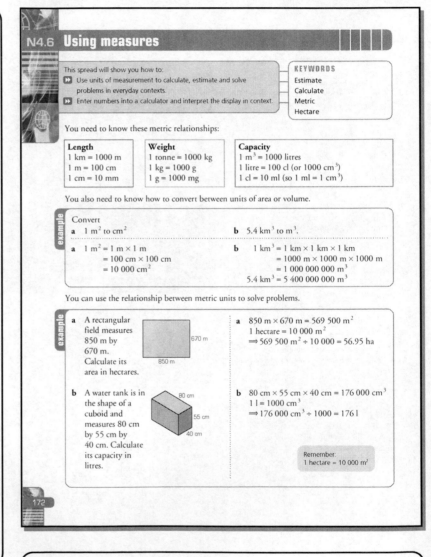

N4.6 Using measures

This spread will show you how to:
- Use units of measurement to calculate, estimate and solve problems in everyday contexts.
- Enter numbers into a calculator and interpret the display in context.

KEYWORDS
Estimate
Calculate
Metric
Hectare

You need to know these metric relationships:

Length	Weight	Capacity
1 km = 1000 m	1 tonne = 1000 kg	1 m³ = 1000 litres
1 m = 100 cm	1 kg = 1000 g	1 litre = 100 cl (or 1000 cm³)
1 cm = 10 mm	1 g = 1000 mg	1 cl = 10 ml (so 1 ml = 1 cm³)

You also need to know how to convert between units of area or volume.

example

Convert
a   1 m² to cm²   b   5.4 km³ to m³.

a   1 m² = 1 m × 1 m   b   1 km³ = 1 km × 1 km × 1 km
    = 100 cm × 100 cm   = 1000 m × 1000 m × 1000 m
    = 10 000 cm²   = 1 000 000 000 m³
                                          5.4 km³ = 5 400 000 000 m³

You can use the relationship between metric units to solve problems.

example

a   A rectangular field measures 850 m by 670 m. Calculate its area in hectares.

670 m

850 m

a   850 m × 670 m = 569 500 m²
    1 hectare = 10 000 m²
    ⇒ 569 500 m² ÷ 10 000 = 56.95 ha

b   A water tank is in the shape of a cuboid and measures 80 cm by 55 cm by 40 cm. Calculate its capacity in litres.

80 cm
55 cm
40 cm

b   80 cm × 55 cm × 40 cm = 176 000 cm³
    1 l = 1000 cm³
    ⇒ 176 000 cm³ ÷ 1000 = 176 l

Remember:
1 hectare = 10 000 m²

172

Plenary

Encourage students to suggest how they would solve this problem and demonstrate their methods.
A field 940 m long has an area of 49.35 ha.
It takes a snail 2.5 minutes to travel 1 metre.
How long would it take the snail to travel all round the perimeter of the field?
Discuss the most sensible units of time for the answer.

Further activities

Students work in pairs on this adaptation of question 4.
One student draws rectangles on a piece of paper and the other student should estimate the area of each one. Swap over, and compare accuracy ratios.

Differentiation

Extension questions:
▶ Questions 1 and 2 focus on converting between metric measures.
▶ Question 3 focuses on converting between metric and imperial measures.
▶ Question 4 extends to estimation and ratio.

Core tier: focuses on metric measures of length, weight and capacity, and includes using a calculator to solve problems connected with time.

Exercise N4.6

1 Convert each of these measurements into the units given in brackets.
  a 450 cm (m)   **b** 63 234 g (kg)   **c** 0.03 m (mm)
  d 1 000 000 cl (litres)   **e** 6.89 tonnes (kg)   **f** 0.276 m (cm)

2 Calculate the volume of this cuboid
  a in cubic metres   **b** in cubic centimetres.

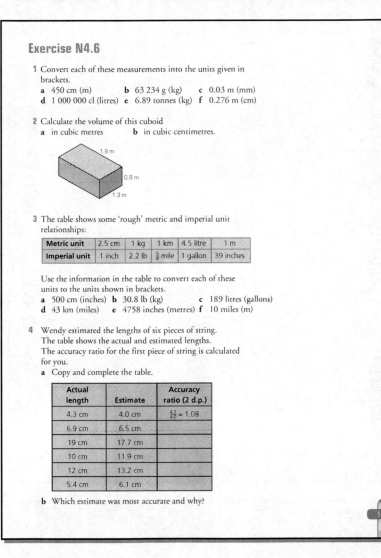

3 The table shows some 'rough' metric and imperial unit relationships:

Metric unit	2.5 cm	1 kg	1 km	4.5 litre	1 m
Imperial unit	1 inch	2.2 lb	$\frac{5}{8}$ mile	1 gallon	39 inches

Use the information in the table to convert each of these units to the units shown in brackets.
  a 500 cm (inches)   **b** 30.8 lb (kg)   **c** 189 litres (gallons)
  d 43 km (miles)   **e** 4758 inches (metres)   **f** 10 miles (m)

4 Wendy estimated the lengths of six pieces of string.
The table shows the actual and estimated lengths.
The accuracy ratio for the first piece of string is calculated for you.
  a Copy and complete the table.

Actual length	Estimate	Accuracy ratio (2 d.p.)
4.3 cm	4.0 cm	$\frac{4.3}{4.0} = 1.08$
6.9 cm	6.5 cm	
19 cm	17.7 cm	
10 cm	11.9 cm	
12 cm	13.2 cm	
5.4 cm	6.1 cm	

  b Which estimate was most accurate and why?

173

Exercise commentary

The questions assess the objectives on Framework Pages 228–31.

Problem solving
Question 4 provides an opportunity to assess Framework Pages 21. The whole exercise assesses Page 31.

Group work
The further activity should be tackled in pairs.

Misconceptions
In calculations involving time, students may interpret 1 hour 20 minutes as 1.2 hours, and 2.4 hours as 2 hours 40 minutes.
Remind them that 0.2 hours = 0.2 × 60 minutes, and so on.
Students generally find it harder to learn metric/imperial conversions than to just convert between metric units.
Reinforce the main conversions and encourage appropriate accuracy.

Links
Mental recall of measurement facts: Framework Page 91.
Area and volume: Pages 235–239.

Homework

N4.6HW involves the relationship between linear units and units of area.

Answers

1 a 4.5 m **b** 63.234 kg **c** 30 mm **d** 10 000 litres **e** 6890 kg
  f 27.6 cm
2 a 1.976 m³ **b** 1 976 000 cm³
3 a 200 in **b** 14 kg **c** 42 gal **d** 27 miles **e** 122 yards **f** 16 000 m
4 a 1.06, 1.07, 0.84, 0.91, 0.89 **b** 4.3 cm length, ratio closest to 1

Summary

The key objectives for this unit are:
▸ Make and justify estimates and approximations of calculations.
▸ Use efficient methods to add, subtract, multiply and divide fractions.
▸ Give solutions to problems to an appropriate degree of accuracy.

Plenary activity

Write on the board:

$$\frac{\square}{\square} \boxed{\text{operation}} \frac{\square}{\square} =$$

Ask students to use the numbers 2, 3, 4 and 5, and any of the operations $\times$, $\div$, $+$ or $-$ to produce the largest possible answer.

Check out commentary

1 Students need to think about what their estimate will be used for.
They should be encouraged to think about the actual problem and not just the numbers in the question.
Encourage discussion centred around these problems.

2 Students are required to show that they are able to calculate with fractions using all four operations.
The questions also include integers and mixed numbers to ensure that students can deal with these when they occur.

3 Encourage students to solve this problem mentally first.
They should then produce an exact answer.
Students should consider the accuracy to which their answer should be given, bearing in mind the accuracy given in the question.

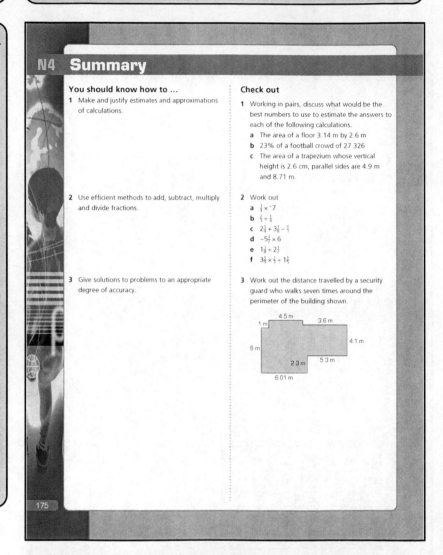

N4 **Summary**

You should know how to ...

1 Make and justify estimates and approximations of calculations.

2 Use efficient methods to add, subtract, multiply and divide fractions.

3 Give solutions to problems to an appropriate degree of accuracy.

Check out

1 Working in pairs, discuss what would be the best numbers to use to estimate the answers to each of the following calculations.
 a The area of a floor 3.14 m by 2.6 m
 b 23% of a football crowd of 27 326
 c The area of a trapezium whose vertical height is 2.6 cm, parallel sides are 4.9 m and 8.71 m.

2 Work out
 a $\frac{1}{3} \times {}^-7$
 b $\frac{4}{7} \div \frac{1}{3}$
 c $2\frac{1}{4} + 3\frac{1}{8} - \frac{1}{7}$
 d $-5\frac{1}{2} \times 6$
 e $1\frac{1}{8} \div 2\frac{1}{2}$
 f $3\frac{1}{10} \times \frac{1}{7} \div 1\frac{1}{5}$

3 Work out the distance travelled by a security guard who walks seven times around the perimeter of the building shown.

175

Development

Converting between units is used in the context of scales in S4.

Links

Addition, subtraction, multiplication and division are used throughout mathematics.
Converting between units is also used in science and geography.

Mental starters

Objectives covered in this unit:
▸ Know and use squares, cubes, and roots
▸ Add several small numbers
▸ Use factors to multiply mentally
▸ Use metric units of length
▸ Discuss and interpret graphs
▸ Apply mental skills to solve simple problems

Resources needed

* means class set needed
Essential:
A5.1OHP – sequence of L-shapes
A5.2OHP – sequence of straws
A5.3OHP – sequence of squares
A5.7OHP – real-life graphs
A5.8OHP – implicit graphs
Counting stick, Tape measures
Useful:
Mini-whiteboards*
R6 number lines
A5.3F – factorization
A5.6ICT – graph plotting

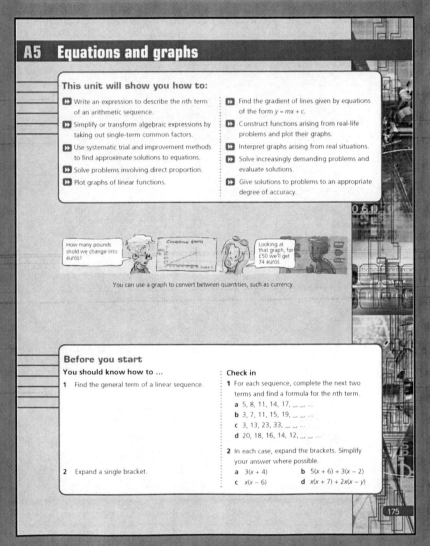

A5 Equations and graphs

This unit will show you how to:

▸▸ Write an expression to describe the *n*th term of an arithmetic sequence.

▸▸ Simplify or transform algebraic expressions by taking out single-term common factors.

▸▸ Use systematic trial and improvement methods to find approximate solutions to equations.

▸▸ Solve problems involving direct proportion.

▸▸ Plot graphs of linear functions.

▸▸ Find the gradient of lines given by equations of the form $y = mx + c$.

▸▸ Construct functions arising from real-life problems and plot their graphs.

▸▸ Interpret graphs arising from real situations.

▸▸ Solve increasingly demanding problems and evaluate solutions.

▸▸ Give solutions to problems to an appropriate degree of accuracy.

How many pounds shold we change into euros?

Looking at that graph, for £50 we'll get 74 euros.

You can use a graph to convert between quantities, such as currency.

Before you start

You should know how to ...

1 Find the general term of a linear sequence.

2 Expand a single bracket.

Check in

1 For each sequence, complete the next two terms and find a formula for the *n*th term.
  a 5, 8, 11, 14, 17, _, _, ...
  b 3, 7, 11, 15, 19, _, _, ...
  c 3, 13, 23, 33, _, _, ...
  d 20, 18, 16, 14, 12, _, _, ...

2 In each case, expand the brackets. Simplify your answer where possible.
  a $3(x + 4)$ b $5(x + 6) + 3(x - 2)$
  c $x(x - 6)$ d $x(x + 7) + 2x(x - y)$

175

Unit commentary

Aim of the unit
The unit consolidates and extends previous algebra, and provides an opportunity to further students' reasoning skills.
The unit starts by encouraging students to justify formulae and explore ways to extend investigative tasks. Trial and improvement, proportion and implicit functions are also covered in this unit.

Introduction
Discuss the extent to which students have developed their algebra in just two years.

Framework references
This unit focuses on:
▸ Teaching objectives pages:
  117, 133–137, 143, 157, 165, 173–177
▸ Problem solving objectives pages:
  7, 9, 27–35

Differentiation

Core tier
Focuses on extending algebraic manipulation techniques in order to solve increasingly complex linear equations, and exploring graphs in a real-life context.

Check in activity

Students sit in a circle on chairs, with one student standing in the centre.
The student in the centre gives an algebra statement (for example, if $x = 6$ then $2x^2 = 72$). If it is true then everyone remains seated. If false, everyone swaps chairs. This should generate a new student standing, to start the process again.

Mental starter

Ask students to imagine that they are at a party.
As they greet each other, the partygoers give one another a pat on the back. Demonstrate with five or six students.
Ask how many pats there will be in total if they all attend the party.

Students should write down their response, which can be checked later in the lesson.

Useful resources

A5.1OHP shows the sequence of L-shapes in the student book.

Introductory activity

Outline the key points in problem-solving by referring to the starter.

1 Break the problem into small steps.
You could make the party problem easier by considering smaller parties first. Select volunteers to demonstrate what happens with 1, 2, 3 and 4 people.

2 Gain results and look for a pattern.
Draw a table of results:

People (n)	1	2	3	4
Pats (p)	0	2	6	12

Look for patterns (it goes up in 2, then 4, then 6, ...).

3 Write a formula or generalisation.
Encourage students to think **how** the problem works:
3 people $\Rightarrow$ 2 pats each $\Rightarrow$ 3 × 2 = 6
10 people $\Rightarrow$ 9 pats each $\Rightarrow$ 10 × 9 = 90
n people $\Rightarrow$ $(n-1)$ pats each $\Rightarrow$ $n \times (n-1)$
So $p = n(n-1)$

4 Explain why the formula works.
n lots of $(n-1)$ pats makes $p = n(n-1)$

Now refer back to the starter.
What should the total number of pats be for the class?

Discuss the problem (L-shape patterns) in the student book.
A5.1OHP illustrates the sequence.

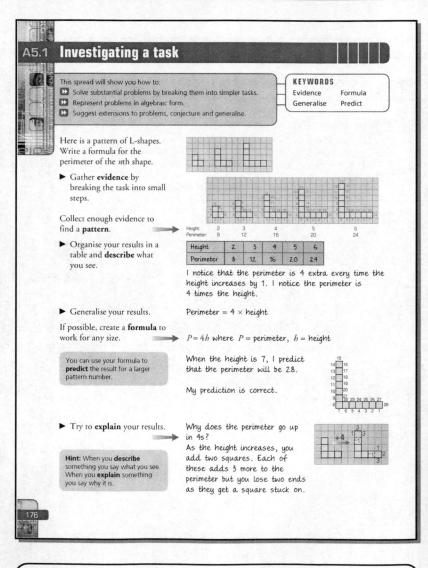

A5.1 Investigating a task

This spread will show you how to:
- Solve substantial problems by breaking them into simpler tasks.
- Represent problems in algebraic form.
- Suggest extensions to problems, conjecture and generalise.

KEYWORDS
Evidence   Formula
Generalise   Predict

Here is a pattern of L-shapes. Write a formula for the perimeter of the mth shape.

► Gather **evidence** by breaking the task into small steps.

Collect enough evidence to find a **pattern**.

► Organise your results in a table and **describe** what you see.

Height	2	3	4	5	6
Perimeter	8	12	16	20	24

I notice that the perimeter is 4 extra every time the height increases by 1. I notice the perimeter is 4 times the height.

► Generalise your results. If possible, create a **formula** to work for any size.

Perimeter = 4 × height

$P = 4h$ where P = perimeter, h = height

You can use your formula to **predict** the result for a larger pattern number.

When the height is 7, I predict that the perimeter will be 28.

My prediction is correct.

► Try to **explain** your results.

Hint: When you **describe** something you say what you see. When you **explain** something you say why it is.

Why does the perimeter go up in 4s?
As the height increases, you add two squares. Each of these adds 3 more to the perimeter but you lose two ends as they get a square stuck on.

176

Plenary

Return to the idea in the starter.
How would it affect the problem if the elder partygoer patted the younger partygoer only?

Students are unlikely to complete all the investigations, as they can be taken to different levels.

Students could be encouraged to extend the given tasks, and devise similar problems of their own for the class.

Differentiation

Extension questions:

▸ Question 1 is a task involving a linear formula with no justification required.
▸ Questions 2–4 are more complex tasks that need to be broken down.
▸ Questions 5 and 6 involve concepts that will not be formalised until later work.

Core tier: focuses on expanding brackets in linear expressions.

Exercise A5.1

Investigate these tasks using the step-by-step approach suggested. Try to generalise and explain your results.

1 In a square grid of varying sizes, the key counter ● must move from the top left square to the empty space in the bottom right.
All counters can be moved horizontally or vertically only, and only into an empty space.
Investigate the minimum total number of counter moves to do this.

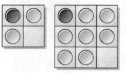

2 Investigate the total number of diagonals that can be drawn from one vertex of a polygon.

3 a Investigate the total number of diagonals that can be drawn in a polygon.

b Use your answer to give the number of diagonals in a 'Mystic Rose'.
This is a 36-sided polygon with all diagonals drawn in. (You might like to construct one yourself.)

4 Investigate the number of leads needed to connect different numbers of computers, if all computers must be attached to each other.

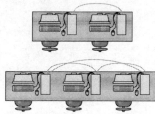

5 a Investigate the number of different arrangements of a family lining up for a photograph.
b Can you find a button on your calculator that generates these results? If so, how many different arrangements of a family of 20 would there be?

6 a Investigate the number of different ways you can pay for a phone call in a phone box if the call costs a multiple of 10 pence and you have 10 pence coins and 20 pence coins available.
b What would happen if the slot also took 5 pence coins?

Exercise commentary

The questions assess the objectives on Framework Pages 27 and 29.

Problem solving

All the questions in this exercise focus on problem solving, and particularly assess the objectives on Pages 7–9 and 27–29.

Group work

All of the tasks are ideally suited to small group work and discussion.

Misconceptions

Students may lose count of the diagonals in question 3. Encourage students to think of all vertices having the same number of diagonals, and decide what to do with the ones already counted at the end.
Encourage use of formulae, not expressions – students may find it easier to write these in words rather than symbols.

Links

This topic links to sequences (Framework Pages 149–151).

Homework

A5.1HW involves similar types of tasks with a similar structure to those in the exercise.

Answers

1 $8 \times$ grid length $- 11$
2 $n - 3$ for an n-sided polygon
3 a $\frac{1}{2}n(n - 3)$ **b** 594
4 $\frac{1}{2}n(n - 1)$ leads for n computers
5 a $n \times (n - 1) \times (n - 2) \dots 2 \times 1$ arrangements for n people.
  b There are 2.4×10^{18} arrangements for 20 people.

Mental starter

Ask students to keep adding odd numbers mentally $(1 + 3 + 5 + ...)$ until you say 'stop'. Students should then write down the number they reached on a piece of paper or a mini-whiteboard, and hold it up.
Who got the highest number?
Write down all the different numbers and ask students what they notice.

Repeat for consecutive integers $(1 + 2 + 3 + ...)$.

Useful resources

Mini-whiteboards may be useful for the mental starter and the introductory activity.
A5.2OHP shows the example in the student book.

Introductory activity

Draw this sequence of diagrams on the board.

Draw the patterns in a definite order:
▶ Top and bottom row first
▶ Then the counters on each edge

Ask how many counters there will be in the square with length 100.
Discuss ideas – students can illustrate their ideas on mini-whiteboards, for example:

```
        100
      _____
  98 |      | 98
     |_____|
        100
```

Encourage the response that there will be 100 across the top and bottom, then 98 up the two sides.

Extend to a square of length *n*.
Encourage a formula, not an expression, for example:
$$C = n + n + (n - 2) + (n - 2)$$
$$= 2n + 2(n - 2)$$

Ask students to justify the formula.
Encourage responses like:
'You need two strips as long as the square for top and bottom, but for the edges you can subtract 2 because of the corners'.
Discuss the sequence of patterns in the student book.
A5.2OHP shows the sequence, and illustrates the justification of the formula.

A5.2 Generalising and justifying findings

This spread will show you how to:
▶ Solve substantial problems by breaking them into simpler tasks.
▶ Present a concise, reasoned argument.
▶ Suggest extensions to problems, conjecture and generalise.

KEYWORDS
Explain   Conclusion
Justify   Generate

If you look at how a pattern builds systematically, it can often help you to construct and justify your formula.

example

Here is a pattern of straws, with increasing heights.
a   Draw the fourth and fifth patterns, and hence draw a table of results.
b   By considering how the diagrams are built, find a general formula.

a

Height, h	1	2	3	4	5
Number of straws, s	4	12	24	40	60

b Look at the diagram with $h = 3$:

In the third diagram:
$$s = (4 \times 3) + (4 \times 3)$$

In general:
$$s = (h + 1) \times h + (h + 1) \times h$$
$$\Rightarrow \quad s = 2h(h + 1)$$
Check with $h = 5$:
$$s = 2 \times 5 \times 6 = 60$$

Note: You could alternatively use the method of **differences** to find a formula.
However, the formula you get $(s = 2h^2 + 2h)$ would be difficult to justify.

You can always **extend** an investigation.

Sanjiv has been investigating the number of corner, middle and edge pieces in a square jigsaw.

corner   edge   middle
C   E   M

Sanjiv could extend the problem by ...
▶ moving on to rectangular puzzles
▶ moving on to 3-D puzzles.

178

Plenary

Recap the sequence explored in the introductory activity, and discuss how it could be extended.
For example:
▶ Try rectangles instead of squares
▶ Try a sequence of 3-D cubes made up of beads:

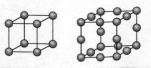

Further activities

All of these tasks can be extended and investigated.
Students could suggest their own investigative task for the rest of the class.

Differentiation

Extension questions:
▶ Questions 1 and 2 provide a straightforward context, with no extension.
▶ Questions 3 and 4 have an accessible lead-in, but with possibilities for extension.
▶ Question 5 is a task involving a number of increasingly complex formulae, best obtained by a structured approach.

Core tier: focuses on constructing and solving linear equations.

Exercise A5.2

Investigate each problem in this exercise.
Focus on generalising your results and justifying your findings by thinking about the structure of the problem.

1 T-shapes
Investigate the relationship between:
▶ height and perimeter
▶ height and number of squares.

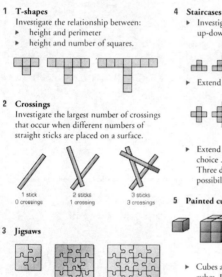

2 Crossings
Investigate the largest number of crossings that occur when different numbers of straight sticks are placed on a surface.

1 stick   2 sticks   3 sticks
0 crossings   1 crossing   3 crossings

3 Jigsaws

corner — edge — middle

▶ Investigate how the number of corner, edge and middle pieces in a square jigsaw is related to its size.
▶ **Prove** that your three results add together give the total number of jigsaw pieces.
▶ Extend to rectangular puzzles.

4 Staircases
▶ Investigate the number of blocks in up-down staircases of differing heights:

▶ Extend to symmetrical stairways:

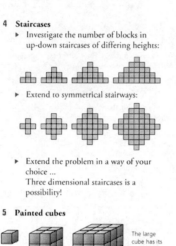

▶ Extend the problem in a way of your choice ...
Three dimensional staircases is a possibility!

5 Painted cubes

The large cube has its outside painted silver.

▶ Cubes are made up of small $1 \times 1 \times 1$ cubes. If the large cube is now broken down, investigate the number of small cubes with:
▶ 0 faces painted
▶ 1 face painted
▶ 2 faces painted ...
▶ **Prove** that all of your formulae add to give the total number of cubes.
▶ Extend your problem to cuboids.

179

Exercise commentary

The questions assess the objectives on Framework Pages 29–35.

Problem solving

All the questions in this exercise focus on problem solving, and particularly assess the objectives on Pages 7–9 and 29–35.

Group work

All of the tasks are ideally suited to small group work and discussion. Encourage students to share their ideas and discuss their justifications.

Misconceptions

There may be a tendency for students to describe, rather than explain, their formulae. You will need to emphasise the difference, referring back to the introductory activity if necessary.

Links

The concepts develop from A5.1, where formulae were found but not explained. This topic links to deriving and using formulae (Framework Pages 139–143).

Homework

A5.2HW provides an opportunity to justify and construct formulae by adopting a structured approach.

Answers

1 $p = 6h - 2$, $n = 3h - 2$ where p = perimeter, h = height, n = no. of squares
2 $c = \frac{1}{2}n(n - 1)$ where c = no. of crossings, n = no. of sticks
3 $c = 4$, $e = 4(n - 2)$, $m = (n - 1)^2$ where c = corner, e = edge, m = middle, n = size
4 $b = h^2$ where b = no. of blocks, h = height; symmetrical: $h^2 + (h - 1)^2$
5 $(n - 2)^3$ have 0, $6(n - 2)^2$ have 1, $12(n - 2)$ have 2, 8 have 3 faces painted.

Mental starter

Demonstrate that $72 + 48$ is identical to $8(9 + 6)$ or 8×15.
Students should find similar ways of writing other sums, for example:

▸ $8 + 32$
▸ $99 + 54$
▸ $5a + 10b$

Useful resources

A5.3OHP shows the sequence of patterns in the introductory activity, and illustrates each of the formulae used to describe it.

Introductory activity

Recap the sequence from the introductory activity in A5.2.

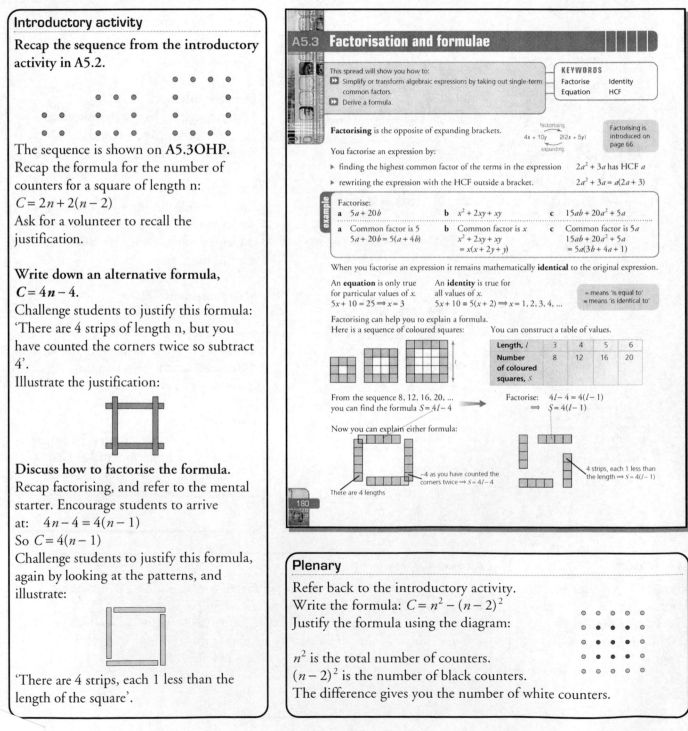

The sequence is shown on **A5.3OHP**. Recap the formula for the number of counters for a square of length n:
$C = 2n + 2(n - 2)$
Ask for a volunteer to recall the justification.

Write down an alternative formula, $C = 4n - 4$.
Challenge students to justify this formula:
'There are 4 strips of length n, but you have counted the corners twice so subtract 4'.
Illustrate the justification:

Discuss how to factorise the formula.
Recap factorising, and refer to the mental starter. Encourage students to arrive at: $4n - 4 = 4(n - 1)$
So $C = 4(n - 1)$
Challenge students to justify this formula, again by looking at the patterns, and illustrate:

'There are 4 strips, each 1 less than the length of the square'.

A5.3 Factorisation and formulae

This spread will show you how to:
▶▶ Simplify or transform algebraic expressions by taking out single-term common factors.
▶▶ Derive a formula.

KEYWORDS
Factorise   Identity
Equation   HCF

Factorising is the opposite of expanding brackets.

$$4x + 10y \quad 2(2x + 5y)$$

factorising / *expanding*

Factorising is introduced on page 66.

You factorise an expression by:
▶ finding the highest common factor of the terms in the expression $2a^2 + 3a$ has HCF a
▶ rewriting the expression with the HCF outside a bracket. $2a^2 + 3a = a(2a + 3)$

example

Factorise:
a $5a + 20b$ **b** $x^2 + 2xy + xy$ **c** $15ab + 20a^2 + 5a$

a Common factor is 5
$5a + 20b = 5(a + 4b)$

b Common factor is x
$x^2 + 2xy + xy$
$= x(x + 2y + y)$

c Common factor is $5a$
$15ab + 20a^2 + 5a$
$= 5a(3b + 4a + 1)$

When you factorise an expression it remains mathematically **identical** to the original expression.

An **equation** is only true for particular values of x.
$5x + 10 = 25 \Rightarrow x = 3$

An **identity** is true for all values of x.
$5x + 10 \equiv 5(x + 2) \Rightarrow x = 1, 2, 3, 4, ...$

= means 'is equal to'
≡ means 'is identical to'

Factorising can help you to explain a formula.
Here is a sequence of coloured squares:

You can construct a table of values.

Length, l	3	4	5	6
Number of coloured squares, S	8	12	16	20

From the sequence 8, 12, 16, 20, ... you can find the formula $S = 4l - 4$

Factorise: $4l - 4 = 4(l - 1)$
$\Rightarrow S = 4(l - 1)$

Now you can explain either formula:

−4 as you have counted the corners twice $\Rightarrow S = 4l - 4$
There are 4 lengths

4 strips, each 1 less than the length $\Rightarrow S = 4(l - 1)$

190

Plenary

Refer back to the introductory activity.
Write the formula: $C = n^2 - (n - 2)^2$
Justify the formula using the diagram:

n^2 is the total number of counters.
$(n - 2)^2$ is the number of black counters.
The difference gives you the number of white counters.

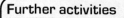

Further activities

The tasks involved in questions 3–5 can be extended by students who have finished early.
Which tasks can be extended into 3-D?

A5.3F contains further practice at using factorisation in context.

Differentiation

Extension questions:
▸ Questions 1 and 2 provide practice at simple factorisation.
▸ Questions 3 and 4 require factorisation, but require students to justify a formula by referring to structure.
▸ Question 5 offers no guidance, and requires a variety of tasks involving factorising and justifying to be carried through from beginning to end.

Core tier: focuses on solving equations with brackets.

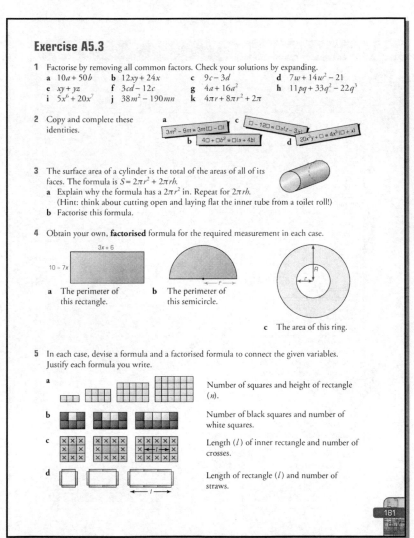

Exercise A5.3

1 Factorise by removing all common factors. Check your solutions by expanding.
  a $10a + 50b$ **b** $12xy + 24x$ **c** $9c - 3d$ **d** $7w + 14w^2 - 21$
  e $xy + yz$ **f** $3cd - 12c$ **g** $4a + 16a^2$ **h** $11pq + 33q^2 - 22q^3$
  i $5x^6 + 20x^7$ **j** $38m^2 - 190mn$ **k** $4\pi r + 8\pi r^2 + 2\pi$

2 Copy and complete these identities.
  a $3m^2 - 9m \equiv 3m(\Box - \Box)$
  b $4\Box + \Box b^2 \equiv \Box(a + 4b)$
  c $\Box - 12\Box \equiv \Box z(z - 3\Box)$
  d $20x^3y + \Box \equiv 4x^3(\Box + x)$

3 The surface area of a cylinder is the total of the areas of all of its faces. The formula is $S = 2\pi r^2 + 2\pi rh$.
  a Explain why the formula has a $2\pi r^2$ in. Repeat for $2\pi rh$.
  (Hint: think about cutting open and laying flat the inner tube from a toilet roll!)
  b Factorise this formula.

4 Obtain your own, **factorised** formula for the required measurement in each case.

  $3x + 6$
  $10 - 7x$

  a The perimeter of this rectangle.
  b The perimeter of this semicircle.
  c The area of this ring.

5 In each case, devise a formula and a factorised formula to connect the given variables. Justify each formula you write.
  a Number of squares and height of rectangle (n).
  b Number of black squares and number of white squares.
  c Length (l) of inner rectangle and number of crosses.
  d Length of rectangle (l) and number of straws.

181

Exercise commentary

The questions assess the objectives on Framework Pages 117, 143 and 157.

Problem solving
Questions 3–5 assess the objectives on Framework Pages 27–35.

Group work
All the investigative tasks lend themselves ideally to pair work or group discussion, but questions 1 and 2 should be attempted individually to consolidate the skills.

Misconceptions
Emphasise that the focus is on explaining a formula.
For example in question 5b ($B = W + 4$), a **description** is: 'You add 4 to the number of whites to get the number of blacks'. An **explanation** is: 'The middle strip has equivalent numbers of blacks and whites but there are 4 extra blacks at the ends.

Links
The work links strongly to problem solving, and can be applied throughout the curriculum.

Homework

A5.3HW consolidates the work of the lesson with short justification tasks, including square and triangular numbers.

Answers

1 **a** $10(a + 5b)$ **b** $12x(y + 2)$ **c** $3(3c - d)$ **d** $7(w + 2w^2 - 3)$
  e $y(x + z)$ **f** $3c(d - 4)$ **g** $4a(1 + 4a)$ **h** $11q(p + 3q - 2q^2)$
  i $5x^6(1 + 4x)$ **j** $38m(m - 5n)$ **k** $2\pi(2r + 4r^2 + 1)$
2 **a** $3m(m - 3)$ **b** $4ab + 16b^2 \equiv 4b(a + 4b)$ **c** $4\pi z - 12\pi^2 \equiv 4\pi(z - 3\pi)$
  d $20x^3y + 4x^4 \equiv 4x^3(5y + x)$
3 **b** $2\pi r(r + h)$ 4 **a** $P = 8(4 - x)$ **b** $P = r(\pi + 2)$ **c** $A = \pi(R^2 - r^2)$
5 **a** $S = n^2 + 2n$, $S = n(n + 2)$ **b** $B = w + 4$, doesn't factorise
  c $C = 2l + 6$, $C = 2(l + 3)$ **d** $S = 2l + 2$, $S = 2(l + 1)$

Mental starter

Say that you are thinking of a number.

When you square it you get 1369.

Challenge students in pairs to find your integer without using a calculator.

Repeat for a number that, when cubed, gives you 4913.

Useful resources

R6 number lines may be useful for illustrating the trial and improvement method.

Introductory activity

Refer to the mental starter.

Discuss how the problem could be solved with a calculator, and formalise the strategy on the board.

$x^2 = 1369 \Rightarrow x = 37$

Present the equation $x^2 + x = 529$.

Discuss strategies for solving it, particularly the method used in the starter.

Use a table, similar to that used in the student book, to solve $x^2 + x = 529$.

Invite students to suggest values of x, and encourage systematic trials.

Use number lines to reinforce how you are getting closer to the solution, $x = 23$.

Describe this as an example of the **trial and improvement** method.

Emphasise that it should be used when an equation is too difficult to be solved directly using algebra.

Refer to the examples in the student book, which include:

▸ an integer solution
▸ a terminating decimal solution
▸ a solution that needs to be rounded to a specified accuracy

The third example is set in the context of shape, and requires the equation to be constructed before it can be solved.

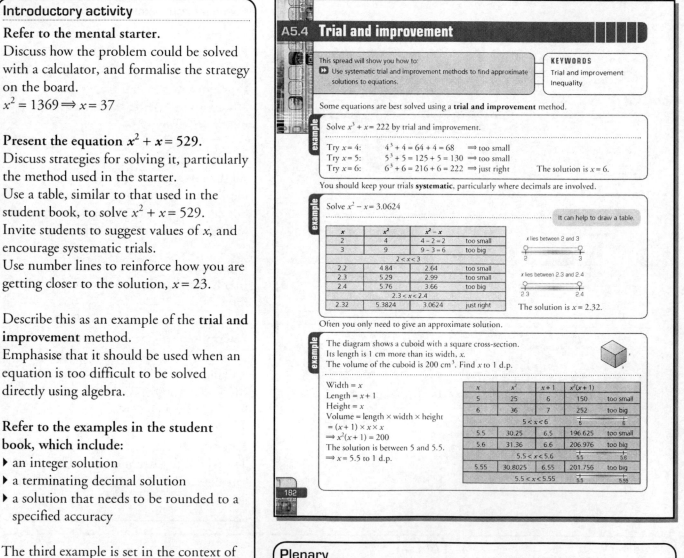

A5.4 Trial and improvement

This spread will show you how to:
▶ Use systematic trial and improvement methods to find approximate solutions to equations.

KEYWORDS
Trial and improvement
Inequality

Some equations are best solved using a **trial and improvement** method.

example

Solve $x^3 + x = 222$ by trial and improvement.

Try $x = 4$: $4^3 + 4 = 64 + 4 = 68 \Rightarrow$ too small
Try $x = 5$: $5^3 + 5 = 125 + 5 = 130 \Rightarrow$ too small
Try $x = 6$: $6^3 + 6 = 216 + 6 = 222 \Rightarrow$ just right   The solution is $x = 6$.

You should keep your trials **systematic**, particularly where decimals are involved.

example

Solve $x^2 - x = 3.0624$

It can help to draw a table.

x	x^2	$x^2 - x$	
2	4	$4 - 2 = 2$	too small
3	9	$9 - 3 = 6$	too big
	$2 < x < 3$		
2.2	4.84	2.64	too small
2.3	5.29	2.99	too small
2.4	5.76	3.66	too big
	$2.3 < x < 2.4$		
2.32	5.3824	3.0624	just right

x lies between 2 and 3

x lies between 2.3 and 2.4

The solution is $x = 2.32$.

Often you only need to give an approximate solution.

example

The diagram shows a cuboid with a square cross-section.
Its length is 1 cm more than its width, x.
The volume of the cuboid is 200 cm³. Find x to 1 d.p.

Width = x
Length = $x + 1$
Height = x
Volume = length × width × height
= $(x + 1) \times x \times x$
$\Rightarrow x^2(x + 1) = 200$
The solution is between 5 and 5.5.
$\Rightarrow x = 5.5$ to 1 d.p.

x	x^2	$x + 1$	$x^2(x + 1)$	
5	25	6	150	too small
6	36	7	252	too big
	$5 < x < 6$			
5.5	30.25	6.5	196.625	too small
5.6	31.36	6.6	206.976	too big
	$5.5 < x < 5.6$			
5.55	30.8025	6.55	201.756	too big
	$5.5 < x < 5.55$			

182

Plenary

Set the problem $3^m = 59\ 049$.

Encourage students to use their calculators to find m.

Repeat for (realistic) problems set by students, encouraging those that will practise a variety of calculator keys.

Further activities

Students can work in pairs, setting equations for a partner to solve using trial and improvement.

Differentiation

Extension questions:

▸ Questions 1 and 2 have exact solutions and provide easily accessible opportunities to form an equation.
▸ Questions 3–6 require decimal solutions and more equations to construct.
▸ Questions 7–11 require the same skills but involve further mathematical knowledge, and use of a calculator in complex cases.

Core tier: focuses on linear equations involving fractions.

Exercise A5.4

1 Solve these equations to 1 d.p. using trial and improvement.
  a $x^2 + x = 132$ **b** $x^3 + 2x = 186.816$
  c $2x^2 + 7 = 20.52$ **d** $2^x = 4096$
  e $\sqrt{x} = 2.5$

2 Write an equation to solve each problem. Using trial and improvement, find the required quantity.

  a Area = 38.44 cm^2.
  What is the length?
  b Area = 63.75 m^2.
  Length is 1 m more than width.
  What is the length?
  c Volume = 769.488 m^3.
  Length is 1 m more than the width.
  The width is 1 m more than the height.
  What is the height?
  d Volume = 178.9555 mm^3.
  What is the base length?

3 These equations have inexact solutions. Find each solution to the required number of decimal places, using trial and improvement.
  a $m(m + 1) = 100$ (1 d.p.)
  b $p(p - 4) = 63$ (2 d.p.)
  c $y^3 - 2y = 70$ (1 d.p.)
  d $x^5 = 5000$ (2 d.p.)
  e $\frac{56}{q} + q = 19$
  (2 solutions ... both to 1 d.p.).

4 The shaded area is 17 mm^2.
Construct an equation and use trial and improvement to find the exact length of the large square.

5 Solve the equation $4^x = 50$, giving your answer correct to 2 d.p.

6 The cube of a number is 200 more than the number itself.
  a Show that this statement can be written as $x(x^2 - 1) = 200$.
  b Find x correct to 1 d.p.

7 $\square^3 = 30 \times \square^2$ ... use trial and improvement to find the number whose cube is thirty times as much as its square.

8 This cuboid has a volume of 150 cm^3. It has a square cross-section and its length is 2 cm more than its width.
  a Show that $x^3 - 2x^2 = 150$.
  b Find x to 1 d.p.

9 The surface area of a cylinder is given by the formula $A = 2\pi r(r + h)$.
Find the radius of a 10 cm high cylinder with surface area 550 cm^2, to 1 d.p.

10 a Solve $x^x = 200$ to 1 d.p.
  b Solve $\sqrt[x]{64\ \text{million}} = 20$ exactly.

11 The graph shown is $y = x^2 + x$. Use trial and improvement to complete the coordinates of P = ($\square$, 93.84).

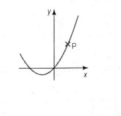

Exercise commentary

The questions assess the objectives on Framework Pages 133 and 135.

Problem solving
The questions in this exercise assess the objectives on Framework Page 29 and 31.

Group work
Students can compare their equations in small groups, to check that they have formed equivalent equations.

Misconceptions
Students at this level tend to have an incomplete knowledge of appropriate accuracy. Encourage working to one more decimal place than is required in inexact solutions.
Equations may be incorrectly simplified, for example $x^2 + x = 50 \Rightarrow x^2 = 50x$.
You may need to recap how to collect like terms.

Links
This lesson links to constructing and solving equations (Framework Page 123).

Homework

A5.4HW provides further practice at trial and improvement, establishing equations and using mathematics from a variety of contexts.

Answers

1 a $x = 11$ **b** $x = 5.6$ **c** $x = 2.6$ **d** $x = 12$ **e** $x = 6.3$
2 a 6.2 cm **b** 8.5 cm **c** 8.2 cm **d** 7.1 mm
3 a $m = 9.5$ **b** $p = 10.18$ **c** $y = 4.3$ **d** $x = 5.49$
  e $q = 3.6$ or 15.4
4 5.25 mm **5** $x = 2.82$ **6 b** 5.9 **7** 30 **8 b** 6.1
9 5.6 cm **10 a** $x = 3.9$ **b** 6 **11** 9.2

Mental starter

You will need a counting stick like this:

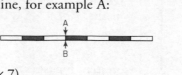

Say that the top line represents the 3 times table, and the bottom line represents the 7 times table. Touch a point on the top line, for example A:

A
↓
B

Say 12 (that is, 4 × 3), then point to B.
Students respond by chanting 28 (that is, 4 × 7).

Useful resources

A **counting stick** can be used for the mental starter.
You can buy one, or make one easily from a metre rule.

Introductory activity

Refer to the mental starter.

Describe how the two sets of multiples are in **direct proportion** – each time the top row increases by 3, the bottom row increases by 7.

Discuss other pairs of quantities that are directly proportional.

For example, describe a cake that needs 500 g of butter for 4 people.

Ask students to complete the table:

Number of people	1	2	3	4
Grams of butter				

Discuss what it would look like if you plotted a graph.

Demonstrate on the board, and encourage students to notice that it is a straight line.

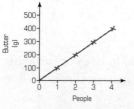

Discuss how to interpret the gradient: each time you get an extra person, you need an extra 125 g of butter.

Link to $y = mx + c$, and discuss the equation of the graph ($y = 125x$).

Refer to the examples in the student book.

Use them to highlight that any two quantities that are in direct proportion will produce a linear graph.

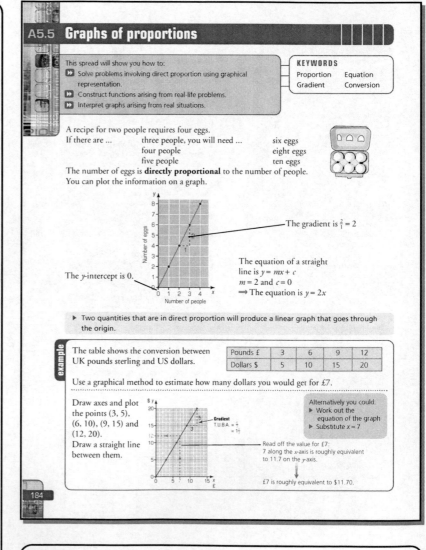

A5.5 Graphs of proportions

This spread will show you how to:
- Solve problems involving direct proportion using graphical representation.
- Construct functions arising from real-life problems.
- Interpret graphs arising from real situations.

KEYWORDS
Proportion Equation
Gradient Conversion

A recipe for two people requires four eggs.

If there are ...	you will need ...
three people,	six eggs
four people	eight eggs
five people	ten eggs

The number of eggs is **directly proportional** to the number of people. You can plot the information on a graph.

The gradient is $\frac{2}{1} = 2$

The y-intercept is 0.

The equation of a straight line is $y = mx + c$
$m = 2$ and $c = 0$
$\Rightarrow$ The equation is $y = 2x$

▶ Two quantities that are in direct proportion will produce a linear graph that goes through the origin.

The table shows the conversion between UK pounds sterling and US dollars.

Pounds £	3	6	9	12
Dollars $	5	10	15	20

Use a graphical method to estimate how many dollars you would get for £7.

Draw axes and plot the points (3, 5), (6, 10), (9, 15) and (12, 20).
Draw a straight line between them.

Gradient
T.U.B.A = $\frac{5}{3}$ = $1\frac{2}{3}$

Alternatively you could:
▶ Work out the equation of the graph
▶ Substitute $x = 7$

Read off the value for £7:
7 along the x-axis is roughly equivalent to 11.7 on the y-axis.

£7 is roughly equivalent to $11.70.

184

Plenary

Try an inverse proportion problem:
It takes one person seven days to dig a plot of land.
How long will it take 2, 3, 4, ... people?
What will this look like on a graph?
What will its equation be?

Differentiation

Extension questions:

▸ Questions 1 and 2 should familiarise students with the idea of proportion.
▸ Questions 3 and 4 guide students towards establishing a proportion statement by linking to gradient.
▸ Question 5 introduces the concept of indirect proportion.

Core tier: focuses on simple problems involving direct proportion graphs.

Exercise A5.5

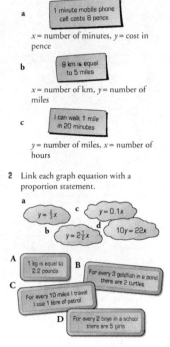

1 Construct a graph to represent each of these direct proportion relationships. Find the equation of your graph.

a
> 1 minute mobile phone call costs 8 pence

x = number of minutes, y = cost in pence

b
> 8 km is equal to 5 miles

x = number of km, y = number of miles

c
> I can walk 1 mile in 20 minutes

y = number of miles, x = number of hours

2 Link each graph equation with a proportion statement.

a
> $y = \frac{2}{3}x$

c
> $y = 0.1x$

b
> $y = 2\frac{1}{2}x$

d
> $10y = 22x$

A
> 1 kg is equal to 2.2 pounds

B
> For every 3 goldfish in a pond there are 2 turtles

C
> For every 10 miles I travel I use 1 litre of petrol

D
> For every 2 boys in a school there are 5 girls

3 a A proportion graph has equation $y = \frac{3}{5}x$. Write a statement that could be modelled using this graph.
   b Repeat for $y = 3\frac{1}{3}x$.

4 This graph allows you to convert your weight in stones into your weight in kg.

a Use the graph to find the weight, in kg, of an 11-stone female.
b Find the gradient of the line and, hence, its equation.
c Use your equation to find the weight, in kg, of a 15-stone male.
d Use your equation to find the weight, in stone, of a 200-kg elephant.
e Show, using the graph and your equation, that 6 stones is less than 40 kg.
f Which is heavier, 13 stones or 80 kg?

5 Two quantities are in inverse proportion if, as one increases by a certain amount, the other decreases by a certain amount. For example the number of men needed to dig a hole is inversely proportional to the number of days taken to dig the hole.
a Copy and complete this table:

Number of men, x	1	2	3	4	5
Number of days to dig, y		6			

b Suggest an equation to connect x and y.
c Plot a graph of this equation.

185

Exercise commentary

The questions assess the objectives on Framework Page 137.

Problem solving
The questions in this exercise assess the objectives on Framework Page 27.

Group work
Questions 2 and 3 should be discussed in pairs or in small groups.

Misconceptions
When constructing a proportion statement, students may incorrectly include a + or − operation. Emphasise the multiplicative nature of variables that are in proportion (for example, if one quantity increases from 20 to 24 you don't add 4 to the other quantity).
Students may need the link with gradient reinforcing.

Links
This topic links to gradient of a linear graph and $y = mx + c$. The idea of proportionality is also used in science.

Homework

A5.5HW provides further practice at direct proportion, and extends into simple cases of indirect proportion.

Answers

1 a $y = 8x$ b $y = \frac{5}{8}x$ c $y = 3x$
2 a B   b D   c C   d A
4 a 70 kg   b $y = 6\frac{4}{11}x$ c 95 kg
   d 31 st   e 6 st ≈ 38 kg   f 13 st
5 a $y = 12, 6, 4, 3, 2.4$ b $y = \frac{12}{x}$

Mental starter

Split students into small groups, each with a tape measure.
Instruct students to measure their height and the height of their navel above the ground.
Students should calculate their height: navel ratio, and compare with their friends.

Briefly discuss the Golden Ratio $\left(\dfrac{\sqrt{5}+1}{2} = 1.618\ldots\right)$

This is meant to be the height: navel ratio of a perfectly proportioned body!

Introductory activity

Discuss the meaning of proportion.
Refer to the starter and the examples from A5.5 to highlight that proportion is linked with division.
Recap that two quantities are in direct proportion if they increase or decrease at the same rate.

Present a proportion problem.
Here are three chocolate bars:

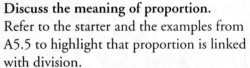

- Show, using the first two bars, that price is proportional to the amount of chocolate:
  $\dfrac{32}{80} = \dfrac{52}{130}$ (or reciprocals)
- Form an equation to find x:
  $\dfrac{32}{80} = \dfrac{x}{290}$
- Ask students to solve the equation:
  $x = \dfrac{32}{80} \times 290 = 116$

So the large bar costs £1.16

In pairs, students should find the weight and price of another Venus bar.

Discuss the examples in the student book.
- Example **a** involves scaling up a recipe.
- Example **b** involves currency conversion.

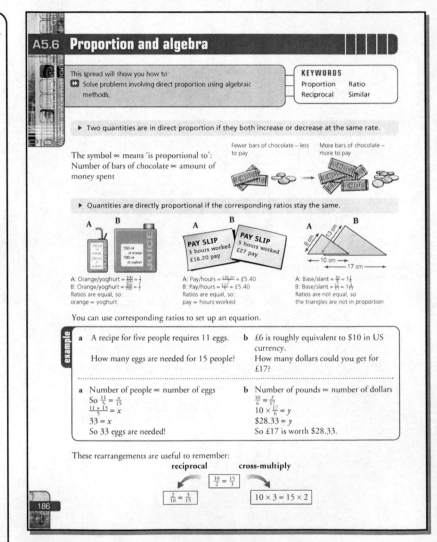

Plenary

Finish with a 'best buy' question, where quantities are not in direct proportion.
A 330 ml can of cola costs 60p.
A 550 ml can of cola costs £1.10.
Which is the better value for money?
Discuss this problem, in particular:
- Is price proportional to the amount of cola?
- How many ml/p (or p/ml) for each can?

Further activities

Are the corresponding lengths of a cube in proportion as it increases in size?
Are the corresponding surface areas and volumes in proportion?
How can these relationships be represented algebraically?

Differentiation

Extension questions:
▶ Question 1 recaps the meaning of direct proportion.
▶ Questions 2 and 3 require an algebraic approach to proportion and some explanation of the meaning of the solution.
▶ Question 4 presents a real-life proportion experiment for students to explain and discuss.

Core tier: focuses on graphs of linear functions.

Exercise A5.6

1 Decide if these quantities are in direct proportion.
 a The speed you travel and the distance you cover.
 b The number of people in a room and the amount of space each person has.
 c The amount you talk on your mobile and the price of your calls.
 d The amount of cereal you eat and the amount of protein you receive.
 e The number of people digging and the time taken to dig a 2m³ trench.
 f Your age and the number of presents you receive on your birthday.

2 The stated quantities are in direct proportion. Set up an algebra equation and solve it to find the missing value in each case.
 a
 Amount of nuts and weight of bar.
 b
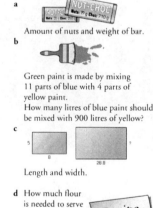
 Green paint is made by mixing 11 parts of blue with 4 parts of yellow paint.
 How many litres of blue paint should be mixed with 900 litres of yellow?
 c
 Length and width.
 d How much flour is needed to serve 5 people?
 Recipe
 SERVES 6
 340 g flour

3 Ben has a small tube of tasty 'Dummies' sweets. Matt has a large one. They both love blue sweets so they counted how many they had:

	Total	Blue
Ben	37	7
Matt	108	23

 a Is the number of blue sweets in proportion to the total number of sweets?
 b Using an equation to help you, decide how many blue sweets Ben should have so that there is direct proportion between the number of blues and the total number.

4 Ayeesha performed a scientific experiment. She measured the height of a plant in the first few days of its life:

Age (days)	1	2	3	4	5	6	7
Height (cm)	2.8	5.6	8.4				

 a Is age in direct proportion to height?
 b Copy and complete the table.
 c Write a formula connecting age and height.
 d How tall should the plant be at the end of a fortnight?
 e After 3 weeks, Ayeesha found the plant to be 50 cm high. Do age and height remain in proportion after this time? Explain your answer.

187

Exercise commentary

The questions assess the objectives on Framework Page 137.

Problem solving
The questions in this exercise require students to represent problems in algebraic form, assessing the objectives on Framework Page 27.

Group work
Questions 1 and 4 lend themselves well to small group or paired discussion.

Misconceptions
Continue to reinforce the involvement of multiplication and division in proportion, rather than addition or subtraction. Students may show reluctance in writing an equation, and may prefer to try various numbers until an answer emerges. Encourage the use of algebra as an elegant maths tool.

Links
The skills learned here will be used later with similar shapes (Framework Page 193). Question 4 highlights the link with experimental science.

Homework

A5.6HW includes proportion problems in a variety of contexts. It also includes 'best buy' type analysis where quantities are not in proportion.

Answers

1 a Yes b No c Depends on tariff
 d Yes e No f No
2 a 61 g b 2475 litres c 18 d 283 g
3 a No b 8
4 a Yes b 11.2, 14, 16.8, 19.6
 c Height = 2.8 × Age d 39.2 cm
 e No

Mental starter

Students will need a blank sheet of A4 paper, or a mini-whiteboard. Instruct students to sketch scatter diagrams with:

▸ Positive correlation
▸ Negative correlation
▸ No correlation

Students should suggest possible labels for each axis.

Useful resources

A5.7OHP contains the graphs shown in the introductory activity and in the student book.

Introductory activity

Briefly recap $y = mx + c$.
Discuss the meaning of m and c, in terms of a linear graph.

Present a real-life scenario.
The graphs show the monthly charges for two mobile phone companies, 'Chatalot' and 'Kallme'. Draw them on the board, or use **A5.7OHP**.

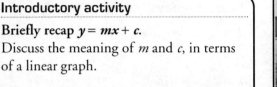

Discuss:
▸ what each graph shows
▸ which company is the better value.
  (no correct answer, because it depends on call length)

Find the equation of each line.
Recap how to find m and c, and give students a few minutes to find:
$y = 2x + 5$ (Chatalot)
$y = \frac{5}{7}x + 15$ (Kallme)
When would the bills be equal in price?

Discuss interpreting m and c.
Chatalot: Each hour costs £2, and you pay £5 standing charge each month.
Kallme: Every seven hours cost £5, but you pay £15 standing charge.

Discuss the graphs in the student book.
You will need to recap scatter graphs from data handling, and briefly discuss a line of best fit.

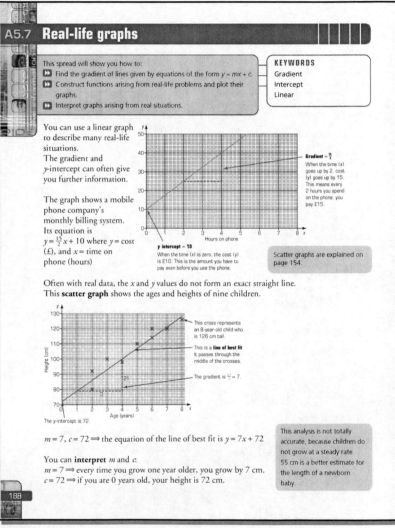

A5.7   Real-life graphs

This spread will show you how to:
▸ Find the gradient of lines given by equations of the form $y = mx + c$.
▸ Construct functions arising from real-life problems and plot their graphs.
▸ Interpret graphs arising from real situations.

KEYWORDS
Gradient
Intercept
Linear

You can use a linear graph to describe many real-life situations.
The gradient and y-intercept can often give you further information.

The graph shows a mobile phone company's monthly billing system.
Its equation is
$y = \frac{15}{2}x + 10$ where $y =$ cost (£), and $x =$ time on phone (hours)

Gradient = $\frac{15}{2}$
When the time (x) goes up by 2, cost (y) goes up by 15. This means every 2 hours you spend on the phone, you pay £15.

y intercept = 10
When the time (x) is zero, the cost (y) is £10. This is the amount you have to pay even before you use the phone.

Scatter graphs are explained on page 154.

Often with real data, the x and y values do not form an exact straight line. This **scatter graph** shows the ages and heights of nine children.

This cross represents an 8-year-old child who is 126 cm tall.

This is a **line of best fit**. It passes through the middle of the crosses.

The gradient is $\frac{y}{x} = 7$.

The y-intercept is 72.

$m = 7$, $c = 72 \Rightarrow$ the equation of the line of best fit is $y = 7x + 72$

You can **interpret** m and c:
$m = 7 \Rightarrow$ every time you grow one year older, you grow by 7 cm.
$c = 72 \Rightarrow$ if you are 0 years old, your height is 72 cm.

This analysis is not totally accurate, because children do not grow at a steady rate. 55 cm is a better estimate for the length of a newborn baby.

188

Plenary

Discuss the **limitations** of the linear models presented.
▸ Mobile phone graphs – why may they not be linear? (different rates for different call lengths)
▸ Age and height scatter graph – definitely not linear because of growth spurts at different ages (particularly inaccurate at the lower end of the graph).

Further activities

Students working in pairs could think of two variables which they feel may be correlated (for example, height and length of stride).
By collecting data, students can derive their own relationship for these variables. Encourage discussion of the limitations of the relationship.

Differentiation

Extension questions:

▸ In question 1, equations and their interpretation are already given and students simply have to match them.
▸ Question 2 requires an equation to be found and analysed.
▸ Questions 3 and 4 explore limitations of real-life equations.

Core tier: focuses on distance-time graphs.

Exercise A5.7

1 Match these statements with the line graph equations that would represent them.

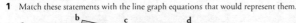

A new born baby weighs 6 pounds and gains 1 pound every 10 days

Electricity is charged at 4 pence per unit with a cost of £5 per month for service.

0°C is 32°F. Every rise of 5°C is equivalent to a rise of 9°F.

Eight kilometres is equivalent to 5 miles.

A	B	C	D
$y = 4x + 500$	$5y = 9x + 160$	$y = 1\frac{1}{2}x$	$y = \frac{1}{10}x + 6$

2 For each graph, find its equation and explain what the equation tells you.

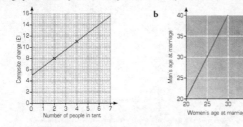

a (Campsite charge (£) vs Number of people in tent)

b (Man's age at marriage vs Women's age at marriage)

3 a Draw scatter graphs for each set of data and find the equation of the line of best fit.

i

Maths	14	37	35	23	24	29	46	21	44	48
Physics	7	29	34	17	27	28	37	9	40	39

... 10 students in a maths test and physics test.

ii

Temperature °C of day	24	32	20	27	25	29	19	23	28	28
Number of ice-creams sold in one hour	8	20	2	15	11	19	1	5	15	18

... ice-creams sold on 10 days in summer.

b In each case, explain what the line shows you. Discuss possible limitations of each line.

4 Mr Edwards came up with this equation to represent a person's weight in stones(x) and their pulse rate each minute(y).

$$y = 8x + 20$$

Interpret the equation and discuss its limitations.

Exercise commentary

The questions assess the objectives on Framework Pages 173–177.

Problem solving

The questions in this exercise assess the objectives on Framework Page 27 and 33.

Group work

All questions are suitable for pair work or small group discussion.

Misconceptions

You may need to recap gradient – emphasise that it is most easily interpreted in fractional form.
You can use the mnemonic 'TUBA' (top up, bottom across). Ensure in questions 2 and 3, however, students are clear which variable is up and which is across.

Links

The topic links to gradient and $y = mx + c$ (Framework Page 167). It also links to limitations of assumptions in statistics (Pages 273 and 275).

Homework

A5.7HW provides further practice at constructing and interpreting real-life graphs.

Answers

1 a D b A c B d C
2 a $y = 1.5x + 5$; Campsite charge is £5 plus £1.50 per person
  b $y = 2x - 20$; At marriage, the man's age is double the woman's minus 20
3 a Equations close to: i $y = 0.9x + 3.7$
   ii $y = 1.7x - 31.6$

Mental starter

This is a 'guess the number' activity, where you give students two clues for each number. For example:

▶ It is a square number but also a multiple of 5
▶ It is a cube number and a square number
▶ It has a factor of 6 and its square root is greater than 10.

Useful resources

A5.8OHP shows the graph connected with the scenario in the introductory activity.

Introductory activity

Present this 'Mensa' style problem to the class.

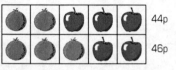

44p

46p

Only show the top row at first (hint that there is a hidden row).
Ask: How much do the apples and oranges cost each?
Encourage students to suggest possible prices. For example: oranges 4p, apples 12p.
Discuss the use of algebra to represent all the possible options in a single equation: $2x + 3y = 44$
Emphasise the importance of specifying the meaning of x and y.
Introduce this as an example of an **implicit equation.**
Discuss strategies for plotting an implicit graph:

▶ Make y the subject (**explicit** form), then draw a table of two or three values, then a graph
▶ Keep in implicit form, and draw a table with $x = 0$ and $y = 0$, then draw a graph.

Stress that for a linear graph you only really need two coordinates, but it is best to have a third one to check.
Plot the graph and discuss what the points show.
Repeat for the second line of the puzzle.
Draw a second line on the same axes and discuss the meaning of the intersection (shown on **A5.8OHP**).
Introduce the term **simultaneous equations.**

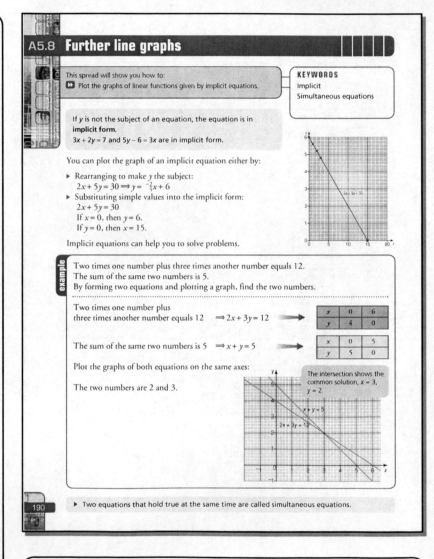

A5.8 Further line graphs

This spread will show you how to:
▶ Plot the graphs of linear functions given by implicit equations.

KEYWORDS
Implicit
Simultaneous equations

If y is not the subject of an equation, the equation is in **implicit form.**
$3x + 2y = 7$ and $5y - 6 = 3x$ are in implicit form.

You can plot the graph of an implicit equation either by:

▶ Rearranging to make y the subject:
$2x + 5y = 30 \Rightarrow y = -\frac{2}{5}x + 6$
▶ Substituting simple values into the implicit form:
$2x + 5y = 30$
If $x = 0$, then $y = 6$.
If $y = 0$, then $x = 15$.

Implicit equations can help you to solve problems.

example

Two times one number plus three times another number equals 12.
The sum of the same two numbers is 5.
By forming two equations and plotting a graph, find the two numbers.

Two times one number plus three times another number equals 12 $\Rightarrow 2x + 3y = 12$

| x | 0 | 6 |
| y | 4 | 0 |

The sum of the same two numbers is 5 $\Rightarrow x + y = 5$

| x | 0 | 5 |
| y | 5 | 0 |

Plot the graphs of both equations on the same axes:

The two numbers are 2 and 3.

The intersection shows the common solution, $x = 3$, $y = 2$.

▶ Two equations that hold true at the same time are called simultaneous equations.

190

Plenary

Discuss the exercise, particularly the later questions that focus on simultaneous equations.

Discuss the different scenarios that students may have invented for question 8.

Further activities

Students can devise a problem similar to the 'apples and oranges' problem in the introductory activity.

The problem can then be presented to the class as a challenge (stress that it must be solvable).

A5.6ICT uses a graph-plotting package to explore equations of the form $y = mx + c$.

Differentiation

Extension questions:

▸ Questions 1 and 2 provide practice at rearranging an equation and plotting the graph of an implicit equation.

▸ Questions 3 −5 continue the theme of plotting implicit equations but introduce simultaneous equations.

▸ Questions 6−8 focus on constructing, solving and interpreting simultaneous equations.

Core tier: focuses on algebraic manipulation set in the context of shape.

Exercise A5.8

1 Copy and complete the table of values for each equation. Hence, on a single set of axes labelled ⁻8 to ⁺8, plot all of the graphs.

a $x + y = 6$

x	0	□
y	□	0

b $3x + 4y = 12$

x	0	□
y	□	0

c $2x + 5y = 15$

x	0	□
y	□	0

2 Match the explicit equations in Sheet A with the implicit equations in Sheet B.

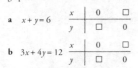

$y = -\frac{3}{8}x + 3$
$y = \frac{1}{4}x + 6$
$y = 2\frac{1}{2} - 4x$
$y = 7 - 2x$

Sheet A

$2x + y = 7$ $3x + 8y = 24$
$8x + 2y = 5$
$3y - x = 18$

Sheet B

3 a Write an explicit equation that would pair with $5y + 2x = 10$.
  b Write an implicit equation that would pair with $y = \frac{2}{3}x + 7$.

4 a Plot the graphs of $x + 2y = 8$ and $y = x - 2$ on a single pair of axes.
  b Use your graphs to solve these equations simultaneously.
    $\left. \begin{array}{l} x + 2y = 8 \\ y = x - 2 \end{array} \right\}$

5 Repeat question 4 for these pairs of equations.
  a $x + y = 9$
    $y = 2x - 3$
  b $x + y = 9$
    $5y - 4x = 18$
  c $3x + 4y = 24$
    $3x + 2y = 18$

6 Why would these pairs of simultaneous equations have no solutions? (Hint: think about their graphs, drawing them if necessary.)
  a $y = 3x - 3$
    $y = 3x + 3$
  b $9x + 3y = 12$
    $y = 4 - 3x$
  c $x + 3y = 11$
    $x + 3y = 20$.

7 In each problem, **i** write a pair of equations to represent the information **ii** plot the equations on a graph and **iii** find the solution to the equations using your graph.

a Two numbers have a sum of 8 and a difference of 2.
b The sum of two numbers is 6. The sum of one of the numbers and 3 times the other is 14.
c A bar of chocolate and a can of coke cost 85p. Three bars and two cokes costs £2.05.

8 Invent a problem that could be represented by these simultaneous equations. Solve the problem using a graph.
  $x + 3y = 10$
  $2x + y = 5$

191

Exercise commentary

The questions assess the objectives on Framework Page 165.

Problem solving
This exercise assesses the objectives on Framework Page 27.
Question 6 also assesses Page 33.

Group work
Students would benefit from discussing questions 6−8, though they should be encouraged to practice graph plotting individually.

Misconceptions
Students may prefer to rearrange an equation into explicit form. However the numbers generated may be difficult to manipulate. Encourage substituting $x = 0$ and $y = 0$ into the implicit form. Emphasise that solutions for both x and y should be presented in a simultaneous equation.

Links
The topic of implicit equations is used here as a bridge to simultaneous equations, which will be covered thoroughly in year 9.

Homework

A5.8HW practices the techniques involved in dealing with basic implicit and simultaneous equations.

Answers

1 a $x = 0, 6$; $y = 6, 0$ b $x = 0, 4$; $y = 3, 0$ c $x = 0, 7.5$; $y = 3, 0$
2 $y = -\frac{3}{8}x + 3$, $3x + 8y = 24$; $y = 7 - 2x$, $2x + y = 7$; $y = 2\frac{1}{2} - 4x$, $8x + 2y = 5$; $y = \frac{1}{3}x + 6$, $3y - x = 18$
3 a $y = -\frac{2}{5}x + 2$ b $3y - 2x = 21$ or equivalent   4 b $x = 4$, $y = 2$
5 a $x = 4$, $y = 5$ b $x = 3$, $y = 6$ c $x = 4$, $y = 3$ 6 They are all parallel
7 a i $x + y = 8$, $x - y = 2$ iii 5, 3   b i $x + y = 6$, $x + 3y = 14$ iii 2, 4
  c i $x + y = 85$, $3x + 2y = 205$ iii Chocolate 35p, Coke 50p
8 $x = 1$, $y = 3$

Summary

The key objectives for this unit are:

▸ Write an expression to describe the nth term of an arithmetic sequence
▸ Given values for m and c, find the gradient of lines given by equations of the form $y = mx + c$
▸ Construct functions arising from real-life problems and plot their corresponding graphs
▸ Give solutions to problems to an appropriate degree of accuracy

Check out commentary

1 Students should not write $m = s + 3$ even though you add three matches each time. Remind students of the difference between describing and explaining.

2 Students can use the mnemonic TUBA to help find gradient.
Encourage small group work, particularly where students find difficulty in interpreting the equation.

3 Encourage students to recognise that they are working with direct proportion. Link to the realisation that when one quantity is divided by the other one you always get the same answer.
This should help to establish the equation $\frac{13}{5} = \frac{75}{x}$ or equivalent.

4 Reinforce the use of trial and improvement in cases where the algebra is too difficult to solve the equation directly.
Emphasise that working is always needed to one decimal place more than the final answer, so that you can round back accurately.

Plenary activity

Draw this sequence:
For the sequence, ask students to produce:
▸ A formula
▸ A graph
▸ A justification.
This activity should link most skills covered in the unit.

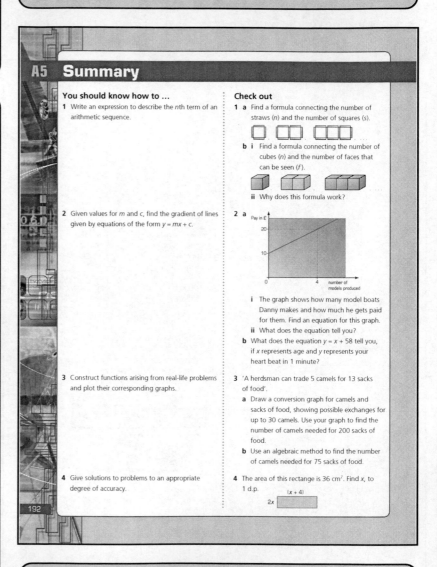

A5 Summary

You should know how to ...

1 Write an expression to describe the nth term of an arithmetic sequence.

2 Given values for m and c, find the gradient of lines given by equations of the form $y = mx + c$.

3 Construct functions arising from real-life problems and plot their corresponding graphs.

4 Give solutions to problems to an appropriate degree of accuracy.

Check out

1 a Find a formula connecting the number of straws (n) and the number of squares (s).

  b i Find a formula connecting the number of cubes (n) and the number of faces that can be seen (f).

    ii Why does this formula work?

2 a

  i The graph shows how many model boats Danny makes and how much he gets paid for them. Find an equation for this graph.
  ii What does the equation tell you?
  b What does the equation $y = x + 58$ tell you, if x represents age and y represents your heart beat in 1 minute?

3 'A herdsman can trade 5 camels for 13 sacks of food'.
  a Draw a conversion graph for camels and sacks of food, showing possible exchanges for up to 30 camels. Use your graph to find the number of camels needed for 200 sacks of food.
  b Use an algebraic method to find the number of camels needed for 75 sacks of food.

4 The area of this rectangle is 36 cm². Find x, to 1 d.p.

192

Development

All of these skills are revisited and developed in Year 9 and beyond. In particular, algebraic methods of solving simultaneous equations will be explored.

Links

Proportion and graphs are used commonly in science.
The interpretation of real-life graphs extends to many subjects.

Mental starters

Objectives covered in this unit:
▸ Order, add, subtract, multiply and divide integers.
▸ Apply mental skills to solve simple problems.
▸ Find fractions and percentages of quantities.

Resources needed

* means class set needed
Essential:
Squared paper
Useful:
Mini-whiteboards
Isometric paper
Envelopes in which films are sent for processing
Different sizes of metric paper, including A2 and A1
P1.1OHP – steps diagrams
P1.2OHP – graph and table for steps results
P1.3OHP – steps arrangements
P1.4OHP – sizes of photographic prints
P1.6OHP – table of paper sizes

P1 Solving problems

This unit will show you how to:

▸ Solve increasingly demanding problems and evaluate solutions.
▸ Present a concise, reasoned argument, using symbols, diagrams, graphs and related explanatory text.
▸ Represent problems and synthesise information in algebraic or graphical form.
▸ Move from one form to another to gain a different perspective on the problem.
▸ Solve substantial problems by breaking them into simpler tasks.

▸ Suggest extensions to problems, conjecture and generalise.
▸ Identify exceptional cases or counter-examples, explaining why.
▸ Compare two ratios.
▸ Interpret and use ratio in a range of contexts.
▸ Use proportional reasoning to solve a problem.
▸ Explore connections in mathematics across a range of contexts.
▸ Give solutions to problems to an appropriate degree of accuracy.

Last week they were both £1 for 1 litre. Which shall we buy?

We could compare the amount of cola per penny.

You can use the unitary method to find the best buy.

Before you start

You should know how to ...

1 Generate and describe non-linear sequences.

2 Draw a graph from a table of results.

3 Write and simplify an integer ratio.

4 Increase or decrease an amount by a given percentage.

Check in

1 i Write down the next two terms in these sequences.
   ii Describe how you would find the nth term.
   a 1 4 9 16   b 2 8 18 32
   c 2 5 10 17

2 Draw a graph to show this sequence.

Term (T)	1	2	3	4
Sequence (S)	1	4	9	16

3 a Simplify these ratios   i 4:12   ii 15:24
   b Write these ratios in the form 1:n
   i 3:18   ii 5:16

4 a Increase £30 by 20%
   b Decrease £40 by 15%

Unit commentary

Aim of the unit

This unit develops problem-solving skills, beginning with identifying the problem and breaking it into simpler tasks. There are investigations of sequences, which involve finding a general rule, and of real-life problems using ratio and proportion. Throughout the emphasis is on working systematically, making and justifying predictions, and presenting results clearly.

Introduction

Discuss the terminology used in sequences: term (1st term, nth term), expression, formula.
Ask students for examples.
Emphasise the importance of correct terminology.

Framework references

This unit focuses on:
▸ Teaching objectives pages: 77, 79, 81, 155–157
▸ Problem-solving objectives pages: 5, 9, 15, 27, 29, 31

Differentiation

Core tier

Focuses on refining techniques in problem-solving, using the context of paving slabs.

Check in activity

Present the problem $1 + 2 + 3 + \ldots + 99 + 100 = ?$
Allow students two minutes to think about it, then invite discussion.
Work towards a strategy by making the problem simpler:
$$1 + 2 + 3 + 4 = T \text{(Total)}$$
$$4 + 3 + 2 + 1 = T$$
$$\overline{5 + 5 + 5 + 5 = 2T}\quad \text{Discuss further.}$$

Mental starter

Write sequences on the board:

3	6	9	12
3	6	10	15
3	6	12	24
3	6	12	21

Ask students to predict the next one or two values in each sequence.

Useful resources

Mini-whiteboards may be useful for the mental starter.
Squared paper
P1.1OHP – step diagrams

Introductory activity

Discuss the task outlined in the student book.

How could you find the total number of different ways to climb 8 steps, 20 steps or 50 steps?

Emphasise that you should start with the simplest case, building up and looking for patterns in the results.
Then you can use your results to predict what will happen in more complicated cases.
You could refer back to the Check in activity described on page 193.

Emphasise that for each case you study, listing solutions or results in a logical order will help to ensure that no errors are made.

Emphasise that you need at least three results, but possibly four or five, to be sure that your prediction of future results is based upon sound reasoning.
Use the sequences in the mental starter to reinforce this.

P1.1OHP contains the step diagrams in the student book for class discussion.

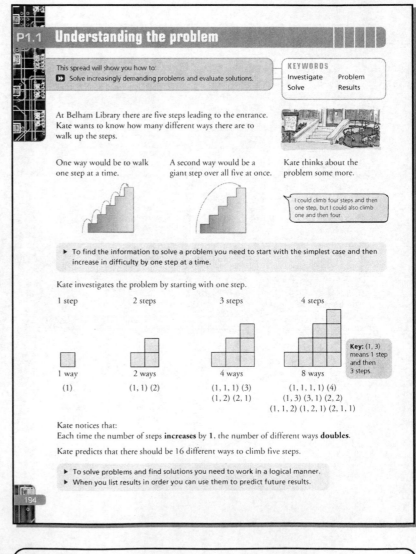

P1.1 **Understanding the problem**

This spread will show you how to:
▶ Solve increasingly demanding problems and evaluate solutions.

KEYWORDS
Investigate   Problem
Solve   Results

At Belham Library there are five steps leading to the entrance. Kate wants to know how many different ways there are to walk up the steps.

One way would be to walk one step at a time.

A second way would be a giant step over all five at once.

Kate thinks about the problem some more.

> I could climb four steps and then one step, but I could also climb one and then four

▶ To find the information to solve a problem you need to start with the simplest case and then increase in difficulty by one step at a time.

Kate investigates the problem by starting with one step.

1 step	2 steps	3 steps	4 steps
1 way	2 ways	4 ways	8 ways
(1)	(1, 1) (2)	(1, 1, 1) (3) (1, 2) (2, 1)	(1, 1, 1, 1) (4) (1, 3) (3, 1) (2, 2) (1, 1, 2) (1, 2, 1) (2, 1, 1)

Key: (1, 3) means 1 step and then 3 steps.

Kate notices that:
Each time the number of steps **increases** by **1**, the number of different ways **doubles**.

Kate predicts that there should be 16 different ways to climb five steps.

▶ To solve problems and find solutions you need to work in a logical manner.
▶ When you list results in order you can use them to predict future results.

194

Plenary

Discuss students' results for the city lights investigation, in particular the patterns they noticed in question 3.

Further activities

Students can look at the patterns they have found and the positions of each type of lamp in the square arrangement of blocks. They can write about any connections they see.

Differentiation

Extension questions:
▸ Question 1 provides an entry into the investigation by asking a closed question.
▸ Question 2 requires students to extend the task in complexity.
▸ Question 3 requires students to identify general patterns.

Core tier: focuses on identifying the information needed to solve a problem in context.

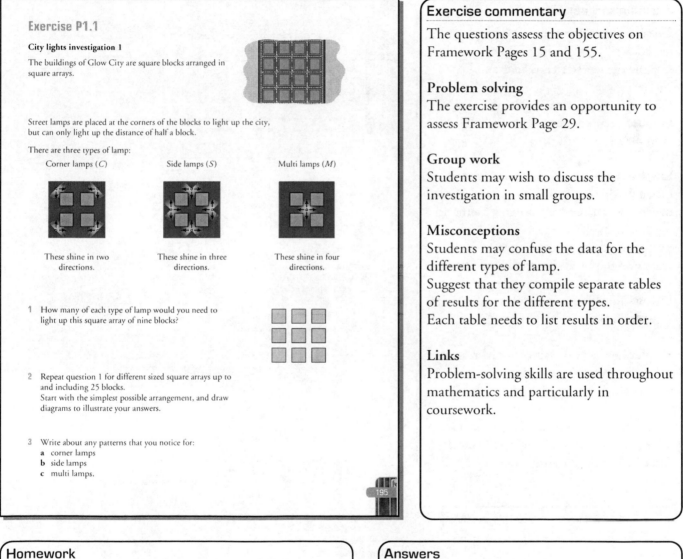

Exercise P1.1

City lights investigation 1

The buildings of Glow City are square blocks arranged in square arrays.

Street lamps are placed at the corners of the blocks to light up the city, but can only light up the distance of half a block.

There are three types of lamp:

Corner lamps (*C*)   Side lamps (*S*)   Multi lamps (*M*)

These shine in two   These shine in three   These shine in four
directions.   directions.   directions.

1 How many of each type of lamp would you need to light up this square array of nine blocks?

2 Repeat question 1 for different sized square arrays up to and including 25 blocks.
   Start with the simplest possible arrangement, and draw diagrams to illustrate your answers.

3 Write about any patterns that you notice for:
   a corner lamps
   b side lamps
   c multi lamps.

195

Exercise commentary

The questions assess the objectives on Framework Pages 15 and 155.

Problem solving
The exercise provides an opportunity to assess Framework Page 29.

Group work
Students may wish to discuss the investigation in small groups.

Misconceptions
Students may confuse the data for the different types of lamp.
Suggest that they compile separate tables of results for the different types.
Each table needs to list results in order.

Links
Problem-solving skills are used throughout mathematics and particularly in coursework.

Homework

P1.1HW is an investigative task based on number patterns within a calendar.

Answers

1 $C - 4$; $S - 8$; $M - 4$
2 $C - 4\ 4\ 4\ 4\ 4$; $S - 0\ 4\ 8\ 12\ 16$; $M - 0\ 1\ 4\ 9\ 16$
3 **a** Always 4
   b Multiples of 4
   c Square numbers

Mental starter

What is the rule?

One student leaves the room. Tell the other students a rule to generate values in a sequence, for example $2n + 3$.

The student comes back in and tries to guess the rule by giving different input numbers to the other students, who give the output number according to the rule.

Write the input number with its output number on the board, in order, until the student guesses the rule.

Useful resources

Squared paper

P1.2OHP – graph and table for steps results

Introductory activity

Discuss the importance of writing results in order, to help generate a rule.

Use the mental starter as an example.

Discuss how you can predict the next number in a sequence, referring to the example in the student book.

Emphasise that you should always test that a prediction is correct by working out the solution using your original method.

Emphasise that you can represent a sequence graphically to see how it changes.

Discuss how you can use algebra to write a general rule.

Emphasise that you can write the rule in words first, and this may help you to write it in algebra.

P1.2OHP illustrates the techniques used to organise results, containing the table and graph from the student book.

This spread will show you how to:
▶ Present a concise reasoned argument.
▶ Represent problems and synthesise information in algebraic and graphical form.

KEYWORDS
Explain   Generalize
Predict   Symbol

Kate wants to know the number of ways to climb a staircase with 20 steps. She draws a table of the results she has found so far:

Number of steps	1	2	3	4	5
Number of ways	1	2	4	8	16

Kate uses colour in her table to show that this is a **prediction**.

Kate tests her prediction by listing the number of ways to climb five steps.
(1, 1, 1, 1, 1) (1, 1, 1, 2) (1, 1, 2, 1) (1, 2, 1, 1) (2, 1, 1, 1) (1, 1, 3)
(1, 3, 1) (3, 1, 1) (1, 4) (4, 1) (1, 2, 2) (2, 1, 2) (2, 2, 1) (2, 3) (3, 2) (5)

There are 16 ways in total. My prediction is correct.

Kate draws a graph of her results.

Different ways to climb steps

The number of ways does not steadily increase, but grows quickly.

To find a general result, Kate needs to use algebra. She thinks that multiplying by 2 should be part of the general rule.

Kate adds an extra row to her table.

Number of steps (n)	1	2	3	4	5
Number of ways (W)	1	2	4	8	16
	$1 = 2^0$	$2 = 2^1$	$2 \times 2 = 2^2$	$2 \times 2 \times 2 = 2^3$	$2 \times 2 \times 2 \times 2 = 2^4$

Kate notices that:

The power of 2 is 1 less than the number of steps (n).
So the number of ways, $W = 2^{n-1}$.
For $n = 20$, $2^{19} = 524\ 288$.
There are 524 288 ways to climb 20 steps.

▶ Tabulate and graph results and use these to predict future results.
▶ Test any predictions where possible.
▶ Explain your results and, if possible, use symbols to represent the nth term.

Plenary

Recap how to find a rule for a sequence of numbers, using the method of differences.

Further activities

Students can draw graphs and diagrams to prove all their predicted results, and justify their predictions.

Differentiation

Extension questions:
▶ Questions 1 and 2 link back to exercise P1.1.
▶ Questions 3–5 focus on the investigative process, including prediction and generalisation.
▶ Question 6 extends the task and requires a general formula.

Core tier: focuses on organising results into tables and representing solutions in graphical form.

Exercise P1.2

City lights investigation 2

1 Look at your results from Exercise P1.1 on page 195.
Draw a graph to show the number of corner lamps needed for different sized square blocks. Use the axes shown.

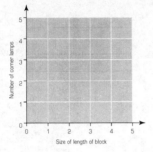

2 Repeat question 1 for side lamps and multi lamps using your results from Exercise P1.1.

3 Write one or two sentences for each graph in questions 1 and 2, describing how the number of lamps needed changes as the number of blocks increases.

4 Predict how many of each type of lamp you would need for:
   a 36 blocks
   b 49 blocks.

   Test your predictions in each case by drawing the square arrangement.

5 Find a general result for each type of lamp needed for an $n \times n$ arrangement of blocks.
   Explain how you found your general results, and try to justify each rule by looking back at the square arrangement.

Hint: Try to use algebra for questions 5 and 6.

6 Investigate the **total** number of lamps required for different square arrays of blocks.
   Try to find a general formula for the total number of lamps needed to light up an $n \times n$ array of blocks.

Exercise commentary

The questions assess the objectives on Framework Pages 155–157.

Problem solving
The exercise provides an opportunity to assess Framework Pages 27 and 33.

Group work
Students may discuss the investigation, in small groups.

Misconceptions
Students may focus only on their calculated results. Emphasise that a test to show a prediction works must be for a new case and should be shown by using the same method(s) as used to find the original results.

Links
Problem-solving skills are used throughout mathematics and particularly in coursework.
Plotting and interpreting graphs: Framework Page 173.

Homework

P1.2HW continues the calendar number investigation from **P1.1HW**.
Students are required to organise results, and to predict and test feature results.

Answers

3 Graph for C is horizontal line as always need the same number. Graph for S is linear as number needed increases by the same amount each time. Graph for M is a curve as the extra amount needed increases each time.
4 **a** $C - 4$; $S - 20$; $M - 25$ **b** $C - 4$; $S - 24$; $M - 36$
5 $C = 4$; $S = 4(n - 1)$; $M = (n - 1)^2$
6 $(n + 1)^2$

Mental starter

Use **What is the rule?** outlined in P1.2, but extend to two rules, for example $3n - 1$ and $2n + 1$.
Assign half the students to the first rule.
They should sit with arms folded.
The rest should sit up straight, and answer according to the second rule.

Useful resources

Squared paper
Isometric paper
P1.3OHP – step arrangement

Introductory activity

Discuss ways in which an investigation could naturally be extended, referring to the example in the student book.

Emphasise that when you extend an investigation you should only change one feature at a time and work systematically as before.

Use the two rules suggested for the mental starter to emphasise that three results is usually insufficient to make predictions. (The value 2 will give the response 5 for both rules.)

Use the method of differences to show how the next result for Kate's table of results in the student book could be 37.

Use the consecutive sum problem to introduce the idea of counter-example.

Emphasise that you only need to find one counter example to prove that a rule does not work, but if you have not found a counter example it does not necessarily prove that the rule will always work – you just haven't found an example to disprove the rule.

P1.3OHP shows the step arrangement in the student book.

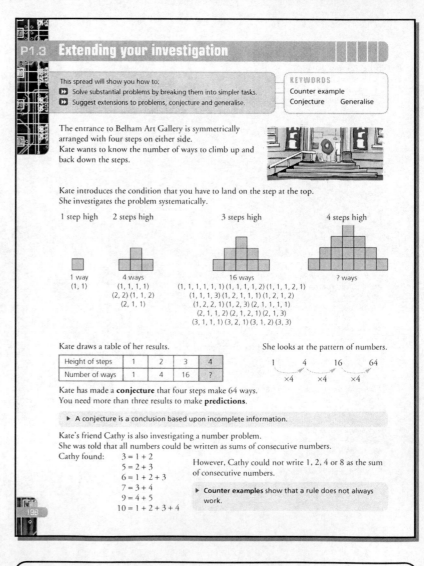

P1.3 Extending your investigation

This spread will show you how to:
▶▶ Solve substantial problems by breaking them into simpler tasks.
▶▶ Suggest extensions to problems, conjecture and generalise.

KEYWORDS
Counter example
Conjecture   Generalise

The entrance to Belham Art Gallery is symmetrically arranged with four steps on either side.
Kate wants to know the number of ways to climb up and back down the steps.

Kate introduces the condition that you have to land on the step at the top.
She investigates the problem systematically.

| 1 step high | 2 steps high | 3 steps high | 4 steps high |

1 way
(1, 1)

4 ways
(1, 1, 1, 1)
(2, 2) (1, 1, 2)
(2, 1, 1)

16 ways
(1, 1, 1, 1, 1, 1) (1, 1, 1, 1, 2) (1, 1, 1, 2, 1)
(1, 1, 1, 3) (1, 2, 1, 1, 1) (1, 2, 2, 1)
(1, 2, 2, 1) (1, 2, 3) (2, 1, 1, 1, 1)
(2, 1, 1, 2) (2, 1, 2, 1) (2, 1, 3)
(3, 1, 1, 1) (3, 2, 1) (3, 1, 2) (3, 3)

? ways

Kate draws a table of her results.

Height of steps	1	2	3	4
Number of ways	1	4	16	?

She looks at the pattern of numbers.

1 ⤳ 4 ⤳ 16 ⤳ 64
  ×4   ×4   ×4

Kate has made a **conjecture** that four steps make 64 ways.
You need more than three results to make **predictions**.

▶ A conjecture is a conclusion based upon incomplete information.

Kate's friend Cathy is also investigating a number problem.
She was told that all numbers could be written as sums of consecutive numbers.
Cathy found:
$3 = 1 + 2$
$5 = 2 + 3$
$6 = 1 + 2 + 3$
$7 = 3 + 4$
$9 = 4 + 5$
$10 = 1 + 2 + 3 + 4$

However, Cathy could not write 1, 2, 4 or 8 as the sum of consecutive numbers.

▶ Counter examples show that a rule does not always work.

198

Plenary

Discuss the results for a rectangular array and link them to the results for a square arrangement.

Further activities

Students could link their results for rectangular arrangements with original results for square arrangements.

Differentiation

Extension questions:
The focus of this exercise is to extend the investigation, making it into a more substantial problem that can be investigated using the methods of the previous lesson.

Core tier: focuses on representing problems and interpreting solutions on algebraic form. Independent and dependent variables are introduced.

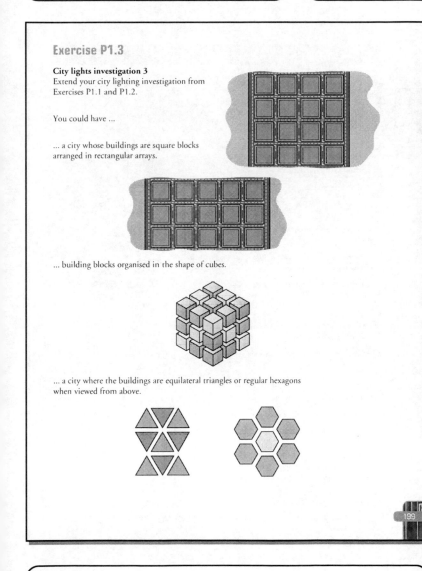

Exercise P1.3

City lights investigation 3
Extend your city lighting investigation from Exercises P1.1 and P1.2.

You could have ...

... a city whose buildings are square blocks arranged in rectangular arrays.

... building blocks organised in the shape of cubes.

... a city where the buildings are equilateral triangles or regular hexagons when viewed from above.

199

Exercise commentary

The questions assess the objectives on Framework Pages 155–157.

Problem solving
The exercise provides an opportunity to assess Framework Pages 27 and 33.

Group work
Students could work in small groups and pool results. For example, if extending to rectangular arrays, one student could look at $2 \times n$ arrays, another $3 \times n$, and so on.

Misconceptions
Students sometimes change more than one feature at a time and try to generalise for all cases at once. Emphasise that you should extend an investigation in only one way at a time.

Links
Problem-solving skills are used throughout mathematics and particularly in coursework.

Homework

P1.3HW allows students to extend the calendar number investigation in their own way.

Answers

For rectangular array $m \times n$
$C = 4$
$S = 2(m - 1) + 2(n - 1)$
$M = (m - 1)(n - 1)$

P1.4 Investigating ratios

Mental starter

Write a range of ratios on the board and ask students to express them in their simplest form and in unitary form.

Useful resources

Mini-whiteboards may be useful for the mental starter.

Envelopes in which films are sent for processing

P1.4OHP – sizes of photographic prints

Introductory activity

Discuss photographic processing, and the choices of print size that are commonly on offer.

Emphasise that the dimensions and details of photograph sizes given in the student book use the terminology given on envelopes for posting films for processing.

Discuss what the terms 30% bigger and 100% bigger might actually refer to.

Emphasise that expressing ratios in unitary form makes it easier to compare them.

Emphasise that when you use ratio for comparisons, it is easier if the units are the same.

P1.4OHP shows the diagram from the student book, giving the dimensions of each size of print.

Use this to discuss:
▸ whether the rectangles are similar
▸ the ratio of the areas.

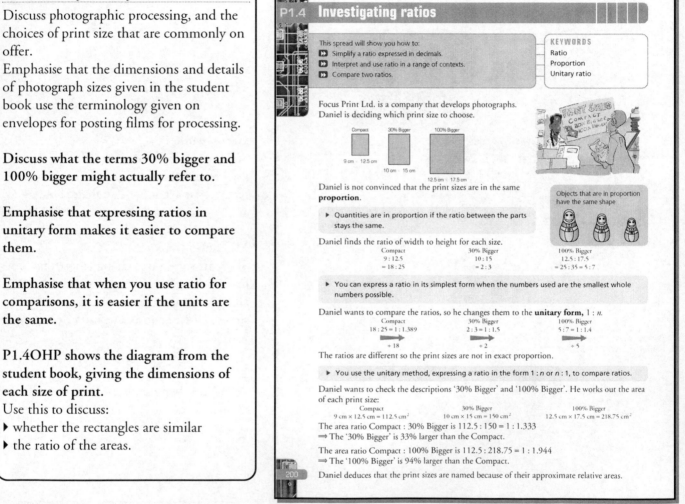

P1.4 Investigating ratios

This spread will show you how to:
▶ Simplify a ratio expressed in decimals.
▶ Interpret and use ratio in a range of contexts.
▶ Compare two ratios.

KEYWORDS
Ratio
Proportion
Unitary ratio

Focus Print Ltd. is a company that develops photographs. Daniel is deciding which print size to choose.

Compact   30% Bigger   100% Bigger

9 cm – 12.5 cm   10 cm – 15 cm   12.5 cm – 17.5 cm

Daniel is not convinced that the print sizes are in the same **proportion**.

Objects that are in proportion have the same shape

▸ Quantities are in proportion if the ratio between the parts stays the same.

Daniel finds the ratio of width to height for each size.

Compact	30% Bigger	100% Bigger
9 : 12.5	10 : 15	12.5 : 17.5
= 18 : 25	= 2 : 3	= 25 : 35 = 5 : 7

▸ You can express a ratio in its simplest form when the numbers used are the smallest whole numbers possible.

Daniel wants to compare the ratios, so he changes them to the **unitary form,** $1 : n$.

Compact	30% Bigger	100% Bigger
18 : 25 = 1 : 1.389	2 : 3 = 1 : 1.5	5 : 7 = 1 : 1.4
÷ 18	÷ 2	÷ 5

The ratios are different so the print sizes are not in exact proportion.

▸ You use the unitary method, expressing a ratio in the form $1 : n$ or $n : 1$, to compare ratios.

Daniel wants to check the descriptions '30% Bigger' and '100% Bigger'. He works out the area of each print size:

Compact	30% Bigger	100% Bigger
9 cm × 12.5 cm = 112.5 cm²	10 cm × 15 cm = 150 cm²	12.5 cm × 17.5 cm = 218.75 cm²

The area ratio Compact : 30% Bigger is 112.5 : 150 = 1 : 1.333
⟹ The '30% Bigger' is 33% larger than the Compact.

The area ratio Compact : 100% Bigger is 112.5 : 218.75 = 1 : 1.944
⟹ The '100% Bigger' is 94% larger than the Compact.

Daniel deduces that the print sizes are named because of their approximate relative areas.

200

Plenary

Discuss the students' results for the investigation and write a general rule.

Discuss ways in which the investigation could be extended.

Write a general rule for the investigation in the exercise.

Extend the investigation.

Extension questions:
The focus of this exercise is an investigative task using ratio.

Core tier: focuses on using ratio to solve a problem relating to types of paving slabs.

Exercise P1.4

Investigation

Consider the digits: 0 1 2 3 4 5 6 7 8 9

For example:

Step 1 Choose any three of these digits. Choose 2, 8, 5

Step 2 Add them up. Call the answer A. $A = 2 + 8 + 5 = 15$

Step 3 Make all the possible two-digit numbers The possible two-digit numbers are 28, 25, 82, 85, 52, 58
from your chosen three digits in step 1. $B = 28 + 25 + 82 + 85 + 52 + 58 = 330$
Add them up. Call the answer B.

Step 4 Find the ratio A : B in its simplest $A : B = 15 : 330$
form. $= 1 : 22$

1 Choose your own set of three numbers and find the ratio A : B using steps 1 to 4.

2 Choose a different set of three numbers and find the ratio A : B.

3 What do you notice about the example and your answers to questions 1 and 2?
Try to explain why this might always happen.

> **Hint:** It may help if you use algebra.

4 Choose four of the digits from the list.
Work through Steps 1 to 4 using these numbers and find the ratio A : B.
(Remember to work systematically to find all possible two-digit numbers in Step 3).

5 Choose a different set of four numbers and find the ratio A : B.

6 Choose a third set of four numbers and find the ratio A : B.

7 What do you notice about your answers to questions 4, 5 and 6?
Try to use algebra to explain your answer.

8 Choose a set of five numbers from the list and use these to find the ratio A : B.

9 Compare and comment on your answers to questions 1, 4 and 8.

201

Exercise commentary

The questions assess the objectives on Framework Pages 77 and 81.

Problem solving
Questions 3, 7 and 9 provide an opportunity to assess Framework Page 31.

Group work
Students may discuss the investigation, although ideally they should work alone.

Misconceptions
Students may have difficulties finding all the 2-digit numbers, especially when four or more digits are chosen.
Emphasise the need to work systematically.
Students often find it hard to justify results, as in question 3. Encourage the use of algebra and provide initial guidance.

Links
Enlargement and scale: Framework Pages 212–217
Ratio: Page 81.

Homework

P1.4HW is an investigative task using ratio to explore Fibonacci numbers.

Answers

1 $1 : 22$ 2 $1 : 22$ 3 Same ratio
4 $1 : 33$ 5 $1 : 33$ 6 $1 : 33$
7 Same ratio   8 $1 : 44$
9 For a set of n numbers the ratio is
$1 : 11(n - 1)$

Mental starter

Write on the board:

Percentages: 1%, 10%, 5%, 15%, 50%, 25%

Amounts: 10 kg, 1000 g, 1 litre, 500 ml, £5

Ask students to choose one from each list and use mental methods to calculate the percentage of the amount.

Useful resources

Mini-whiteboards may be useful for the mental starter.

Introductory activity

Discuss the problem outlined in the student book.

Emphasise that you need to compare the **same things in each case,** that is the volume for 1p, or the cost for 250 ml.

Emphasise the importance of using the **same units in the ratios you are comparing.**

Discuss why it is not necessary to have an **amount per penny** for a comparison. Why would it be equally valid to find what 10p or 50p would buy?

Emphasise that the important point is that one of the quantities used in the comparison is the same in each case.

Discuss how supermarkets nowadays print the cost per unit quantity (metric) on the actual shelves, as a guide to encourage them to buy the supermarket's own brand.

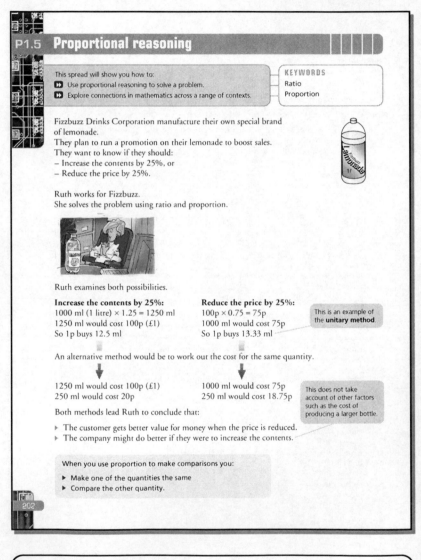

P1.5 Proportional reasoning

This spread will show you how to:

▶ Use proportional reasoning to solve a problem.

▶ Explore connections in mathematics across a range of contexts.

KEYWORDS
Ratio
Proportion

Fizzbuzz Drinks Corporation manufacture their own special brand of lemonade.

They plan to run a promotion on their lemonade to boost sales.
They want to know if they should:
– Increase the contents by 25%, or
– Reduce the price by 25%.

Ruth works for Fizzbuzz.
She solves the problem using ratio and proportion.

Ruth examines both possibilities.

Increase the contents by 25%:
1000 ml (1 litre) × 1.25 = 1250 ml
1250 ml would cost 100p (£1)
So 1p buys 12.5 ml

Reduce the price by 25%:
100p × 0.75 = 75p
1000 ml would cost 75p
So 1p buys 13.33 ml

This is an example of the **unitary method.**

An alternative method would be to work out the cost for the same quantity.

1250 ml would cost 100p (£1)
250 ml would cost 20p

1000 ml would cost 75p
250 ml would cost 18.75p

This does not take account of other factors such as the cost of producing a larger bottle.

Both methods lead Ruth to conclude that:

▶ The customer gets better value for money when the price is reduced.
▶ The company might do better if they were to increase the contents.

When you use proportion to make comparisons you:

▶ Make one of the quantities the same
▶ Compare the other quantity.

202

Plenary

Discuss students' methods for finding solutions to questions 2 and 3.

Students could extend question 3 and find ratios for other shades of brown.

They could research the ratios of colours used in mixing paints.

Differentiation

Extension questions:
- Question 1 presents a problem that is very similar to the scenario on page 202.
- Question 2 is a ratio problem in the context of area.
- Question 3 is a more complex ratio problem in the context of mixing colours.

Core tier: focuses on extending an investigation in a systematic way.

Exercise P1.5

1. A manufacturer wants to run a promotion on a bottle of cola. They need to decide whether to increase the contents or reduce the price.
   The usual size bottle is 1 litre and costs £1.
   a. Compare the effect of increasing the contents by 10% or reducing the price by 10%.
   b. Repeat the comparison using
      i. 20%
      ii. 40%
      iii. 50%.
   c. Comment on your answers to **a** and **b**.
   d. If the contents were increased by 25%, what price reduction on the usual size bottle would give the same value for money?

2. A square ABCD has a triangle DEF inside.

   The ratio of lengths AE : EB = 1 : 1
   The ratio of lengths BF : FC = 1 : 1

   Find the ratio
   shaded area : unshaded area for this shape.

3. Primary colours are red (R), blue (B) and yellow (Y).
   Different shades of brown can be made mixing any two of green, orange and violet.

You can make green (G) by mixing blue and yellow in the ratio B : Y = 2 : 5	You can make orange (O) by mixing red and yellow in the ratio R : Y = 2 : 3	You can make violet (V) by mixing red and blue in the ratio R : B = 3 : 5

   a. Yellowy-brown is made mixing green and orange in the ratio 2 : 3.
      What is the ratio of red to blue to yellow in yellowy-brown?
   b. Reddish-brown is made mixing violet and orange in the ratio 3 : 2.
      What is the ratio of red to blue to yellow in reddish-brown?

203

Exercise commentary

The questions assess the objectives on Framework Pages 79 and 81.

Problem solving
Question 2 provides an opportunity to assess Framework Page 29.
The exercise assesses Page 5.

Group work
The further activity is suitable for group work.

Misconceptions
Students may get confused about which quantities they are comparing.
Emphasise that they must make one of the quantities in each case the same.
You may need to recap percentage increase and decrease, and the fact that 1000 ml = 1 ℓ.

Links
Problem-solving skills are used throughout mathematics and particularly in coursework.

Homework

P1.5HW uses proportional reasoning to solve a problem relating to profit.

Answers

1. c. The customer gets better value for money when the price is reduced.
   d. 20%
2. 3 : 5
3. a. $R : B : Y = 6 : 4 : 19$
   b. $R : B : Y = 13 : 15 : 6$

Mental starter

Write a selection of numbers on the board. Ask students to find the square roots.
Some should give exact answers – these should be done mentally.
For non-exact square roots ask students to estimate first and then use a calculator to find the answer. Include the square root of 2.

Useful resources

Mini-whiteboards may be useful for the mental starter.
Different sizes of metric paper, including A2 and A1

P1.6OHP – table of paper sizes

Introductory activity

Discuss the problem outlined in the student book.

Ask students to check, by measuring, the dimensions of the paper sizes given.
Then check the ratios given in the table.

Discuss why the ratios given in the table are not identical.
How close to each other they need to be for you to make a prediction?

Emphasise that unitary ratios are used for comparison, and that when comparing amounts the units should be the same.

Discuss how Joshua should check his predictions.

P1.6OHP contains the table of paper sizes as shown in the student book, but extended to include A2 and A1 which you can complete by calculation.
It also includes an extra column for area, which can be completed by calculation.
Highlight the fact that the area of A1 is $1m^2$ and this is why the paper sizes are metric.

P1.6 Comparing ratios

This spread will show you how to:
▶ Compare ratios.
▶ Interpret and use ratio in a range of contexts.

KEYWORDS
Compare   Interpret
Ratio

Joshua gets through a lot of paper!

He starts to wonder how paper sizes got their names.

Hey Josh, you got a spare sheet of A4?

Why is A4 paper called A4?

Joshua investigates with the paper he can find in his classroom.

A sheet of A3 paper folded in half is the size of A4 paper.

Fold A4 paper in half and the paper size is A5.

width of A3 = length of A4

width of A4 = length of A5

Joshua measures the length and width of a sheet of A3, A4, A5 and A6 paper. He then calculates the ratio of width (W) to length (L).

Paper sizes A0 to A6 are known as **metric paper sizes**.

Paper size	A3	A4	A5	A6
Width (mm)	297	210	148	105
Length (mm)	420	297	210	148
Ratio $W : L$	1 : 1.41414	1 : 1.41429	1 : 1.41892	1 : 1.40952

Joshua notices that the ratios are all roughly equal to $1 : \sqrt{2}$ (1.41421...), and he deduces that:

▶ The width of one paper size is the length of paper the next size smaller
▶ Metric paper sizes have width : length ratio = $1 : \sqrt{2}$.

Joshua predicts the dimensions of A2, A1 and A0 paper:

A2 paper has width 420 mm (this is the length of A3).
A2 length should be $420 \times \sqrt{2} = 594$ mm.
A1 width = 594 mm, length = $594 \times \sqrt{2} = 840$ mm.
A0 width = 840 mm, length = $840 \times \sqrt{2} = 1188$ mm.

Joshua checks his prediction by measuring.

▶ You can compare ratios and interpret the results to solve problems.

Plenary

Discuss students' results for the investigations in the exercise.

Further activities

Students can extend question 5 to other groups of letters, or passages of writing of different styles.

Differentiation

Extension questions:
The focus of this exercise is to use ratio in practical contexts and to discover how Morse assigned codes to letters.

Core tier: focuses on the open-ended nature of an investigation, and how it can be extended further.

Exercise P1.6

Investigation
Morse code uses patterns of dots and dashes to represent letters of the alphabet.

A .−	I ..	Q −−.−	Y −.−−
B −...	J .−−−	R .−.	Z −−..
C −.−.	K −.−	S ...	
D −..	L .−..	T −	
E .	M −−	U ..−	
F ..−.	N −.	V ...−	
G −−.	O −−−	W .−−	
H	P .−−.	X −..−	

When you transmit a letter in Morse code:

▹ A dash takes three times longer than a dot to transmit (so if a dot is 1 unit, a dash is 3 units).
▹ The gap between dots and dashes is 1 unit.

> The time (in units) to transmit the letter A is: $1 + 1 + \overline{3}$ = 5
>
> The time (in units) to transmit the letter D is: $\overline{3} + 1 + 1 + 1 + 1$ = 7
>
> The transmitting ratio of A : D is 5 : 7.

1   Find the transmitting ratio of:
   a E : T   **b** A : N   **c** S : O
   d E : Q   **e** K : X

2   Choose a passage of writing in any novel (about ten lines of writing).
   a Count the number of times the letter E appears and the number of times the letter T appears.
   b Work out the ratio of the number of times E appears to the number of times T appears, E : T. (You may want to express this as a unitary ratio).
   c Compare your ratio with your answer to question 1a and comment on what you notice.

3   In a standard popular game set there are 12 letter Es and 6 letter Ts.
   a Write down the ratio E : T.
   b Compare with your answer to question 2 and comment.

4   **a** Combine your total number of Es and Ts from question 2 with the totals from one or more people.
   b Work out the new ratio E : T.
   c Comment on your answer.

5   Choose another pair of letters to compare and work out their Morse transmitting ratio.
   Choose a passage of writing, count the number of times the letters appear and work out their ratio. Compare your two ratios.

205

Exercise commentary

The questions assess the objectives on Framework Pages 79 and 81.

Problem solving
Questions 2–5 provide an opportunity to assess Framework Page 31.

Group work
Questions 4 and 5 are suitable for group work.

Misconceptions
The exercise involves Morse Code, which may be unfamiliar to many students. When calculating transmitting times for sentences, students should know that the gap between words is equivalent to seven dots.

Links
Problem-solving skills are used throughout mathematics and particularly in coursework.

Homework

P1.6HW requires students to research Samuel Morse and the Morse Code for numbers.

Answers

1 **a** 1 : 3
  b 5 : 5 = 1 : 1
  c 5 : 11
  d 1 : 13
  e 9 : 11
3 **a** 2 : 1

Summary

The key objectives for this unit are:

▶ Solve substantial problems by breaking them into simpler tasks, using a range of efficient techniques, methods and resources.

▶ Present a concise, reasoned argument, using symbols, diagrams, graphs and related explanatory text.

▶ Use proportional reasoning to solve a problem, choosing the correct numbers to take as 100%, or as a whole.

▶ Give solutions to problems to an appropriate degree of accuracy.

Check out commentary

1 The obvious answer to this question is 64, so ensure that students actually understand what the question is asking. You could discuss the question as a whole-class exercise in the first instance, and encourage students to identify other squares within the array (such as the large square encompassing the board).

Discuss strategies, and encourage simplification of the problem.

Ask: what if the board was only

▶ One square big

▶ 2 by 2

▶ 3 by 3, and so on.

Encourage students to organise their results, and find a pattern that can be extended to the 8 × 8 case.

2 Students often apply the unitary method correctly, but then misinterpret their answers.

For example, dividing ml by pence will yield the results 208.3 and 210.5. Students may incorrectly deduce that the 208.3 figure is smaller so it must represent better value.

Emphasise that there are two ways of tackling 'best buy' problems: by comparing amounts per unit cost, or by comparing cost per unit amount.

Plenary activity

As a whole class activity, discuss different approaches to question 1 in the Check out exercise.

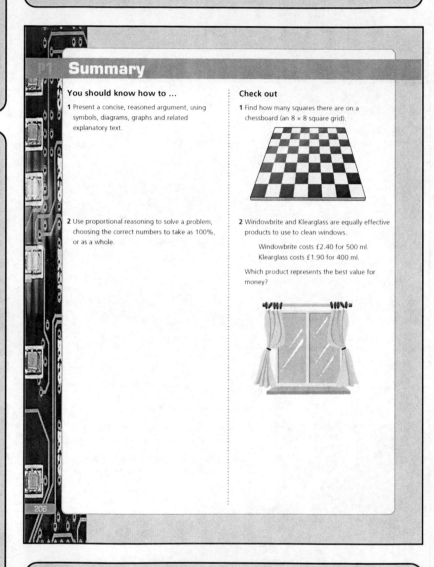

Summary

You should know how to ...

1 Present a concise, reasoned argument, using symbols, diagrams, graphs and related explanatory text.

2 Use proportional reasoning to solve a problem, choosing the correct numbers to take as 100%, or as a whole.

Check out

1 Find how many squares there are on a chessboard (an 8 × 8 square grid).

2 Windowbrite and Klearglass are equally effective products to use to clean windows.

   Windowbrite costs £2.40 for 500 ml.
   Klearglass costs £1.90 for 400 ml.

   Which product represents the best value for money?

Development

Proportional reasoning is developed in Year 9, particularly in connection with trigonometry.

Links

Problem-solving skills are used throughout mathematics and particularly in coursework.

Mental starters

Objectives covered in this unit:

▶ Visualise, describe and sketch 2-D shapes, 3-D shapes and simple loci.
▶ Use metric units for conversion.
▶ Apply mental skills to solve simple problems.

Resources needed

* means class set needed

Essential:
S4.1OHP – 3-D solids
S4.2OHP – plans and elevations
S4.3OHP – volume
S4.6OHP – midpoint
S4.7OHP – loci
S4.8OHP – more loci
S4.9OHP – RHS triangles
Protractors*

Useful:
Box of solids
R6 Number lines
Counting stick
R28 Protractor
R8 Coordinate grid
R1 Digit cards
Coins
Compass
S4.9ICT – interactive geometry

S4 Dimensions and scales

This unit will show you how to:

▶▶ Visualise and use 2-D representations of 3-D objects.
▶▶ Analyse 3-D shapes through 2-D projections, including plans and elevations.
▶▶ Use and interpret maps and scale drawings.
▶▶ Use straight edge and compasses to construct a triangle, given right angle, hypotenuse and side (RHS).
▶ Find, by reasoning, the locus of a point that moves according to a simple rule.

▶▶ Calculate the surface area and volume of right prisms.
▶▶ Use bearings to specify direction and solve problems.
▶▶ Given the coordinates of points A and B, find the midpoint of the line segment AB.
▶▶ Solve increasingly demanding problems and evaluate solutions.
▶▶ Solve substantial problems by breaking them into simpler tasks.

> Looking at the map scale, it's 2 km away. Come on!

> Yes, but it's on a bearing of 045° – that's this way!

You can use a map and a compass to find distance and direction

Before you start

You should know how to ...

1 Identify planes of symmetry in 3-D shapes.

2 Calculate the volume of a simple prism.

3 Convert between metric units.

4 Read and plot coordinates in all four quadrants.

Check in

1 How many planes of symmetry does this cuboid have?

2 a Find the volume of this shape.
  b What is the area of A?

  volume = 80 mm³

3 Copy and complete:
  a 4 cm = ___ mm   b 42 mm = ___ cm
  c 52 cm = ___ m   d 1000 g = ___ kg
  e 3.5 l = ___ ml   f 2000 kg = ___ g

4 Plot on a grid:
  A(4, 2)   B(‾3, 2)   C(‾2, 1)
  D(5, ‾1)   E(0, 5)   F(‾3, 0)

Unit commentary

Aim of the unit

This unit begins by exploring 3-D shapes and progresses into the applications of geometrical construction, including scale drawings, bearings and loci.

The unit ends by reviewing the techniques used to construct triangles, including RHS triangles.

Introduction

Discuss the 3-D nature of the world in which we live, and how students can use their knowledge of 2-D shapes to describe 3-D surfaces, and to make representations such as nets and elevations.

Framework references

This unit focuses on:
▶ Teaching objectives pages:
  199, 201, 217-225, 233, 239-241.
▶ Problem solving objectives pages:
  15, 17, 19, 27, 31, 33, 35.

Check in activity

Discuss all the different shapes that students know, and classify them into broad groups (starting with '2-D' and '3-D').

Differentiation

Core tier

Focuses on 3-D shapes, scale drawings and loci, using simpler contexts.

Mental starter

Write these words on the board: *face, edge, vertex, parallel, perpendicular, plane of symmetry.*

Organise students in pairs. Students should describe a solid shape and their partner should try to identify the shape correctly.

Encourage students to use the words on the board.

Afterwards discuss other words that students may have used.

Useful resources

S4.1OHP shows standard 3-D solids.

A **box of solids** is useful for the introductory activity.

Introductory activity

Refer to the mental starter.

Discuss the vocabulary associated with 3-D shapes and illustrate its meaning with examples.

You can use a box of solids to help, and you can also use **S4.1OHP**, which contains a variety of standard solids. Recap their names and meanings.

Discuss cross-sections through a 3-D shape.

Use the example of a triangular prism. Ask students to visualise the cross-sections when the prism is cut in a particular direction. Encourage students to name the shapes produced.

Link to the definition of a **prism** by showing that the cross-section along its length will yield congruent triangles.

Extend to planes of symmetry.

Show with an example that a plane of symmetry is a cross-section that divides a shape into two identical and reversed halves.

Discuss more complex solids.

Using the box of shapes, combine two solids to make a more complex solid. The examples in the student book show a pyramid on top of a cube, and an octahedron made of two pyramids. Explore other shapes that can be made, and invite the class to describe them.

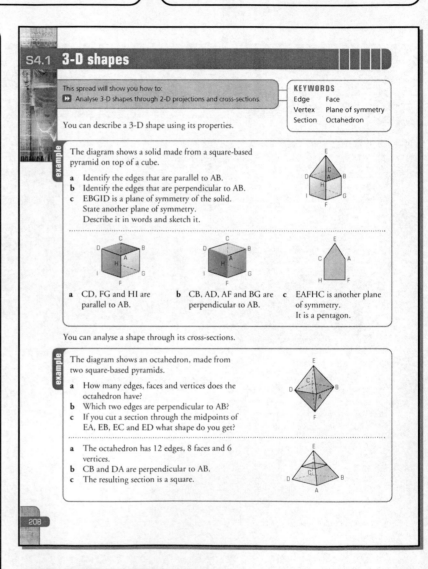

This spread will show you how to:
▶ Analyse 3-D shapes through 2-D projections and cross-sections.

KEYWORDS
Edge   Face
Vertex   Plane of symmetry
Section   Octahedron

You can describe a 3-D shape using its properties.

The diagram shows a solid made from a square-based pyramid on top of a cube.

a   Identify the edges that are parallel to AB.
b   Identify the edges that are perpendicular to AB.
c   EBGID is a plane of symmetry of the solid. State another plane of symmetry. Describe it in words and sketch it.

a   CD, FG and HI are parallel to AB.
b   CB, AD, AF and BG are perpendicular to AB.
c   EAFHC is another plane of symmetry. It is a pentagon.

You can analyse a shape through its cross-sections.

The diagram shows an octahedron, made from two square-based pyramids.

a   How many edges, faces and vertices does the octahedron have?
b   Which two edges are perpendicular to AB?
c   If you cut a section through the midpoints of EA, EB, EC and ED what shape do you get?

a   The octahedron has 12 edges, 8 faces and 6 vertices.
b   CB and DA are perpendicular to AB.
c   The resulting section is a square.

Plenary

Discuss shapes with curved faces: spheres, ellipsoids, cones and cylinders. Use the box of shapes to illustrate.

Ask students to visualise sections cut at different angles through the shapes, such as sections of a cone.

Further activities

Students can work in pairs on this activity. Student 1 draws a simple solid shape, ensuring Student 2 cannot see it. Student 2 asks questions, such as 'How many vertices does it have?' Student 2 should try to draw the shape.

This can be made into a game, with a maximum of 10 questions, and points scored.

Differentiation

Extension questions:
- Question 1 allows students to practise using terminology connected with 3-D shapes.
- Question 2 focuses on sketching solids.
- Questions 3 and 4 require students to visualise solid shapes and their sections.

Core tier: focuses on cubes and cuboids.

Exercise S4.1

1 Here are two triangular prisms joined to make a new solid.
   a Use words like *parallel, perpendicular, faces, edges* and *vertices* to describe the new solid.
   b Sketch another solid using the two triangular prisms. Draw it and write a description of it below your diagram.
   c Repeat part **b** to make another solid.

2 Use the shapes shown to sketch as many different solids as you can. The faces shaded blue are identical.
   a
   b

3 This cube has been sliced to give a rectangular cross-section. Is it possible to slice a cube so that the cross-section is:
   a a square
   b a triangle
   c a pentagon
   d a hexagon?

   If so, describe how it can be done.
   Which of these cross-sections are planes of symmetry?

4 Can you make a solid with:
   a only two faces
   b only three faces
   c four faces
   d five faces
   e six edges
   f five vertices
   g six vertices
   h three vertices?

   In each case, explain your answer and sketch the solid where appropriate.

209

Exercise commentary

The questions assess the objectives on Framework Pages 199 and 201.

Problem solving

The questions in this exercise require visualisation of 3-D shapes to solve problems, assessing the objectives on Framework Page 15. Questions 3 and 4 also assess Pages 33 and 35.

Group work

Questions 2, 3 and 4 can be explored in pairs, with students comparing answers.

Misconceptions

Many students find visualising 3-D shapes difficult. Making a box of solids available will help, and you should encourage sketching.

Emphasise the distinction between edges, faces and vertices, and ensure that students are confident in identifying these on curved surfaces.

Links

This work links to plane symmetry (Framework Page 207).

Homework

S4.1HW gives students practice at analysing everyday 3-D shapes.

Answers

3 a Yes b Yes c No d No
4 a No b No c Yes d Yes
  e Yes f Yes g Yes h No

Mental starter

Sketch the net of a cube on the board.
Ask students to identify it.
Repeat for a cuboid, square-based pyramid, triangular prism, and a tetrahedron.

Useful resources

S4.2OHP shows the plans and elevations in the student book.
A **box of solids** is useful for the introductory activity.

Introductory activity

Recap the ideas of the previous lesson.
Recall that you can describe a solid in terms of its:
▸ Faces, edges and vertices
▸ Cross-sections
▸ Planes of symmetry

Extend to the idea contained in the mental starter.
Describe how you can illustrate a solid by its net, and show that nets are not unique. Use the example of a cube.

Extend to projections of a solid.
Recall from work in year 7 that you can illustrate a 3-D shape by its:
▸ Plan
▸ Front elevation
▸ Side elevations

Illustrate with the examples in the student book.
These are shown on S4.2OHP.
Show how you can sketch the views of a solid if you know its shape. However, if measurements are specified you can construct the views accurately, using techniques already learned.

Explore the elevations of other shapes.
Using a box of solids, invite the class to sketch the plan and elevations of various shapes. Discuss responses.

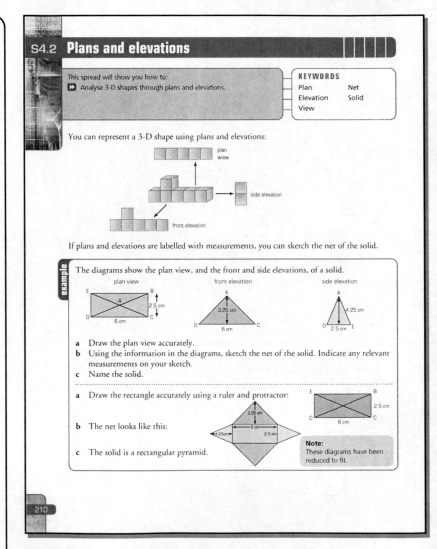

S4.2 Plans and elevations

This spread will show you how to:
▸ Analyse 3-D shapes through plans and elevations.

KEYWORDS
Plan   Net
Elevation   Solid
View

You can represent a 3-D shape using plans and elevations:

plan view
side elevation
front elevation

If plans and elevations are labelled with measurements, you can sketch the net of the solid.

example

The diagrams show the plan view, and the front and side elevations, of a solid.

plan view   front elevation   side elevation

a   Draw the plan view accurately.
b   Using the information in the diagrams, sketch the net of the solid. Indicate any relevant measurements on your sketch.
c   Name the solid.

a   Draw the rectangle accurately using a ruler and protractor:

b   The net looks like this:

c   The solid is a rectangular pyramid.

Note:
These diagrams have been reduced to fit.

Plenary

Discuss question 4, and ask students to justify their responses.

Further activities

Students can work in pairs, inventing plans and elevations for their partner to sketch the solid.

Encourage students to be critical of each other's (and their own) plans and elevations, to ensure that they have been drawn correctly.

Differentiation

Extension questions:

▸ Question 1 provides practice at drawing nets.
▸ Questions 2 and 3 focus on constructing plans and elevations.
▸ Question 4 requires students to solve a problem involving possible and impossible solids.

Core tier: focuses on plans and elevations of shapes made from multilink cubes.

Exercise S4.2

1 By making appropriate measurements draw the nets of the solids shown in these plans and elevations. Sketch each solid first.

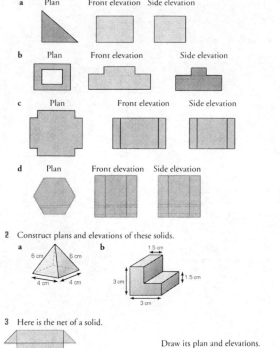

a   Plan   Front elevation   Side elevation

b   Plan   Front elevation   Side elevation

c   Plan   Front elevation   Side elevation

d   Plan   Front elevation   Side elevation

2 Construct plans and elevations of these solids.

a   6 cm   6 cm
    4 cm   4 cm

b   1.5 cm
    3 cm   1.5 cm
       3 cm

3 Here is the net of a solid.

Draw its plan and elevations.
Sketch the solid.

4 For each of these shapes, decide whether it could be the plan of a solid. Justify your answer.

211

Exercise commentary

The questions assess the objectives on Framework Pages 199 and 201.

Problem solving

Question 4 assesses the objectives on Framework Page 33.

Group work

This exercise can be attempted in pairs, with students comparing and discussing solutions.

Misconceptions

Students may find it hard to draw plans and elevations correctly, and may miss out edges. Emphasise that edges should appear as lines within an elevation, as in question 1 parts c and d.

Links

This topic links to constructing nets (Framework Page 223).

Homework

S4.2HW is similar to question 1, requiring students to move from elevations to solids and finally to nets.

Answers

4 All could be plans of solids.

Mental starter

Recap approximate imperial to metric conversions of length.

▶ How many centimetres are there in 4 inches?

▶ How many inches are there in 30 centimetres?

▶ How many feet are there in 3 metres?

Say that students' grandparents may have run the 100 yards race at school.

Discuss which is longer: 100 metres or 100 yards.

Useful resources

S4.3OHP shows the diagrams in the student book.

A **box of solids** may be useful for the introductory activity.

Introductory activity

Recap 3-D shapes from the previous two lessons.

Briefly recall the names, using a box of solids as a visual prompt.

Discuss the meaning of the term volume.

Encourage responses along the lines of 'the amount of space occupied by a solid shape'.

Discuss units of volume, particularly cm^3 and m^3. Ensure that students are familiar with the terms 'centimetres cubed' and cubic centimetres'.

Also mention imperial units of volume: ft^3 and in^3.

Discuss the volume of a prism.

Volume = area of uniform cross-section × length.

Illustrate with the first example in the student book, which is shown on **S4.3OHP**.

Discuss the meaning of the term surface area.

Encourage responses along the lines of 'the amount of surface occupied by the faces of a solid shape'.

Discuss units of surface area – students should be satisfied that they are the same as for the area of a 2-D shape.

Show how you can use the net of a solid to calculate its surface area.

Illustrate with the second example in the student book, which is also shown on **S4.3OHP**.

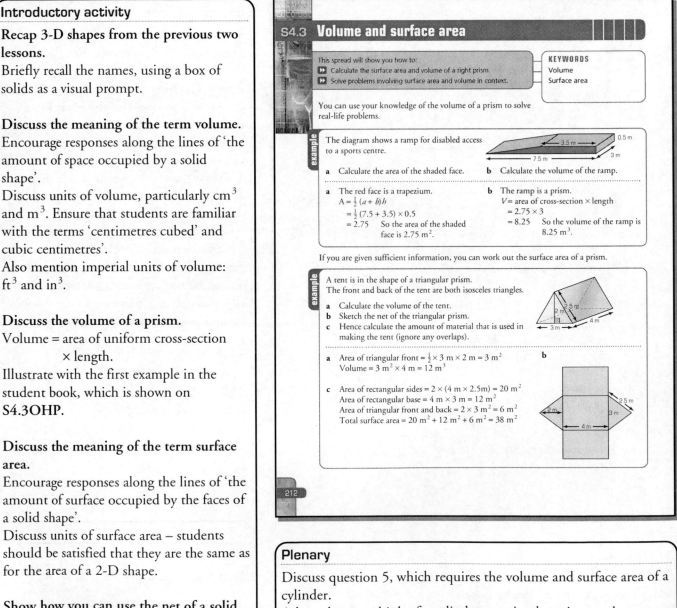

Plenary

Discuss question 5, which requires the volume and surface area of a cylinder.

Ask students to think of a cylinder as a circular prism, and try to construct a formula.

Further activities

In pairs or small groups, students can estimate the volume and surface area of their classroom.
When calculating the surface area only include walls and ceiling, so allow for windows and doors.

Differentiation

Extension questions:
▶ Question 1 involves the volume of a simple cuboid.
▶ Questions 2 – 4 focus on the volume and surface area of shapes made from cuboids.
▶ Question 5 extends to the volume and surface area of a cylinder.

Core tier: focuses on the volume and surface area of cuboids and shapes made from cuboids.

Exercise S4.3

1 Find the volume of this box of cornflakes. The cornflakes have a volume of 120 cm³. What percentage of the box do they fill?

2 This toy house is in the shape of a cuboid with a triangular prism on top. What is the volume of the house?

3

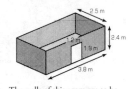

a The walls of this room are to be painted. What area is to be painted? (The door will be painted separately, so don't count it!)
b What area of carpet is required?
c If carpet costs £19.99 a square metre, how much will it cost to carpet the room?
d What is the volume of the room?

4 The diagram shows a rather strange-shaped jug.

a Find the volume of the jug (excluding the handle and the spout).
b The jug is to be filled with water. What capacity of water will it hold? (1 cm³ = 1 ml.)
c The outside (except the handle and spout) is to be painted blue. What is the area that is to be painted? (Include the base of the jug.)

5 a Find the volume of water (in millilitres) that this vase will hold.

b The curved surface area and the base are to be painted pink. What area is to painted?

213

Exercise commentary

The questions assess the objectives on Framework Pages 239 and 241.

Problem solving
The questions in this exercise assess the objectives on Framework Page 19. Questions 1 and 2 assess Page 29.

Group work
Question 5 can be discussed in pairs.

Misconceptions
Students sometimes make mistakes when calculating the surface area of shapes in context, for example by neglecting the difference between boxes with or without lids. Emphasise the importance of reading the question, and encourage students to sketch the net.

Links
This topic links to 2-D representations of 3-D shapes (Framework Pages 199 and 201).

Homework

S4.3HW provides practice at calculating the volume and surface area of a prism, in the context of a swimming pool.

Answers

1 480 cm³, 25%
3 a 28.08 m²   b 9.5 m²   c £189.91
  d 22.8 m³
4 a 2220 ml   b 994 cm²
5 a 84.82 ml   b 120.2 cm²

Mental starter

Draw a number line on the board divided into 10 equal sections (or use **R6**). Alternatively use a counting stick for this activity.

State a scale, for example 1 cm to 5 m. Say that the top scale represents cm and the bottom scale represents m. Point to divisions on the top scale (for example, at 7 cm) and ask the class to recite the number on the bottom scale (35 m). Vary it by pointing to the bottom scale.

Extend to other scales involving units of length.

Useful resources

R6 contains number lines for use in the mental starter.

Alternatively, a **counting stick** may be useful for the mental starter.

Introductory activity

Refer to the mental starter.

Discuss the meaning and use of the term scale. In particular, discuss these questions:

▶ What is scale?
▶ Where is scale used (maps, drawings, axes on a graph)
▶ How are scales expressed?

Discuss the two main written ways of expressing a scale:

▶ As a comparison between units (for example, 2 cm to 5 m).
▶ As a ratio, without units (for example, 1 : 500).

Emphasise that the two forms are interchangeable, and refer to the first example in the student book, which shows the use of scale in drawing a plan.

Discuss the use of scale in maps.

Encourage students to describe the scales that they may have seen:

▶ 1 inch equals 3 miles
▶ 1 : 50 000
▶ the use of a segmented bar.

Discuss the second example in the student book, which shows a section taken from an OS map.

Ensure that students know how to get from map lengths to real lengths and vice versa. You may need to recap metric conversion.

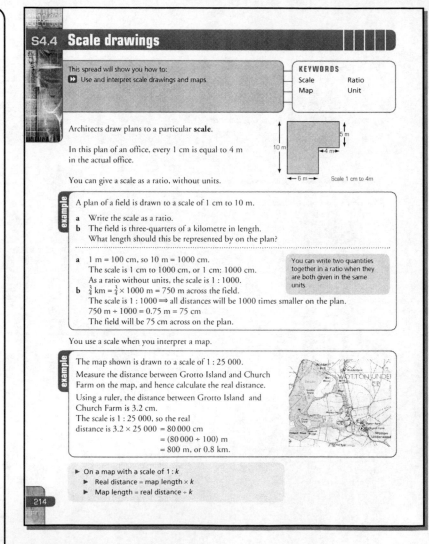

S4.4 Scale drawings

This spread will show you how to:
▶▶ Use and interpret scale drawings and maps.

KEYWORDS
Scale   Ratio
Map   Unit

Architects draw plans to a particular **scale**.

In this plan of an office, every 1 cm is equal to 4 m in the actual office.

You can give a scale as a ratio, without units.

example
A plan of a field is drawn to a scale of 1 cm to 10 m.
a Write the scale as a ratio.
b The field is three-quarters of a kilometre in length.
   What length should this be represented by on the plan?

a 1 m = 100 cm, so 10 m = 1000 cm.
   The scale is 1 cm to 1000 cm, or 1 cm: 1000 cm.
   As a ratio without units, the scale is 1 : 1000.
b $\frac{3}{4}$ km = $\frac{3}{4}$ × 1000 m = 750 m across the field.
   The scale is 1 : 1000 ⟹ all distances will be 1000 times smaller on the plan.
   750 m ÷ 1000 = 0.75 m = 75 cm
   The field will be 75 cm across on the plan.

You can write two quantities together in a ratio when they are both given in the same units.

You use a scale when you interpret a map.

example
The map shown is drawn to a scale of 1 : 25 000.
Measure the distance between Grotto Island and Church Farm on the map, and hence calculate the real distance.
Using a ruler, the distance between Grotto Island and Church Farm is 3.2 cm.
The scale is 1 : 25 000, so the real distance is 3.2 × 25 000 = 80 000 cm
                     = (80 000 ÷ 100) m
                     = 800 m, or 0.8 km.

▶ On a map with a scale of 1 : k
   ▶ Real distance = map length × k
   ▶ Map length = real distance ÷ k

214

Plenary

Ask what scale students would choose if they were to draw a scale plan of the classroom. This will involve first making an estimate of its dimensions, and then thinking of a sensible sized drawing to fit into their exercise books.

Further activities

In pairs, students can make a scale drawing of the classroom. This will involve taking measurements and deciding on a sensible scale.

Differentiation

Extension questions:
▸ Question 1 involves interpreting a simple scale drawing.
▸ Questions 2 and 3 focus on constructing and interpreting drawings to a larger scale, involving kilometres.
▸ Question 4 involves estimating the distance along a road.

Core tier: focuses on making simple scale drawings.

Exercise S4.4

1 Here is a plan of Sally's bedroom.

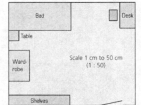

a What are the dimensions of her real bed?
b What are the dimensions of her real wardrobe?
c What are the dimensions of her real table?
d What are the dimensions of her real desk?
e What are the dimensions of her real shelves?
f Sally wants to place a table 2 m by 1 m next to the desk. Will it fit?

2 A boat travels 100 km due South and 80 km due East as shown by the sketch.

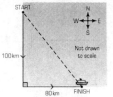

Draw a scale diagram to work out how far the boat is from its starting point. (Scale 1 cm : 10 km)

3 Here is a map of three islands.

Find the distance in kilometres from:
a Stemleaf to Havington
b Havington to Budtown
c Budtown to Stemleaf.
On a copy of the map plan a sea route from Stemleaf to Codleigh via Havington. Estimate how far your journey would be.

4 The map shows a part of a town.

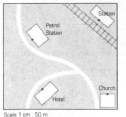
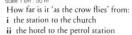

a How far is it 'as the crow flies' from:
i   the station to the church
ii the hotel to the petrol station
iii the petrol station to the church?
b Estimate the distance along the road from:
i   the petrol station to the church
ii the petrol station to the hotel.

215

Exercise commentary

The questions assess the objectives on Framework Page 217.

Problem solving
The questions require confidence in moving between real lengths and scale lengths, assessing the objectives on Framework Page 27.

Group work
Students can work in pairs on the further activity.

Misconceptions
Students often have difficulty in using ratio scales. Emphasise that a ratio scale works with any unit of length, but the units must be the same. Another frequent problem is whether to × or ÷ when converting units.
Large to small ⇒ ×
Small to large ⇒ ÷

Links
Scale drawings link to unit conversion (Framework Page 229), and also ratio (Page 81).

Homework

S4.4HW requires students to make a scale drawing of a room in their house.

Answers

1 a 150 cm by 65 cm   b 75 cm by 50 cm
  c 25 cm by 25 cm   d 35 cm by 60 cm
  e 150 cm by 25 cm; No
2 128 km
3 a 12.5 km   b 10 km   c 9 km
4 a i 175 m   ii 120 m   iii 225 m
  b i About 225 m   ii About 250 m

Mental starter

This activity involves knowing the points of a compass.
Start by saying that you are facing North.
Now turn (verbally) through various multiples of 45°, both
clockwise and anticlockwise and ask for the resulting direction.
For example:
I turn through 135° anticlockwise. Now which direction am I facing?
Extend by introducing combinations of rotations.

Useful resources

A **compass** may be useful as a visual aid.
R28 shows a protractor (full-circle).
You will need a class set of **protractors**.

Introductory activity

Refer to the mental starter.

Discuss the points of the compass, and
how they can be used for specifying
direction.
Refer to the subdivisions, for example
north-north-east (NNE), which effectively
divide the circle into $22\frac{1}{2}°$ sections.

**Explore the need for more accurate
specification.**

Ask the question: What direction will I be
facing if I turn through 10° clockwise?

Define bearings.

Show how bearings can be used to specify
any direction (within a plane).
R28 shows a protractor, on which you can
write the 3-figure bearings and compass
points.

Show how you can specify the position of
one point (for example on a map) from
another point by stating:
▶ Its distance
▶ Its bearing
Illustrate how the bearing of A from B is
not the same as the bearing of B from A.

**Show how you can use bearings to solve
problems.**

The example in the student book involves
a ship's journey, and requires the
construction of an accurate scale drawing.
Encourage students to draw the diagram
for themselves to verify the result.

S4.5 Bearings

This spread will show you how to:
▶▶ Use bearings to specify directions and solve problems.

KEYWORDS
Bearing   Clockwise
Scale drawing

You use **bearings** to specify directions.

▶ A bearing is a clockwise angle measured from North.

You should know the points of
the compass: North, South, East
and West.

To find the bearing of B from A:

1 Draw a North line at A.
2 Measure the angle clockwise from A.
3 Always write three figures, so 062° not 62°.

The bearing of B
from A is 062°.

The bearing of B from A is **not** the same as the bearing of A from B.

Draw a North line at B.   Mark in the corresponding angle.   Extend the angle clockwise to line AB.

Corresponding angle = 62°

62° + 180° = 242°
The bearing of A from B is 242°.

You can use bearings to solve problems involving scale drawings.

example

A ship travels for 25 km on a bearing of 042°, then 35 km on a bearing of 127°.
a Use a scale drawing of 1 cm to 10 km to show the ship's path.
b Find the ship's distance and bearing from its starting point.

a

It is a good idea to draw a
rough sketch first.

b The distance on the scale drawing, using a ruler, is 4.5 cm.

The real distance is (4.5 × 10) km = 45 km to nearest km
The bearing, using a protractor, is 275°.

216

Plenary

Pose this problem:
The town Q is 5 km due East of the town P.
Can you fix the position of the town R if:
▶ Its bearings from P and Q are 020° and 200° respectively
▶ Its bearings from P and Q are 210° and 62° respectively?
Discuss why the latter case is impossible.

Further activities

Students working in pairs can plan a 'round the world' tour using an atlas. They should give distances and bearings. Discuss why their measurements will not be completely accurate.

Differentiation

Extension questions:

▸ Questions 1 and 2 involve simple calculation of bearings without context.

▸ Questions 3–7 focus on constructing a scale drawing form text involving bearings in context.

▸ Question 8 involves completing a scale drawing using clues given in the text.

Core tier: focuses on bearings, using simpler contexts or no context.

Exercise S4.5

1 The bearing of P from Q is 045°.

What is the bearing of Q from P?

2 The bearing of X from Y is 107°.

What is the bearing of Y from X?

3 A ship travels on a bearing of 120° for 100 km and then on a bearing of 230° for a further 200 km.
Represent this with a scale drawing.
What is the ship's distance and bearing from the starting point?

4 A plane flies on a bearing of 060° for 200 km, changes course to fly 300 km on a bearing of 200° and changes course again to fly 250 km on a bearing of 135°. What is its bearing and distance from its original position?

5 A dog walks across a field from C on a bearing of 045° at a steady speed of 1 m/s.

At the same time a cat starts at D at a steady speed of 1 m/s on a bearing of 315°. Do they meet?
Explain your answer.

6 P and Q are two ships 20 km apart. R is a submarine which is on a bearing of 075° from P and 290° from Q.

Using a scale of 1 cm : 2 km draw a scale diagram of P and Q and hence locate R.

7 A ship travels from S on a bearing of 042° for 25 km, then on a bearing of 140° for 25 km.

Draw a scale diagram to show the ship's course. (Use a scale of 1 : 200 000.)

8 A boat is seen on a bearing of 160° from lighthouse B. At the same time, it is seen on a bearing of 300° from lighthouse A.

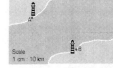

Copy the diagram and mark the position, B, of the boat.

217

Exercise commentary

The questions assess the objectives on Framework Page 233.

Problem solving

All questions require students to give solutions to an appropriate degree of accuracy, assessing the objectives on Framework Page 31. Question 5 assesses Page 33.

Group work

Students can compare the accuracy of their scale drawings in questions 3–8.

Misconceptions

The convention of bearings being measured clockwise often leads to confusion since this runs counter to rotations, in which the anticlockwise direction is positive. Emphasise that compasses run in the same direction as clocks.

Links

Bearings link to angles and lines (Framework Pages 181–183), and also scale drawings (Page 217).

Homework

S4.5HW uses the context of a treasure map to practise bearings.

Answers

1 225°
2 327°
3 200°, 190 km
4 145°, 450 km
5 Yes, after 17 s

Mental starter

Show a coordinate grid (there is one on **R8**).
Plot various points on the grid and ask for their coordinates.
Students can respond using digit cards.

Useful resources

R8 is a blank grid for use in the mental starter.
S4.6OHP shows the second diagram from the student book.
Digit cards may be useful for the starter and plenary.

Introductory activity

Refer to the mental starter.
Ensure that students are confident at drawing and specifying coordinates in all four quadrants.

Discuss the meaning of the term line segment.
Illustrate how you can draw a straight line segment between any two points on a grid.

Discuss the meaning of the term midpoint.
Students should have encountered the midpoint of a line when they were constructing perpendicular bisectors.

Refer to the grid in the student book, which shows the right-angled triangle ABC.
There is a copy of this grid on **S4.6OHP**.
Use the grid to illustrate how you can identify the coordinates of the midpoint by creating two congruent triangles.

You may need to recap congruence and corresponding angles.

Generalise the result to points with coordinates (x_1, y_1) and (x_2, y_2).
You can use a blank grid (like the one on **R8**), and derive the formulae for the coordinates of a midpoint.
Link to finding the mean of two numbers.

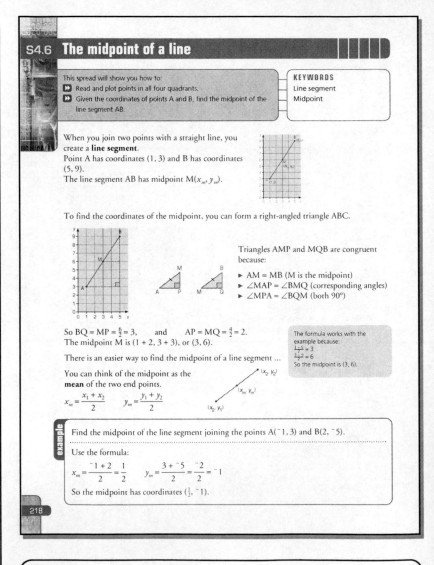

S4.6 **The midpoint of a line**

This spread will show you how to:
▶ Read and plot points in all four quadrants.
▶ Given the coordinates of points A and B, find the midpoint of the line segment AB.

KEYWORDS
Line segment
Midpoint

When you join two points with a straight line, you create a **line segment**.
Point A has coordinates (1, 3) and B has coordinates (5, 9).
The line segment AB has midpoint $M(x_m, y_m)$.

To find the coordinates of the midpoint, you can form a right-angled triangle ABC.

Triangles AMP and MQB are congruent because:
▶ AM = MB (M is the midpoint)
▶ ∠MAP = ∠BMQ (corresponding angles)
▶ ∠MPA = ∠BQM (both 90°)

So BQ = MP = $\frac{6}{2}$ = 3, and AP = MQ = $\frac{4}{2}$ = 2.
The midpoint M is (1 + 2, 3 + 3), or (3, 6).

There is an easier way to find the midpoint of a line segment ...

You can think of the midpoint as the **mean** of the two end points.

$$x_m = \frac{x_1 + x_2}{2} \qquad y_m = \frac{y_1 + y_2}{2}$$

The formula works with the example because:
$\frac{1+5}{2}$ = 3
$\frac{3+9}{2}$ = 6
So the midpoint is (3, 6).

example

Find the midpoint of the line segment joining the points A(⁻1, 3) and B(2, ⁻5).

Use the formula:
$$x_m = \frac{^-1 + 2}{2} = \frac{1}{2} \qquad y_m = \frac{3 + ^-5}{2} = \frac{^-2}{2} = ^-1$$

So the midpoint has coordinates ($\frac{1}{2}$, ⁻1).

218

Plenary

Say that the midpoint of a line segment is (3, 4) and one endpoint is (5, 7). What are the coordinates of the other endpoint?
Students can respond using digit cards to represent the coordinates.
Repeat with other midpoints and endpoints.

Further activities

Challenge students to invent a method to calculate the coordinates of a point that is:
▸ One-quarter of the way along a line
▸ Three-quarters of the way along a line.
Encourage students to generalise and test their results.

Differentiation

Extension questions:
▸ Question 1 provides simple practice at finding the midpoint of a line given two coordinates.
▸ Questions 2–6 focus on finding the midpoint or an endpoint given sufficient information.
▸ Question 7 requires students to generalise a result using algebra.

Core tier: focuses on finding the midpoint of a line given numerical rather than algebraic coordinates.

Exercise S4.6

1 Plot these pairs of points on a grid and calculate the midpoints of the lines joining them.
   a A(6, 3)   B(4, 7)
   b C(4, 2)   D(2, 5)
   c E(1, 2)   F(5, 4)
   d G($^-$1, 2)   H($^-$3, 2)
   e I(0, 4)   J(0, 5)
   f K(2, $^-$1)   L(2, $^-$5)
   g M(3, $^-$2)   N(5, $^-$2)
   h O($^-$1, $^-$3)   P($^-$4, $^-$5)
   i Q(6, 2)   R($^-$3, $^-$7)
   j S($^-$2, $^-$1)   T(4, 2)

2 Look at your results for question 1.
   If two points have coordinates A(x, y) and B(a, b) what are the coordinates M of the midpoint of AB?

3 If A is point (2, 3) and the midpoint, M of AB is (3, 4), what is the point B?

4 If A is (5, 6) and M is (9, 10) what is B?

5 If A is ($^-$1, 4) and M is (2, 3) what is B?

6 a If A(2, 4) and B(3, 5), could M(1, 6) be the midpoint?
      Explain your answer.
   b If A($^-$1, 5) and B(1, $^-$2), could M(1, 3) be the midpoint?
      Explain.

7 The midpoint of a line segment AB is (x_m, y_m).
   The point A has coordinates (x_1, y_1).

   Derive two formulae that would give the coordinates of B(x_2, y_2).

   $x_2 = $ _____
   $y_2 = $ _____

219

Exercise commentary

The questions assess the objectives on Framework Page 219.

Problem solving
Question 2 requires students to generalise a result, assessing the objectives on Framework Page 35. Questions 6 and 7 assess Page 33.

Group work
Students can work in pairs on the further activity.

Misconceptions
Students may find the formula for the midpoint of a line rather daunting. Emphasise the link with the mean of two values.
Encourage the use of correct notation when generalising, for example x_1 rather than $x1$ or x^1.

Links
Calculating the distance between two points links to later work on Pythagoras' theorem (Framework Page 189).

Homework

S4.6HW provides further practice at calculating the midpoint of a line, and extends to calculating the centre of a parallelogram.

Answers

1 a (5, 5)   b (3, 3$\frac{1}{2}$)   c (3, 3)   d ($^-$2, 2)
   e (0, 4$\frac{1}{2}$)   f (2, $^-$3)   g (4, $^-$2)   h ($^-$2$\frac{1}{2}$, $^-$4)
   i (1$\frac{1}{2}$, $^-$2$\frac{1}{2}$)   j (1, $\frac{1}{2}$)
2 ($\frac{x+a}{2}$, $\frac{y+b}{2}$)   3 (4, 5)   4 (13, 14)
5 (5, 2)   6 a No   b No
7 $x_2 = 2x_m - x_1$, $y_2 = 2y_m - y_1$

Ask students to imagine a propeller blade spinning around (you could use the visual image of a wind turbine). What would be the path traced out by a point at the end of one of the blades?

Then ask students to consider the path traced out by the midpoint of one of the blades. How would this shape relate to the previous one?

Finally ask students to consider a point on the central hub of the propeller.

Useful resources

S4.7OHP shows the constructions in the student book.

Introductory activity

Refer to the mental starter.

Describe the path traced out by a moving point as a **locus**.

Show that you can describe a locus by a rule.

Draw a point marked A on the board.

Specify the rule: The point P can move so that it is 5 cm from A.

Ask students to describe the path traced out by P. Encourage the response that it is a circle, and ask for the radius (and diameter).

Refer to the first example in the student book.

The diagrams are shown on **S4.7OHP**.

Encourage students to realise that the locus is an angle bisector, and recap how to construct an angle bisector accurately. Ask what the angle should be in this case (45°).

Refer to the second example in the student book.

Encourage students to realise that the locus is a perpendicular bisector, and recap how to construct this accurately.

As with the previous example, go through the stages of the construction step-by-step, and ask students to construct it in their books.

The technical diagrams from the second example are also shown on **S4.7OHP**.

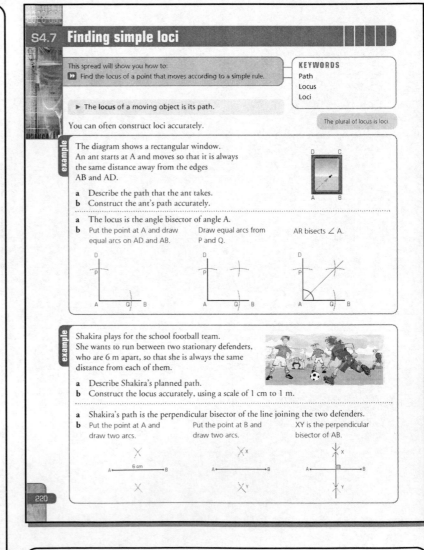

S4.7 Finding simple loci

This spread will show you how to:
▶ Find the locus of a point that moves according to a simple rule.

KEYWORDS
Path
Locus
Loci

The plural of locus is loci.

▶ The **locus** of a moving object is its path.

You can often construct loci accurately.

The diagram shows a rectangular window. An ant starts at A and moves so that it is always the same distance away from the edges AB and AD.

a   Describe the path that the ant takes.
b   Construct the ant's path accurately.

a   The locus is the angle bisector of angle A.
b   Put the point at A and draw equal arcs on AD and AB.   Draw equal arcs from P and Q.   AR bisects ∠ A.

Shakira plays for the school football team. She wants to run between two stationary defenders, who are 6 m apart, so that she is always the same distance from each of them.

a   Describe Shakira's planned path.
b   Construct the locus accurately, using a scale of 1 cm to 1 m.

a   Shakira's path is the perpendicular bisector of the line joining the two defenders.
b   Put the point at A and draw two arcs.   Put the point at B and draw two arcs.   XY is the perpendicular bisector of AB.

Plenary

Discuss question 4, particularly part e.

You may wish to illustrate this motion by rolling a circular object along your desk (like the lid of a jar), with a point marked along its circumference.

Further activities

Challenge students to consider the locus of a point P that is always 3 cm away from a straight line AB, 5 cm long.

Students can work in pairs to investigate this type of locus by changing the numbers.

Differentiation

Extension questions:
▶ Question 1 requires students to construct a perpendicular bisector.
▶ Questions 2–4 focus on real-life loci.
▶ Questions 5 and 6 focus on constructing an angle bisector.

Core tier: focuses on finding simple loci, including a perpendicular bisector.

Exercise S4.7

1 Draw two points A and B, 15 cm apart.
   Draw 20 points that are *equidistant* (the same distance) from A and B.
   Join these points to form the locus of points equidistant from A and B.

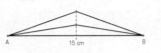

2 Describe your route to school using instructions such as left, right, and straight on.

3 Describe the route from your form room to the school gym or sports hall.

4 Draw a diagram to show:
   a the path traced out by a conker on a string
   b the path of the tip of a windscreen wiper
   c the path of the ball thrown into the air
   d the path of your nose as you walk along a straight road
   e the path of a stone on a car wheel as the car moves.

5 Draw an angle BAC of 70° with arms AB and AC 8 cm long.
   Use a ruler to draw 20 points that are equidistant from AB and AC.
   Join up the points to form a straight line. This line is the angle bisector of BÂC.

6 Construct the angle bisector of angle WXY, which will show the locus of points equidistant from XY and XW.

Exercise commentary

Exercise commentary
The questions assess the objectives on Framework Page 225.

Problem solving
The exercise assesses the objectives on Framework Page 15.

Group work
Questions 3 and 4 can be discussed in pairs or small groups.

Misconceptions
Students often do not see the point in using compasses to construct bisectors, when they could just use a protractor. Discuss the limitations of a protractor - it can only measure angles between lines that are at least 5 cm long, and it is only accurate to the nearest degree.

Links
Loci have strong links with construction (Framework Pages 221 and 223).

Homework

S4.7HW requires students to construct a locus given a simple rule.

Answers

Check students' diagrams.

Mental starter

Ask students to stand up and form loci (you may need to move chairs and desks out of the way).

Students should form the locus of points:

▸ equidistant from two adjacent walls
▸ equidistant from the door and a specified window
▸ 2 m from a specified individual.

Useful resources

S4.8OHP shows the diagram in the second example.

Coins of various denominations would be useful as a visual prop.

Introductory activity

Recap the loci explored last lesson.

These can be categorised by their shapes:

▸ A circle
▸ A perpendicular bisector
▸ An angle bisector

Emphasise the rules that are associated with each of these types of loci. They are illustrated in the student book.

Extend to more complicated loci.

Discuss the locus of a point on the edge of a coin as it is rolled along a table surface. The path is traced out in the second example in the student book, which is also shown on **S4.8OHP**.

Ask how the locus would change if the coin were bigger.

Extend to loci in three dimensions.

Refer to the last example in the student book, and highlight the similarity with a point that moves in two dimensions so that it is always a fixed distance from another point.

You can use this as a definition for a sphere: the locus of points in three dimensions that are a fixed distance from a given point.

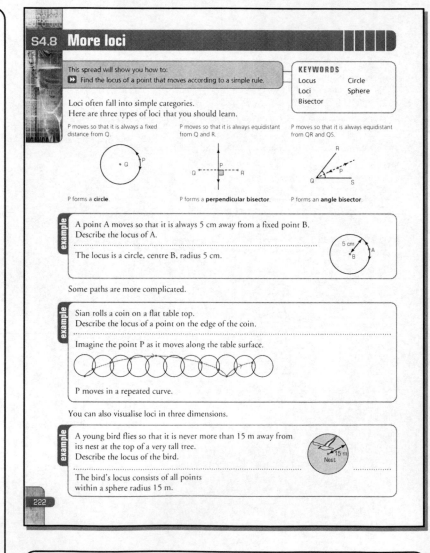

Plenary

Refer back to the coin moving along a table described in the introductory activity.

What would be the locus of a point on a 50p coin, or a 20p coin? Students can discuss this in small groups. If they do not have the correct coinage, they could make a paper cut-out as a visual aid (discuss the correct mathematical names for the shape of these coins).

Students could extend question 4, and explore the loci of other points on the square.
Try using a triangle instead of a square.

Differentiation

Extension questions:
▸ Question 1 involves drawing a simple circular locus.
▸ Questions 2–4 focus on constructing loci subject to various rules.
▸ Questions 5 and 6 are harder problems involving scale drawings in context.

Core tier: focuses on using loci in context to solve problems.

Exercise S4.8

1 Construct the locus of all points 6 cm from a fixed point A.

2 A radio mast can transmit within a 40 km radius. Draw a scale drawing to show the region that can receive the signal. Use a scale of 1 : 1 000 000.

3 Construct the loci of the paths of the points which are equidistant from these pairs of lines.

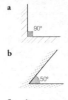

a

b

c

(Copy the diagrams first.)

4 Cut out a 2 cm square ABCD and place it on a straight line. Mark the corner B of the square as shown.
Rotate the square about corner D. Mark the position of B as you rotate the square.

When CD is on the line, continue rotating, now about corner C to keep the square moving along the line and finish the locus of B.

5 ABCD is a plan of a garden on a scale 1 cm : 10 m. Construct the locus of points:

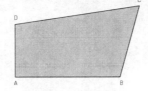

a equidistant from AB and AD
b 30 m from B.
c Trees must be planted nearer to AB than AD, and less than 30 m from B. Show where they could be planted on your diagram.

6 The diagram shows an accurate scale drawing of a field, on a scale of 1 cm to 2 m.

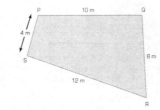

Copy the diagram (you could use tracing paper).
A man walks so that he is always equidistant from P and Q.
A dog is tethered to R on a rope 3 m long. Can the dog reach the man?

223

Exercise commentary

The questions assess the objectives on Framework Page 225.

Problem solving
Questions 2, 5 and 6 assess the objectives on Framework Page 15.

Group work
Question 4 can be attempted in pairs.
Questions 5 and 6 can be discussed in small groups.

Misconceptions
Students may be hasty to assume that a locus has a particular shape, which is not borne out by further examination.
Encourage students to apply the rule a few times to fix the shape.
In question 4, ensure that students mark the position of B a sufficient number of times to reveal the locus.

Links
Construction: Framework Pages 221–223.

Homework

S4.8HW requires students to construct accurate loci, using contexts similar to those in the exercise. Students will need compasses.

Answers

1 Circle centred on A with radius 6 cm
2 Circle centred on the mast with radius 4 cm
6 No

Mental starter

Ask if triangle XYZ can be constructed if:

▸ XY = 4 cm, XZ = 3 cm, YZ = 5 cm
▸ XY = 2 cm, YZ = 2 cm, ZX = 5 cm
▸ X̂ = 60°, Ŷ = 70°, Ẑ = 100°

Emphasise that students should not construct the triangles, but should attempt to sketch them. Encourage justification of responses.

Useful resources

S4.9OHP shows the construction diagrams in the first example of the student book.

Introductory activity

Recap the construction learned in year 7.
In particular, recap how to construct a triangle, given:
▸ Two sides and an angle (SAS)
▸ Two angles and a side (ASA)
▸ Three sides (SSS)

Briefly discuss right-angled triangles.
Illustrate how the longest side is always opposite the right angle, and this is called the **hypotenuse**.

Demonstrate how to construct a right-angled triangle given a right angle, a side and the hypotenuse.
You can use the first example in the student book, which is illustrated on **S4.9OHP**.

Encourage students to work through the steps individually.
Emphasise that they should keep the construction marks, as they show evidence of working.

Use the second example in the student book to show how you can solve real-life problems by constructing a right-angled triangle to scale.

Highlight how real-life problems can often be represented by a simplified diagram.

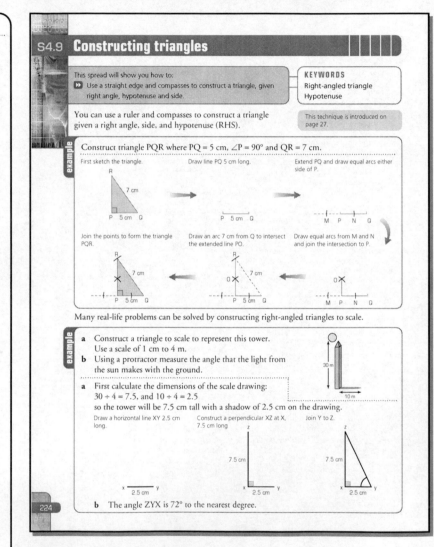

Plenary

Refer back to the mental starter and discuss question 7, which also involves impossible triangles.
Discuss criteria for possible triangles:
▸ Sum of angles should be less than 180°
▸ Longest side should be smaller than the sum of the two shorter sides.

Further activities

Extend question 7 by considering whether there are any instances where two possible triangles can be described by the same information.
Students can explore these possibilities in pairs.

S4.9ICT uses an interactive geometry package to explore transformations.

Differentiation

Extension questions:

▸ Questions 1–3 require students to construct a triangle given information on lengths and angles.
▸ Questions 4–6 require students to solve problems involving constructing triangles.
▸ Question 7 explores possible and impossible triangles.

Core tier: focuses on constructing a triangle given three sides (SSS).

Exercise S4.9

1   Construct △EFG where EF = 6 cm, ∠E = 30° and ∠F = 60°.

2   Construct △ABC where AB = BC = 10 cm and ∠B = 45°.

3   Construct △XYZ where XY = 5.9 cm, YZ = 6.2 cm and ∠XZ = 6.5 cm.

4   A 10 m ladder rests against a wall with its foot 3 m away from the wall.

Construct a scale diagram and use the diagram to find:
   a   how far the ladder reaches up the wall
   b   the angle between the ladder and the ground.

5   The net of a tetrahedron is made from equilateral triangles of side 4 cm. Draw an accurate construction of the net.

6   Boat A can see plane C at an angle of 42°. Boat B can see plane C at angle of 49°, and AB = 85 m.
   Find the height *h* of C above the horizontal.

7   Is it possible to construct △ABC such that
   a   ∠A = 60°, ∠B = 60°, ∠C = 60°
   b   ∠A = 40°, ∠B = 60°, AB = 5 cm
   c   BC = 6 cm, AC = 10 cm, ∠B = 90°
   d   BC = 7 cm, AC = 4.95 cm, ∠B = 45°?
   Give reasons for your answers, and construct the triangle where possible.

225

Exercise commentary

The questions assess the objectives on Framework Page 223.

Problem solving
Questions 4 and 6 assess the objectives on Framework Page 17.
Question 7 assesses the objectives on Pages 33–35.

Group work
Question 7 can be discussed in pairs or small groups.

Misconceptions
When constructing scale drawings, as in question 4, students may be unsure as to an appropriate scale. Suggest that they consider how long the longest length and the shortest length would appear on their diagram with a particular scale. Discourage diagrams that are too small and cramped.

Links
Scale drawings: Framework Page 217

Homework

S4.9HW focuses on constructing scale drawings involving triangles, and includes a consideration of impossible triangles as in question 7.

Answers

4 **a** 9.5 m   **b** 73°
6 43 m
7 **a** Yes   **b** Yes   **c** Yes   **d** Yes

Summary

The key objectives for this unit are:
▶ Solve substantial problems by breaking them into simpler tasks, using a range of efficient techniques, methods and resources.

Plenary activity

Write these measurements on the board:
5.5 cm, 34 mm, 1.75 km, 17.5 cm, 850 m, 70 mm, and 1.375 km.
Write down the scale 1: 25 000.

Ask the class to find pairs of measurements (for example 4 cm would pair with 1 km) and hence identify the odd one out.

Check out commentary

1 Students may not leave sufficient space on their page for the complete drawing. Encourage a rough sketch beforehand, to give an idea of the basic shape and size.
For a fully accurate drawing, students should construct a perpendicular from North to identify the line pointing East, rather than using a protractor.

2 Students should be able to identify basic types of loci from the description of the rule, and it should be clear that this is a question involving the perpendicular bisector of a line.
Students should trace the two points A and B, and draw the line AB between them.
Ensure that students use compasses, not a ruler and protractor, to solve this problem.

3 The purpose of a sketch is rather like a numerical estimate: it gives students a rough idea of the answer. Encourage students not to dwell too long on making the sketch in part **a** too accurate, but instead to focus on accuracy in part **b**.
Students should think of a sensible scale to use, which is neither too small to measure accurately from, nor too large.

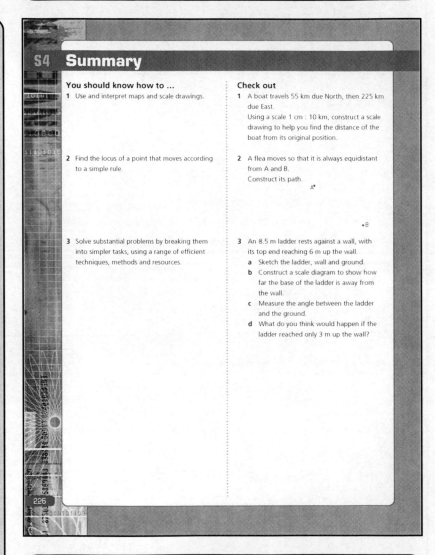

S4 **Summary**

You should know how to ...

1 Use and interpret maps and scale drawings.

2 Find the locus of a point that moves according to a simple rule.

3 Solve substantial problems by breaking them into simpler tasks, using a range of efficient techniques, methods and resources.

Check out

1 A boat travels 55 km due North, then 225 km due East.
Using a scale 1 cm : 10 km, construct a scale drawing to help you find the distance of the boat from its original position.

2 A flea moves so that it is always equidistant from A and B.
Construct its path.

A•

•B

3 An 8.5 m ladder rests against a wall, with its top end reaching 6 m up the wall.
a Sketch the ladder, wall and ground.
b Construct a scale diagram to show how far the base of the ladder is away from the wall.
c Measure the angle between the ladder and the ground.
d What do you think would happen if the ladder reached only 3 m up the wall?

226

Development

All of the themes of this unit are extended in Year 10.
In particular, right-angled triangles are studied in the contexts of Pythagoras and trigonometry.

Links

Geometrical constructions have application in design and technology, particularly in technical drawing.
A knowledge of bearings has a practical use in orienteering.

Mental starters

Objectives covered in this unit:
▸ Order, add, subtract, multiply and divide integers.
▸ Discuss and interpret graphs.

Resources needed

* means class set needed
Essential:
Coins*
Stopwatches*
Graph or squared paper

Useful:
Mini-whiteboards
D3.3OHP – coin spinning data
D3.4OHP – scatter graphs
D3.4ICT – creating charts

D3   Data projects

This unit will show you how to:

▸▸ Suggest a problem to explore using statistical methods, frame questions and raise conjectures.

▸▸ Design a survey or experiment to capture the necessary data from one or more sources.

▸▸ Determine the sample size and degree of accuracy needed.

▸▸ Design, trial and, if necessary, refine data collection sheets.

▸▸ Construct tables for large discrete and continuous sets of raw data, choosing suitable class intervals.

▸▸ Interpret graphs and diagrams and draw inferences to support or cast doubt on initial conjectures.

▸▸ Communicate interpretations and results of a statistical enquiry.

▸▸ Solve increasingly demanding problems and evaluate solutions.

▸▸ Present a concise, reasoned argument, using symbols, diagrams, graphs and related explanatory text.

How long do you think this line is, to the nearest centimetre?

You should always start a survey with clear questions or statements.

Before you start

You should know how to ...

1 Find the range, mean, median and mode for a set of data.

2 Use an assumed mean.

3 Draw frequency diagrams for continuous data.

Check in

1 Find the range, mean, median and mode of this data set.

| 6 | 4 | 5 | 2 | 9 | 4 |

2 Using an assumed mean find the mean of these data sets.

a 2006 2004 2005 2002 2009 2004
b 1.6   1.4   1.5   1.9   1.4   1.2

3 Draw a frequency diagram to represent this data.

Time (seconds)	0–15	15–30	30–45	45–60
Frequency	4	7	8	5

227

Unit commentary

Aim of the unit

This unit focuses on exploring a problem using statistical methods: formulating a hypothesis; experiment design and data collection; interpreting and presenting the results. Students consider sample size, sources of bias, and ways of grouping continuous data.

Introduction

Discuss why it is sometimes necessary to conduct an experiment:
▸ to find possible solutions to problems
▸ to simplify a task
▸ to understand real-life problems.
Discuss examples in context.

Framework references

This unit focuses on:
▸ Teaching objectives Pages: 249–53, 257–9, 263, 267–273

▸ Problem-solving objectives Pages: 25, 27, 29 and 31.

Differentiation

Core tier

Focuses on interpreting and comparing distributions using statistics and diagrams.

Check in activity

Write these seven numbers on the board:
13, 17, 16, 13, 18, 15, 13.
Ask for the mode (13), median (15) and the mean (15).
Ask which average is least representative and why.
Ask how you could add an eighth number to make the mean 16.

Mental starter

Students work in pairs and time how long each can keep a coin spinning.

Discuss strategies for measuring the spin time and for recording the data.

Useful resources

Coins
Stopwatches

Introductory activity

Discuss what data Charlie needs to test her hypothesis. Could she use both primary and secondary data?

Discuss possible sources of bias (refer to the mental starter) for example, table surface not smooth, using different coins, different people timing, difficulty of judging start/stop times of spin, how much practice in spinning may be needed/allowed, and so on.

Discuss Charlie's hypotheses.
Could the same data be used to investigate more than one hypothesis?

Emphasise that when you state a hypothesis, you also need to state the reason behind it.
Discuss the importance of referring to your hypothesis (or hypotheses) throughout your survey, as a focus for your investigation.
You may wish to refer back to the data handling cycle that students should have encountered in Year 7.

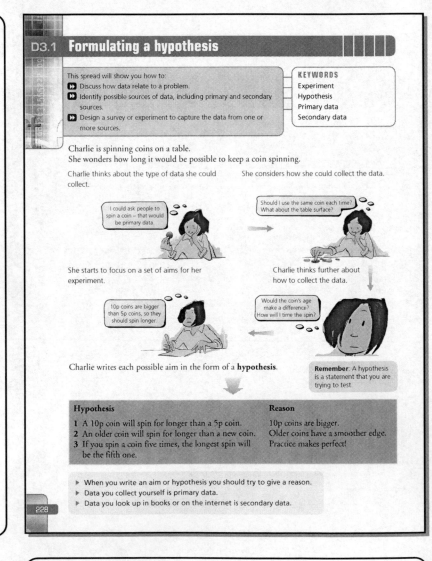

D3.1 Formulating a hypothesis

This spread will show you how to:
▶ Discuss how data relate to a problem.
▶ Identify possible sources of data, including primary and secondary sources.
▶ Design a survey or experiment to capture the data from one or more sources.

KEYWORDS
Experiment
Hypothesis
Primary data
Secondary data

Charlie is spinning coins on a table.
She wonders how long it would be possible to keep a coin spinning.

Charlie thinks about the type of data she could collect.

She considers how she could collect the data.

> I could ask people to spin a coin – that would be primary data.

> Should I use the same coin each time? What about the table surface?

She starts to focus on a set of aims for her experiment.

Charlie thinks further about how to collect the data.

> 10p coins are bigger than 5p coins, so they should spin longer.

> Would the coin's age make a difference? How will I time the spin?

Charlie writes each possible aim in the form of a **hypothesis**.

Remember. A hypothesis is a statement that you are trying to test.

Hypothesis	Reason
1 A 10p coin will spin for longer than a 5p coin.	10p coins are bigger.
2 An older coin will spin for longer than a new coin.	Older coins have a smoother edge.
3 If you spin a coin five times, the longest spin will be the fifth one.	Practice makes perfect!

▶ When you write an aim or hypothesis you should try to give a reason.
▶ Data you collect yourself is primary data.
▶ Data you look up in books or on the internet is secondary data.

228

Plenary

Discuss students' hypotheses for each question in the exercise, and the reasons behind them.

Further activities

For each question, students could identify further related hypotheses, with reasons, that could be investigated.

Differentiation

Extension questions:
The focus of this exercise is to formulate hypotheses and give reasons for their choice.

Core tier: focuses on constructing and using frequency tables and two-way tables.

Exercise D3.1

For each of the situations described in questions 1 to 5:

a Write down a hypothesis that you can test, giving reasons for your choice.
b Identify what data you will need to test your hypothesis.

State if it is primary or secondary data.
(You may be able to collect data from more than one source.)

1 Dave was investigating spinning coins at his school.
   He had two ideas that he wanted to investigate.
   i Does your age make any difference to how long you can spin a coin?
   ii Does it make a difference to how long you can spin a coin if you are left-handed or right-handed?

2 Kylie listens to many different types of music.
   She thinks that the length of a song depends upon what type of music it is.

3 Sally had become very forgetful.
   She wondered if men and women got forgetful as they got older.

4 Helen's mum was always saying, 'I'll do that in a minute' and not doing it for ages.
   Helen wondered whether people actually knew how long a minute was.

5 Evan read in a book that hand span and wrist measurement were related.

He wondered if any other body measurements were related.

229

Exercise commentary

The questions assess the objectives on Framework Pages 249 and 251.

Problem solving

All questions provide an opportunity to assess Framework Page 29.

Group work

All questions are suitable for discussion in pairs or small groups.

Misconceptions

In any single experiment or survey it is possible to collect data and use it to study more than one hypothesis.
If you have to use completely different data to study a second hypothesis then it is a different experiment.

Links

This topic links to practical investigations in science, geography and PSHE.

Homework

D3.1HW focuses on forming a hypothesis to carry out a music survey.

Answers

There are many different possible answers to each question, for example:
1 i The older you are the longer you can spin a coin as older people would have had more practice.
   ii Left-handed people are better at spinning coins than right-handed people as they are more dextrous. Both require primary data.

229

Mental starter

Write these spinning times for a 2p piece on the board:

      Spins 1–10   4, 5, 8, 10, 5, 6, 3, 8, 5

      Mean time for spins 11–50 (40 spins) 5.5 seconds

Ask students to calculate the mean for spins 1–10 and then the mean for spins 1–50.

Useful resources

Coins

Stopwatches

Introductory activity

Discuss how much data you would need (**sample size**) for an experiment to test the length of spin of a coin. (refer to the mental starter).

What degree of accuracy do you need for the data? (Refer to the mental starter for D3.1.)

Discuss the hypotheses outlined on the student's page.

Emphasise that reasons have been given for the hypotheses.

Discuss how you could check that your data collection sheet will actually capture the data you need (pilot survey).

Emphasise that a pilot survey may also uncover problems to do with collecting the data.

For example the need to have all conditions the same for every member of the sample; whether the degree of accuracy is appropriate, and so on.

Discuss the different methods of collecting primary data:

▸ controlled experiment
▸ observation
▸ data logging
▸ questionnaire.

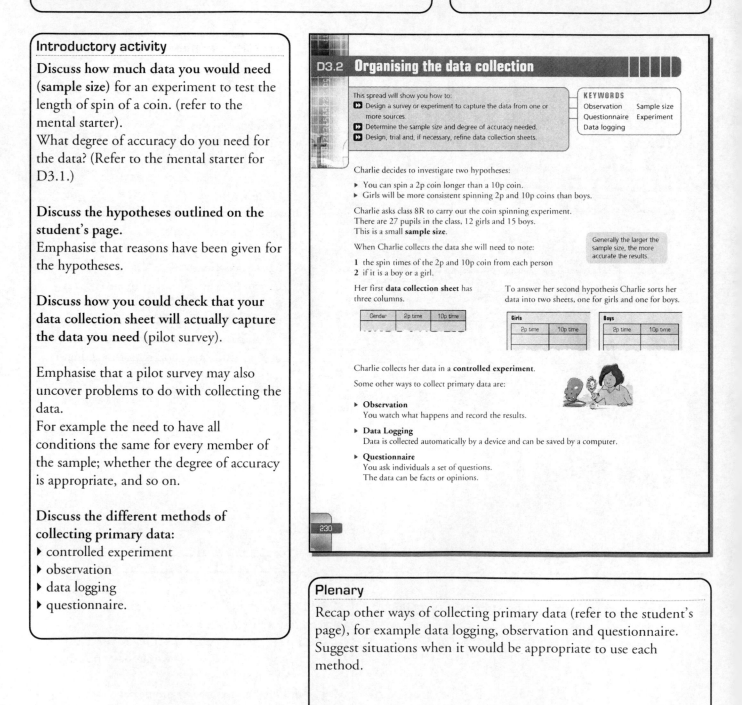

D3.2 Organising the data collection

This spread will show you how to:
▸▸ Design a survey or experiment to capture the data from one or more sources.
▸▸ Determine the sample size and degree of accuracy needed.
▸▸ Design, trial and, if necessary, refine data collection sheets.

KEYWORDS
Observation   Sample size
Questionnaire   Experiment
Data logging

Charlie decides to investigate two hypotheses:
▸ You can spin a 2p coin longer than a 10p coin.
▸ Girls will be more consistent spinning 2p and 10p coins than boys.

Charlie asks class 8R to carry out the coin spinning experiment. There are 27 pupils in the class, 12 girls and 15 boys. This is a small **sample size**.

When Charlie collects the data she will need to note:

1 the spin times of the 2p and 10p coin from each person
2 if it is a boy or a girl.

Generally the larger the sample size, the more accurate the results.

Her first **data collection sheet** has three columns.

Gender	2p time	10p time

To answer her second hypothesis Charlie sorts her data into two sheets, one for girls and one for boys.

Girls

2p time	10p time

Boys

2p time	10p time

Charlie collects her data in a **controlled experiment**.

Some other ways to collect primary data are:

▸ **Observation**
You watch what happens and record the results.

▸ **Data Logging**
Data is collected automatically by a device and can be saved by a computer.

▸ **Questionnaire**
You ask individuals a set of questions. The data can be facts or opinions.

230

Plenary

Recap other ways of collecting primary data (refer to the student's page), for example data logging, observation and questionnaire. Suggest situations when it would be appropriate to use each method.

Students could choose one hypothesis from the answers to question 1 and carry out a pilot survey, using the data collection sheet they have designed.

Differentiation

Extension questions:
The focus of this exercise is to design an experiment and data collection sheet.
It is not intended that both questions 1 and 2 are completed during lesson time – see group work commentary and exercise D3.3.

Core tier: focuses on choosing the most appropriate statistics for a distribution.

Exercise D3.2

1   For each of the hypotheses you chose for each question in exercise D3.1 on page 229:

   a   describe how you would carry out a survey or experiment to collect this data

   b   design a data collection sheet to collect this data.

2   You are going to carry out a coin spinning experiment.

You can work in small groups on this question.

   a   Write out one or more hypotheses that you would like to test.
      Give reasons for choosing these hypotheses.

   b   Write down how you are going to carry out the experiment.
      i   Include the sample size that you will need.
      ii   Say how you are going to time each spin, and how accurate your times need to be.
      iii   Write about what you are going to do to ensure that the experiment is fair.

   c   Design the data collection sheet that you will use to collect the data.

You need to ensure that it is easy to use and that it will enable you to answer your hypotheses.
You may want to refine the data collection sheet to sort the data.

   d   Carry out your experiment and record your results in your data collection sheet.
You should aim to collect at least 30 pieces of data.

231

Exercise commentary

The questions assess the objectives on Framework Pages 251 and 253.

Problem solving

Questions 1 and 2 provide an opportunity to assess Framework Page 27.
Question 3 also assesses Page 31.

Group work

Question 1 can be discussed in pairs. Question 2 is a whole class experiment. Once the hypothesis and data collection methods have been agreed, students can work in small groups and pool their data.

Misconceptions

Students are often unclear about sources of bias until they actually carry out an experiment and experience the possible problems for themselves.

Links

The results from question 2 may be used in Exercise D3.3.
This topic links to practical investigations in science, geography and PSHE.

Homework

D3.2HW allows students to plan and design a data collection sheet for their music survey.

Answers

There are no unique answers to this exercise.

Mental starter

Write several small sets of data on the board:

20 012, 20 009, 20 015, 20 007

and ask students to find the mean, using an assumed mean.

You may need to recap how to calculate an assumed mean.

Useful resources

Mini-whiteboards may be useful for the mental starter.

Graph or squared paper

D3.3OHP – the coin spinning data from the student book.

Introductory activity

Recap finding the mean using an assumed mean, using the mental starter.

Discuss how the data from the experiment should be presented.

Emphasise that data should be analysed by: tabulating results; finding an average and range; drawing graphs.

Discuss ways of grouping the data.

Emphasise that the data could be grouped in other ways (e.g. 4–6, 7–9, 10–12, etc or 5–6, 7–8, 8–10, etc), which may lead to different-shaped graphs.

Emphasise that it is important to choose carefully how to present data.

Emphasise that the same data could be used to answer a hypothesis comparing the coin-spinning times of boys and girls Charlie designed her data collection sheet to enable her to do this (see D3.4).

Emphasise the importance of planning an experiment or survey so that it gives you the data you need.

D3.3OHP contains the coin spinning data from the student book.

You can use this to discuss how to present the data.

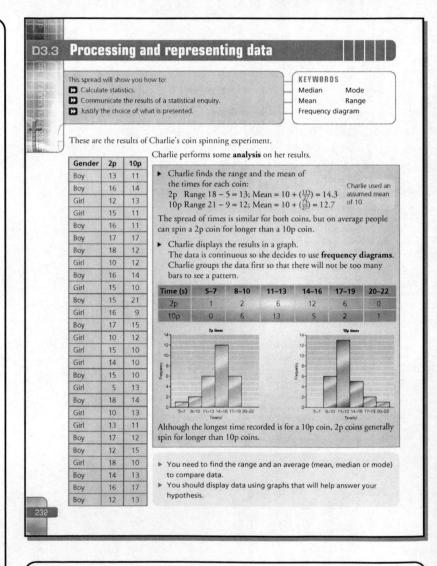

Plenary

Discuss the types of graph that students have used to display their data from question 2.

Recap the different types of graph used for discrete, continuous and qualitative data and discuss why they are used.

From the results of the coin-spinning experiment, students could write down further hypotheses to investigate. What data would they need to collect to explore these hypotheses?

Differentiation

Extension questions:
The focus of this exercise is to sort data, draw graphs and calculate summary statistics. It is not intended that both questions 1 and 2 are completed during lesson time – the choice depends upon the choice made in Exercise D3.2.

Core tier: explores more unusual types of statistical diagram.

Exercise D3.3

1 Use the results of Charlie's experiment, shown in the table on page 232.
  a Calculate the median and mode for **i** 2p coins **ii** 10p coins.

  b Comment on these results and suggest which average (mean, median or mode) is the most appropriate to use these data. Give reasons for your choice.
  c Draw a scatter graph to display the results. Comment on what it shows.

Time to spin 10p coin (s) 15, 10, 5
Time to spin 2p coin (s) 5, 10, 15

  d Group the data differently from Charlie's groupings, and draw frequency diagrams for your grouped data.
  e Compare your graphs with the graphs on page 232, and comment on any differences and similarities in how the data is represented.

2 Use the results of your own coin spinning experiment that you carried out in Exercise D3.2 on page 231.
  a Calculate:
    i the range
    ii the mean
    iii the median
    iv the mode of your data.
    Comment on what these results show and which average is the best to use.
  b Draw graphs to display your results that will help answer your hypothesis. Comment on your graphs.

233

Exercise commentary

The questions assess the objectives on Framework Pages 253, 257, 259, 263 and 267.

Problem solving
Questions 1 and 2 provide an opportunity to assess Framework Page 27.
Question 2d assesses Page 31.

Group work
Question 1 d and e and all of question 2 are suitable for work in pairs or small groups.

Misconceptions
Students need to take care that their choice of graph to display continuous data is appropriate.

Links
Interpreting graphs links to work on interpreting graphs in sections A5.5, A5.7 and A5.8.
This topic links to practical investigations in science, geography and PSHE.

Homework

D3.3HW provides an opportunity for students to represent real data from a music survey.

Answers

1 a **i** 15, 15 and 16 **ii** 12, 10 and 13
  b Median, both bi-modal, both means distorted by outlier
  c No real trend shown
All other questions have no unique answers.

Mental starter

Sketch scatter diagrams on the board, showing positive, negative and no correlation. Ask students to suggest different typical situations that would show these trends.

Useful resources

Graph or squared paper
D3.4OHP – the scatter graphs from the student book

Introductory activity

Emphasise that a statistical investigation is incomplete without an explanation of the diagrams and calculations.

Discuss circumstances when it is sensible to exclude outliers from the data used for analysis.

Emphasise that sorting the data (for example by gender) allows you to make comparisons, and spot similarities and differences.

Use the scatter graphs to emphasise that comments on graphs should refer to the variables they depict.
Discuss the scatter graphs in the student book, which are shown on **D3.4OHP**.

Compare the statistics for the two distributions.
▸ The modes show that boys could spin a coin for longer.
▸ The ranges show that girls have more variable spinning times.

D3.4 Interpreting and discussing data

This spread will show you how to:
▶▶ Interpret tables, graphs and diagrams for continuous data.
▶▶ Compare two distributions using the range and one or more of the mode, median and mean.

KEYWORDS
Interpret
Infer
Scatter graph

Charlie organises results in order of the 2p spinning times.

Girls' times (seconds)

2p	5	10	10	12	12	13	14	15	15	15	16	18
10p	13	12	13	12	13	11	10	10	10	11	9	10

The 5 seconds value for the 2p coin may have been incorrectly recorded, as it does not seem to fit with the rest of the data.

Boys' times (seconds)

2p	12	12	13	14	15	15	16	16	16	16	17	17	17	18	18
10p	13	15	11	13	10	21	11	14	14	17	12	15	17	12	14

By inspection, Charlie notes the range and modal times in seconds.

	Girls		Boys	
	2p	10p	2p	10p
Range (s)	13	4	6	11
Mode (s)	15	10	16	14

If the 5 seconds value is ignored, the range for girls' 2p times is 8 seconds.

The statistics suggest that:

▸ Boys could keep a coin spinning longer than girls.

| Modes (Girls) | 15 | 10 | (Boys) | 16 | 14 |

The mode times for both coins are greater for boys than girls.

▸ Boys show more consistency between coins.

| Range (Girls) | 13 | 4 | (Boys) | 6 | 11 |

The range of spin times for both coins are closer for the boys.

Charlie draws scatter graphs for the results of the girls and boys.

For girls, as the 2p spinning time increases, the 10p spinning time decreases.

For boys, there is no apparent trend in the spinning times.

234

Plenary

Discuss students' answers to the exercise, in particular the reasons they give for their interpretation.

Further activities

Students could find the average number of days absent for the class and compare it with the data in question 3, or with other classes in the school.

D3.4ICT shows how to create charts using both Word and Excel.

Differentiation

Extension questions:
This exercise focuses on using averages and the mode, on interpreting tables and diagrams and making conjectures.

Core tier: focuses on scatter graphs and correlation.

Exercise D3.4

1   The diagram shows a bar chart of the number of deaths per day, and a line graph of the sulphur levels in London during the great smog in December 1952. Compare and comment on the two graphs.

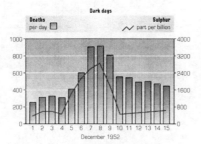

2   The table gives the number of pedestrian injuries (to the nearest 5) among children in Great Britain in 2000.

Age of casualty	0	1	2	3	4	5	6	7	8	9	10	11	12	13	14	15
Number of casualties	25	105	360	565	715	770	925	1035	1185	1335	1430	1765	1920	1565	1365	1125

a   Draw a frequency polygon to represent these data and comment on your graph.
b   Compare and comment on the figures for primary and secondary school children.
c   Calculate the mean, median and mode for this data.
d   Which of the three averages would you choose to represent the data? Give a reason for your answer.

3   The table gives the total number of student days absent for each year group in two schools.

School	Year					Total
	7	8	9	10	11	
Student numbers						
Cowpers	24	31	11	42	56	600
Darwings	57	65	29	84	135	1100

a   Draw graphs to represent these data.
b   Calculate statistics to compare these data.
c   Use your graphs and statistics and compare the numbers of student absences in these two schools.

235

Exercise commentary

The questions assess the objectives on Framework Pages 269, 271 and 273.

Problem solving
Questions 1, 2 and 3 provide an opportunity to assess Framework Pages 25 and 27.

Group work
Question 3 is suitable for group work. Students can collect data for the further activity while they work on this question.

Misconceptions
Students may need reminding how to find averages using a frequency distribution.

Links
Interpreting graphs links to work on interpreting graphs in sections A5.5, A5.7 and A5.8.
This topic links to practical investigations in science, geography and PSHE.

Homework

D3.4HW requires students to comment on their music survey and provide interpretation.

Answers

1   Number of deaths increased with increase in sulphur levels, but stayed high as levels dropped
2   Around age at which walk to school without parents accident rate rises, then falls in mid-teens
    c   Mean 9.67; Median 10; Mode 12; use median

Summary

The key objectives for this unit are:
▸ Design a survey or experiment to capture the necessary data from one or more sources.
▸ Determine the sample size and degree of accuracy needed.
▸ Design, trial and if necessary refine data collection sheets.
▸ Communicate interpretations and results of a statistical enquiry using selected tables, graphs and diagrams in support.
▸ Present a concise, reasoned argument, using symbols, diagrams, graphs and related explanatory text.

Check out commentary

1 Emphasise that if you are to collect meaningful data then it is important to know why you are collecting it – you need an aim or a hypothesis first. Students should describe their survey in question 1, and state clearly whether they are going to use primary or secondary data.

2 Students should appreciate that the move relevant data they collect, the more reliable their results will be. However, encourage them to consider practical factors, in particular the time needed to actually carry out the data collection.

3 You may need to remind students that wrist circumference and hand span are continuous data, and this should help to inform their table layout. You may need to revise inequalities, and discuss sensible intervals.

Plenary activity

Discuss situations where anthropomorphic data (body measurements) are important. For example the design of irons and kettles; determining the variable position of a car seat (also the gear stick and pedals); clothing sizes.

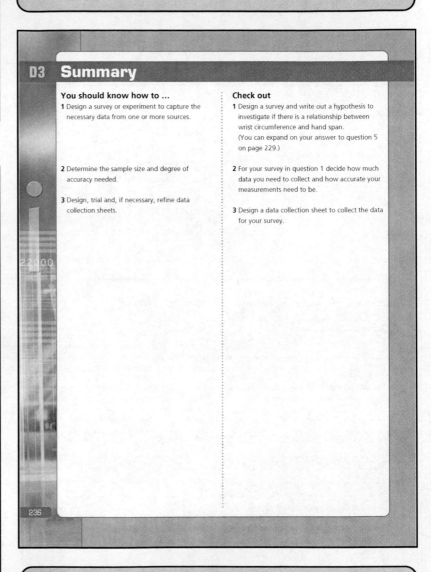

D3 Summary

You should know how to ...

1 Design a survey or experiment to capture the necessary data from one or more sources.

2 Determine the sample size and degree of accuracy needed.

3 Design, trial and, if necessary, refine data collection sheets.

Check out

1 Design a survey and write out a hypothesis to investigate if there is a relationship between wrist circumference and hand span.
(You can expand on your answer to question 5 on page 229.)

2 For your survey in question 1 decide how much data you need to collect and how accurate your measurements need to be.

3 Design a data collection sheet to collect the data for your survey.

236

Development

The themes of this unit are developed further in D4.

Links

The work in this unit links to statistical project work in science, geography and PSHE.
Interpreting graphs links to work on interpreting graphs in sections A5.5, A5.7 and A5.8.

Mental starters

Objectives covered in this unit:
▶ Order, add, subtract, multiply and divide integers.
▶ Round numbers, including to one or two decimal places.

Resources needed

* means class set needed
Essential:
Dice*
Sets of dominoes*
Scientific calculator*
D4.1OHP – possibility space diagram
and bar chart
Useful:
Mini-whiteboards
Packet of sweets
Bag of rice
Bag of marbles
Graph or squared paper
Paint and small paintbrush

D4 Experiments

This unit will show you how to:

▶▶ Design an experiment to capture the necessary data from one or more sources.
▶▶ Determine the sample size and degree of accuracy needed.
▶▶ Gather data from ICT-based sources. Identify all the mutually exclusive outcomes of an experiment.
▶▶ Compare experimental and theoretical probabilities in a range of contexts.

▶▶ Appreciate the difference between mathematical explanation and experimental evidence.
▶▶ Solve increasingly demanding problems and evaluate solutions.
▶▶ Solve substantial problems by breaking them into simpler tasks.
▶▶ Present a concise, reasoned argument, using symbols, diagrams, graphs and related explanatory text.

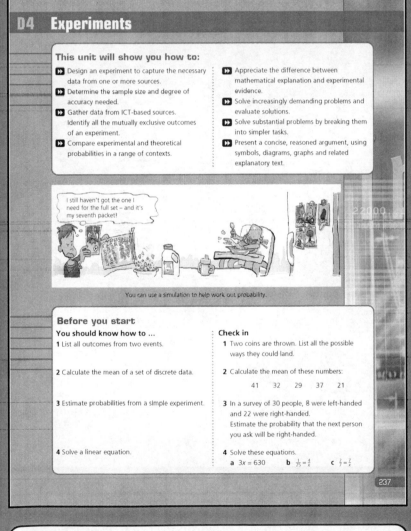

I still haven't got the one I need for the full set – and it's my seventh packet!

You can use a simulation to help work out probability.

Before you start

You should know how to ...
1 List all outcomes from two events.

2 Calculate the mean of a set of discrete data.

3 Estimate probabilities from a simple experiment.

4 Solve a linear equation.

Check in
1 Two coins are thrown. List all the possible ways they could land.

2 Calculate the mean of these numbers:
41   32   29   37   21

3 In a survey of 30 people, 8 were left-handed and 22 were right-handed.
Estimate the probability that the next person you ask will be right-handed.

4 Solve these equations.
a $3x = 630$ b $\frac{1}{25} = \frac{4}{x}$ c $\frac{2}{7} = \frac{7}{x}$

237

Unit commentary

Aim of the unit
This unit sets students practical experiments that enable them to compare experimental and theoretical probability and investigate sample sizes for experiments.

Introduction
Discuss situations in which the expected theoretical result does not always happen. For example, if the first five throws of a dice result in 1, 2, 3, 4 and 5, is the next throw likely to result in a six?

Framework references
This unit focuses on:
▶ Teaching objectives Pages: 283–285

▶ Problem-solving objectives Pages 5, 23, 31, 33.

Differentiation

Core tier
Focuses on determining probabilities in the context of a game of chance.

Check in activity

Ask: How many throws of a dice to you need before you get a six?
Discuss what students think, and then split them into pairs.
Each pair has a dice and a piece of paper to record their results.
Pool the class results and take an average.
Is the answer as expected?

Mental starter

Play the game 'two-dice bingo' (rules outlined in student book).

Useful resources

Mini-whiteboards may be useful for the mental starter.
Dice
Dominoes
D4.1OHP – the possibility space diagram and bar chart from the student book

Introductory activity

Discuss strategies for winning the game 'two-dice bingo' (refer to the mental starter and the student book).

Discuss how you can use theoretical probability to choose which numbers give the best chance of winning.

Emphasise that the theoretical best chance of winning is not necessarily the same as what may happen in a practical situation.

Emphasise that this is especially so when the sample is small, such as the number of turns it would take to win a game of 'two-dice bingo'.

D4.1OHP contains the possibility space diagram and bar chart from the student book.
Use this to illustrate how, by listing all the possible outcomes, you can evaluate the theoretical probabilities.

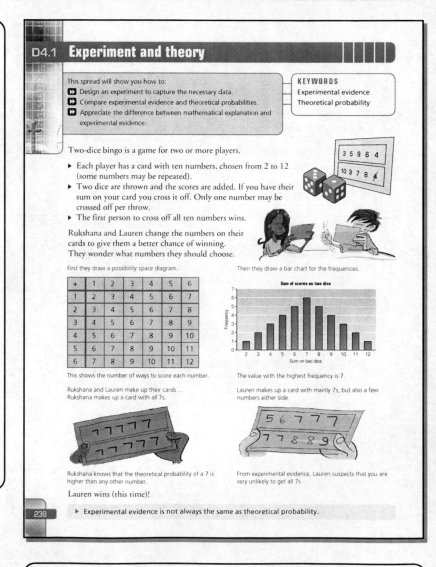

D4.1 **Experiment and theory**

This spread will show you how to:
▶ Design an experiment to capture the necessary data.
▶ Compare experimental evidence and theoretical probabilities.
▶ Appreciate the difference between mathematical explanation and experimental evidence.

KEYWORDS
Experimental evidence
Theoretical probability

Two-dice bingo is a game for two or more players.
▶ Each player has a card with ten numbers, chosen from 2 to 12 (some numbers may be repeated).
▶ Two dice are thrown and the scores are added. If you have their sum on your card you cross it off. Only one number may be crossed off per throw.
▶ The first person to cross off all ten numbers wins.

Rukshana and Lauren change the numbers on their cards to give them a better chance of winning. They wonder what numbers they should choose.

First they draw a possibility space diagram.

+	1	2	3	4	5	6
1	2	3	4	5	6	7
2	3	4	5	6	7	8
3	4	5	6	7	8	9
4	5	6	7	8	9	10
5	6	7	8	9	10	11
6	7	8	9	10	11	12

This shows the number of ways to score each number.

Then they draw a bar chart for the frequencies.

The value with the highest frequency is 7.

Rukshana and Lauren make up their cards ... Rukshana makes up a card with all 7s.

Lauren makes up a card with mainly 7s, but also a few numbers either side.

Rukshana knows that the theoretical probability of a 7 is higher than any other number.

From experimental evidence, Lauren suspects that you are very unlikely to get all 7s.

Lauren wins (this time)!

238

▶ Experimental evidence is not always the same as theoretical probability.

Plenary

Discuss which of the strategies students devised were best at winning games of two-dice bingo. Ask students to explain the reasoning behind their strategies.

Further activities

Students could play domino bingo in small groups, using the strategies devised in question 1.

Differentiation

Extension questions:
▸ Question 1 requires students to draw a possibility space diagram.
▸ Question 2 focuses on comparing theoretical and experimental probabilities.
▸ Question 3 extends the task.

Core tier: focuses on statistical modelling, in the context of a game show.

Exercise D4.1

In the game of dominoes there are 28 pieces.
Each piece is divided into two halves.
Each half is marked with between 0 and 6 dots.

1 The sum of the dots on a domino is from 0 to 12 inclusive.
   a Draw a sample space diagram to show the different ways you can get each sum of dots on a domino.

		1st half					
		0	1	2	3	4	5	6
2nd half	0							
	1							

   b Draw a frequency table to show the total frequency of each score.

Total score	0	1	2
Frequency			

   c Devise, with reasons, a strategy to win a game similar to two-dice bingo, but where the sum of dots on a domino is used in place of two dice.
   d In pairs, use your strategy to play several games of domino bingo.
   e Comment on your strategy and that of your partner.

2 Courtney has a bag of 28 dominoes.
   He chooses a domino at random, makes a note of the total sum, and replaces it in the bag.
   He repeats this 100 times.
   His results are shown in the table.

Sum	0	1	2	3	4	5	6	7	8	9	10	11	12
Frequency	2	3	7	7	11	13	14	12	10	6	5	6	4

   a Use the frequency table in question **1b** to calculate the theoretical probability of each score.
   b Compare the theoretical and experimental probabilities. Comment on your answer.

3 Poppy invents an enlarged domino set in which each half is marked with between 0 and 7 dots.
   a How many pieces will there be?
   b Draw a sample space diagram to show the different total scores.
   c Draw a bar chart to show the frequency of each score.

239

Exercise commentary

The questions assess the objectives on Framework Pages 283 and 285.

Problem solving

The exercise and further activity provide an opportunity to assess Framework Page 23. Questions 1 and 2 also assess Page 31.

Group work

Question 1d requires students to work in pairs.
The suggested further activity is suitable for work in small groups.

Misconceptions

Students may not list all possible outcomes.
You may need to offer guidance in designing a possibility space diagram.
When comparing probabilities, emphasise that they are unlikely to be the same.

Links

Problem solving: Framework Page 23.

Homework

D4.1HW requires students to compare experiment with theory in the context of a charity game.

Answers

1 b Frequency: 1, 1, 2, 2, 3, 3, 4, 3, 3, 2, 2, 1, 1
2 a $\frac{1}{28}, \frac{1}{28}, \frac{1}{14}, \frac{1}{14}, \frac{3}{28}, \frac{3}{28}, \frac{1}{7}, \frac{3}{28}, \frac{3}{28}, \frac{1}{4}, \frac{1}{4}, \frac{1}{28}, \frac{1}{28}$
  b Decimal probabilities are similar. Dominces have been picked at random.
3 a 36
  c Frequency for totals 0–14: 1, 1, 2, 2, 3, 3, 3, 4, 4, 4, 3, 3, 2, 2, 1, 1, 1

Mental starter

Assign the numbers 1–6 to six students.
Use a dice to choose which two students receive a sweet.
Run the simulation several times.

Useful resources

Packet of sweets
Dice
Scientific calculator

Introductory activity

Use the mental starter to introduce the concept of simulation.

Relate the experiment outlined in the student book to current promotions in breakfast cereals.

Refer to the student's page to discuss how you can experiment using a simulation to discover how big a reliable sample size needs to be.

Discuss that the experiment outlined assumes that there are equal numbers of each toy car.

Discuss how you might assign numbers if the proportions of colours were not equal – you may need to introduce random numbers.

Run the simulation in class (or a similar one using a current promotion).

Emphasise that you need to carry out lots of trials to get a reliable result.

Discuss the class simulation.
How many boxes of cereal would be needed?

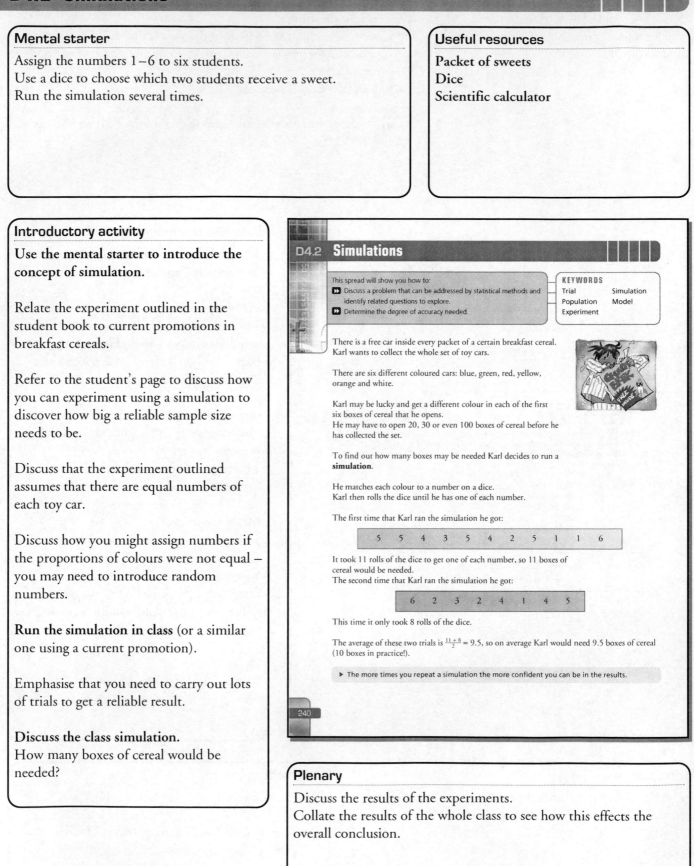

D4.2 Simulations

This spread will show you how to:
▶ Discuss a problem that can be addressed by statistical methods and identify related questions to explore.
▶ Determine the degree of accuracy needed.

KEYWORDS
Trial   Simulation
Population   Model
Experiment

There is a free car inside every packet of a certain breakfast cereal. Karl wants to collect the whole set of toy cars.

There are six different coloured cars: blue, green, red, yellow, orange and white.

Karl may be lucky and get a different colour in each of the first six boxes of cereal that he opens.
He may have to open 20, 30 or even 100 boxes of cereal before he has collected the set.

To find out how many boxes may be needed Karl decides to run a **simulation**.

He matches each colour to a number on a dice.
Karl then rolls the dice until he has one of each number.

The first time that Karl ran the simulation he got:

| 5 | 5 | 4 | 3 | 5 | 4 | 2 | 5 | 1 | 1 | 6 |

It took 11 rolls of the dice to get one of each number, so 11 boxes of cereal would be needed.
The second time that Karl ran the simulation he got:

| 6 | 2 | 3 | 2 | 4 | 1 | 4 | 5 |

This time it only took 8 rolls of the dice.

The average of these two trials is $\frac{11+8}{2} = 9.5$, so on average Karl would need 9.5 boxes of cereal (10 boxes in practice!).

▶ The more times you repeat a simulation the more confident you can be in the results.

240

Plenary

Discuss the results of the experiments.
Collate the results of the whole class to see how this effects the overall conclusion.

Further activities

Pool your results from question 2 with the results of other students in your class and comment on the simulation results when you have a larger amount of data.

Differentiation

Extension questions:
The focus of this exercise is to experiment with simulation and to find out many times you need to run a simulation to get consistent results.
Question 1 requires a dice, and question 2 requires a scientific calculator.

Core tier: focuses on analysing results of a probability experiment.

Exercise D4.2

1 A restaurant offers three starters on its set menu: prawn cocktail, melon and soup.

Menu
Starters
Prawn cocktail
Melon
Soup of the day

a Describe how you could use a dice to simulate the choice of starter that people might choose.
What assumptions do you need to make?
b Suppose that 20 people choose the set menu one evening.
Use a dice to carry out a simulation and find out how many of each starter the chef may need.
c Compare your answers with others in your class.

Over a long period of time the chef notices that, on average out of every six customers, 3 choose prawn cocktail, 2 choose melon and 1 chooses soup.

d Describe how you could use a dice to simulate the choice of starter that people might choose.
e Carry out this simulation for a group of 20 people.

2 Find the function on a scientific calculator that generates random digits.
Each time you press this key a different number between 0 and 1 is shown.
You are going to run a simulation using only the **first** digit after the decimal point that your calculator generates.

0.548

The restaurant offers four desserts.

Desserts
Chocolate fudge cake
Ice-cream sundae
Crème caramel
Fresh fruit salad

Over a long period of time the chef notices that:

40% of customers choose chocolate fudge cake.
30% of customers choose ice-cream sundae.
20% of customers choose creme caramel.
10% of customers choose fresh fruit salad.

a Assign the digits 0 to 9 to the choices for dessert in the proportions given.
b Use your calculator to simulate the choices made by 50 customers at the restaurant.
c Compare your results with other people in your class.

241

Exercise commentary

The questions assess the objectives on Framework Page 285.

Problem solving
All questions in this exercise provide an opportunity to assess Framework Page 23.

Group work
All questions are suitable for work in small groups.

Misconceptions
It does not matter which numbers are assigned to which items as long as the numbers are assigned before numbers are generated.
Students may need help in assigning numbers when items are not in equal proportions.

Links
The topic of simulation links to probability: Framework Pages 277–285.

Homework

D4.2HW is a simulation activity based on the Donkey Derby game at Brighton Pier.

Answers

1 a Assign 2 numbers to each starter, assuming each is equally likely to be chosen.
  d Assign digits in proportion 3 : 2 : 1.
2 a Assign digits in proportion 4 : 3 : 2 : 1.

Mental starter

Rearrange and solve equations of the type
$$\frac{4}{7} = \frac{10}{n}$$
You may need to briefly recap reciprocals and cross-multiplying.

Useful resources

Mini-whiteboards may be useful for the mental starter
Bag of rice
Bag of marbles or **pasta shapes**
Paint and paintbrush (small).

Introductory activity

Ask students to estimate how many grains of rice there are in a bag.

Discuss the capture-recapture method of estimating the size of a population.
Highlight the important elements:
▸ Select an initial sample
▸ Tag the sample in some way
▸ Return the sample to the population
▸ Allow the population to thoroughly mix
▸ Select a second sample (not necessarily the same size)
▸ Count how many are tagged.

Demonstrate, using the example of Alex's bag of rice as outlined in the student book, how the method can be applied in a real scenario.

Referring to the mental starter, show how you can calculate an estimate for the total population.
Try the experiment with a small bag of rice. You will probably need to use about a quarter of 125 g bag of rice, and paint the selected grains.
Allow the paint to dry during the lesson, and wait until the end of the lesson to complete the experiment.

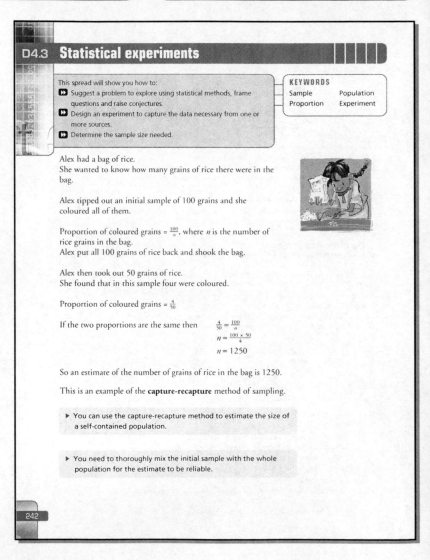

D4.3 Statistical experiments

This spread will show you how to:
▸▸ Suggest a problem to explore using statistical methods, frame questions and raise conjectures.
▸▸ Design an experiment to capture the data necessary from one or more sources.
▸▸ Determine the sample size needed.

KEYWORDS
Sample   Population
Proportion   Experiment

Alex had a bag of rice.
She wanted to know how many grains of rice there were in the bag.

Alex tipped out an initial sample of 100 grains and she coloured all of them.

Proportion of coloured grains = $\frac{100}{n}$, where n is the number of rice grains in the bag.
Alex put all 100 grains of rice back and shook the bag.

Alex then took out 50 grains of rice.
She found that in this sample four were coloured.

Proportion of coloured grains = $\frac{4}{50}$

If the two proportions are the same then
$$\frac{4}{50} = \frac{100}{n}$$
$$n = \frac{100 \times 50}{4}$$
$$n = 1250$$

So an estimate of the number of grains of rice in the bag is 1250.

This is an example of the **capture-recapture** method of sampling.

▸ You can use the capture-recapture method to estimate the size of a self-contained population.

▸ You need to thoroughly mix the initial sample with the whole population for the estimate to be reliable.

242

Plenary

Collate the results of the experiment in question 4 and try to find out how many trials were needed to be reasonably happy that estimates were accurate.

Alternatively
Complete the rice experiment outlined in the introductory activity.

Further activities

Question 4 can be extended to a variety of similar objects.

Differentiation

Extension questions:
The focus of this exercise is to use the capture-recapture method to estimate population size.

▶ Questions 1–3 practise solving problems.
▶ Question 4 is a group activity.

Core tier: focuses on probability simulations in the context of a game show.

Exercise D4.3

1   Lewis wanted to estimate the number of deer living in an area of woodland.
    A sample of 30 of the deer were caught, tagged and set free.
    The next day a second sample of 12 deer were caught.
    Five of this sample were tagged.
    Estimate the number of deer in this area of woodland.

2   At a bird sanctuary, 60 ducks were caught, tagged and returned to the sanctuary.
    In a second sample of 100 ducks, 24 were tagged.
    Find an estimate of the number of ducks in the bird sanctuary.

3   Jenna wanted to estimate how many fish were living in her pond.
    She caught an initial sample of 12 fish, tagged them and returned them to the pond.

    After a short while the fish had all intermingled and Jenna caught a second sample of eight fish. In this sample three were tagged.
    a   Find an estimate of how many fish are living in the pond.

    Later that day Jenna took a sample of ten fish. In this sample four were tagged.
    b   Use this sample to find an estimate of the number of fish in the pond.
    c   How could Jenna obtain a better estimate of the number of fish in her pond?

Experiment

4   This is an experiment to do in small groups.
    a   Take a number of like objects , such as uncooked pasta or marbles, and place them in a large bag.

> Three-coloured pasta (tricolore) is widely available in supermarkets.

   b   Remove some of the objects and mark them in some way to make them different from the rest.
   c   Put the objects back in the bag and shake up the bag.
   d   Each person in the group takes turns to remove a sample of the objects (you decide for yourself how big you want the sample to be), and then finds an estimate for the total number in the bag.
   e   Compare your estimates and decide on the best estimate for your group.
   f   Tip out all the objects and count them to find how close your estimate was to the actual number in the bag.
   g   Compare the estimate of your group with the estimates of other groups in the class.

Exercise commentary

The questions assess the objectives on Framework Pages 79, 81 and 285.

Problem solving
Question 4 provides an opportunity to assess Framework Pages 5 and 23. The exercise assesses Page 33.

Group work
Question 4 is designed for students to work in groups.

Misconceptions
Students may have problems setting up and solving an equation with the unknown as denominator.
As a check the total population will always be greater than the sample taken.
Students should be encouraged to develop their own strategies for ensuring that samples are as random as possible.

Links
Transforming algebraic expressions: Framework Pages 117–119.

Homework

D4.3HW requires students to research Charles Babbage and Ada Lovelace.

Answers

1   72
2   250
3   **a** 32 **b** 30 **c** Find mean average of 2 trials; carry out more trials and find the average of all trials.

Summary

The key objectives for this unit are:
▶ Design a survey or experiment to capture the necessary data from one or more sources.
▶ Determine the sample size and degree of accuracy needed.
▶ Communicate interpretations and results of a statistical enquiry using selected tables, graphs and diagrams in support.
▶ Solve substantial problems by breaking them into simpler tasks.
▶ Present a concise, reasoned argument, using symbols, diagrams, graphs and related explanatory text.

Check out commentary

1a Encourage students to start by fixing the number of possible outcomes. For example, there may be 10 possible choices, only one of which results in a correct guess.

Students can then use the first digit of the random number generator on a calculator to simulate the amount of money made (or lost).

Vary the number of possible outcomes and examine the effect on the profit.

1b Students will need to allocate digits to the three possible outcomes using a weighting indicated by the probabilities. Students often need reminding that zero is a digit.

Encourage students to repeat the simulation a number of times, and take an average.

Plenary activity

Discuss how students might estimate the number of animals (for example birds, squirrels or foxes) that visit the school playground after lunch.

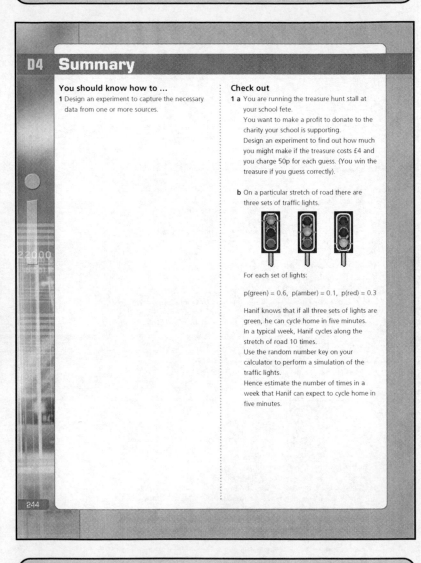

D4 Summary

You should know how to ...

1 Design an experiment to capture the necessary data from one or more sources.

Check out

1 a You are running the treasure hunt stall at your school fete.

You want to make a profit to donate to the charity your school is supporting.

Design an experiment to find out how much you might make if the treasure costs £4 and you charge 50p for each guess. (You win the treasure if you guess correctly).

b On a particular stretch of road there are three sets of traffic lights.

For each set of lights:

p(green) = 0.6, p(amber) = 0.1, p(red) = 0.3

Hanif knows that if all three sets of lights are green, he can cycle home in five minutes.
In a typical week, Hanif cycles along the stretch of road 10 times.
Use the random number key on your calculator to perform a simulation of the traffic lights.
Hence estimate the number of times in a week that Hanif can expect to cycle home in five minutes.

244

Development

The themes of this unit are developed in Year 10 with relative frequency.

Links

The work in this unit links to problem solving (Framework Page 23).
Statistical methods are used in science, geography and PSHE.
Simulation has strong links to ICT.

Glossary

algebra	Algebra is the branch of mathematics where symbols or letters are used to represent numbers.
algebraic expression A2.5	An algebraic expression is a term, or several terms connected by plus and minus signs.
alternate S1.1, S1.2	A pair of alternate angles is formed when a straight line crosses a pair of parallel lines. Alternate angles are equal.

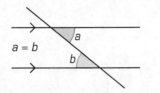

$a = b$

angle: acute, obtuse, right, reflex	An angle is formed when two straight lines cross or meet each other at a point. The size of an angle is measured by the amount one line has been turned in relation to the other.

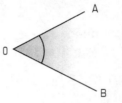

An acute angle is less than 90°.

An obtuse angle is more than 90° but less than 180°.

A right angle is a quarter of a turn, or 90°.

A reflex angle is more than 180° but less than 360°.

angles on a straight line S1.1, S1.2	Angles on a straight line add up to 180°.

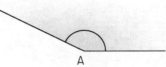

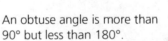

$a + b = 180°$

approximate N3.5, A4.4	To approximate an answer is to work out a rough answer using easier figures.
arc S1.4	An arc is part of a curve.
area S2.1	The area of a surface is a measure of its size. Square millimetre, square centimetre, square metre, square kilometre are all units of area.

Glossary

bar chart
D2.6, D4.2

The heights of the bars on a bar chart represent the frequencies of the data.

base (number)
A2.1

The base is the number which is raised to a power.
For example in 2^3, 2 is the base.

bearing, three-figure bearing
S4.5

A bearing is a clockwise angle measured from the North line giving the direction of a point from a reference point.
A bearing should always have three digits.

N
120°
A
B

The bearing of B from A is 120°.

bias
D1.6

An experiment or selection is biased if not all outcomes are equally likely.

BIDMAS
A2.1, A4.2, N4.3

BIDMAS is a mnemonic to remind you of the correct order of operations: **b**rackets, **i**ndices, **d**ivision or **m**ultiplication, **a**ddition or **s**ubtraction.

bisect, bisector
S4.7

To bisect is to cut in half.
A bisector is a line that cuts something in half.

cancel
N4.5

You cancel a fraction by dividing the numerator and denominator by a common factor.

capacity
S2.2

Capacity is a measure of how much liquid a hollow 3-D shape can hold.

centre
S4.8

The centre of a circle is the point from which all points on the circumference are equidistant.

centre of rotation
S3.1

The centre of rotation is the fixed point about which a rotation takes place.

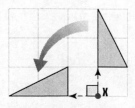

chord
S1.4

A chord is a straight line joining two points on a curve, or a circle.

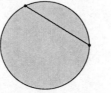

circumference
S1.4, S2.3, S2.4

The circumference is the distance around the edge of a circle.

coefficient
A2.3, A3.2, A3.3,

The coefficient is the number part of an algebraic term.
For example in $3n^5$ the coefficient of n^5 is 3.

collect like terms
A2.3

To collect like terms is to put together terms with the same letter parts. For example $5x + 3x = 8x$ and $4y^2 - y^2 = 3y^2$.

common factor
NA1.3, A2.3

A common factor is a factor of two or more numbers or terms. For example $2p$ is a common factor of $2p^2$ and $6p$.

commutative
N4.1, N4.2

An operation is commutative if the order of combining two terms does not matter.

compasses
S1.5

A pair of compasses is a geometrical instrument used to draw circles or arcs.

compensation
N3.3

The method of compensation is used to make calculations easier. For example to add 99, add 100 and then compensate by subtracting 1.

conclude, conclusion
A5.2

To conclude is to formulate a result or conclusion based on evidence.

congruent
S3.1, S3.2, S3.3

Congruent shapes are exactly the same shape and size.

constant
A3.3

A constant is an algebraic term that remains unchanged.
For example, in the expression $5x + 3$ the constant is 3.

construction lines
S1.5

Construction lines are the arcs drawn when making an accurate diagram.

continuous
D3.3

Continuous data can take any value between given limits, for example height.

coordinates
S4.6

The coordinates of a point give its position in terms of its distance from the origin along the x- and y- axes.

correlation
D2.4

Correlation is a measure of the relationship between two variables.

corresponding
S1.1, S1.2

A pair of corresponding angles is formed when a straight line crosses a pair of parallel lines.
Corresponding angles are equal.

$a = b$

247

Glossary

counter-example
P1.3

A counter-example is an example that shows that a rule does not work.

cross-multiply
A4.1

Cross-multiplying is a method for removing fractions from equations.

cross-section
S2.5

The cross-section of a solid is the shape of its transverse section.

cube root
NA1.4

The cube root of x is the number that when cubed gives you x. For example $\sqrt[3]{64} = 4$, because $4 \times 4 \times 4 = 64$.

cubic
S2.5

A cubic equation contains a term in x^3 as its highest power.

data
D2.1, D3.1

Data are pieces of information.

data collection sheet
D3.2, D3.4

A data collection sheet is a form designed for the systematic collection of data.

data logging
D3.2

Data logging is the automatic collection of data.

decagon
S1.3

A decagon has ten sides.

decimal
N2.1, N2.6, N3.6, N4.1

A decimal number is a number written using base ten notation.

decimal place
N3.2

Each column after the decimal point is called a decimal place.

degree (°)

Angles are measured in degrees. There are 360° in a full turn.

degree of accuracy
N3.2

The degree of accuracy of an answer depends on the accuracy of the figures used in the calculation.

denominator
N2.2, A4.2

The denominator is the bottom number in a fraction. It shows how many parts there are in the whole.

diagonal
A3.1

A diagonal line is one which is neither horizontal nor vertical.

diameter
S1.4, S2.3, S2.4

The diameter is a chord that passes through the centre of a circle.

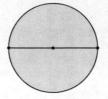

difference pattern
NA1.5, NA1.6

You can find a general rule for a sequence by looking at the pattern of differences between consecutive terms.

digit
N3.1

A digit is any of the numbers 0, 1, 2, 3, 4, 5, 6, 7, 8, 9.

dimension
P1.6

A dimension is a length, width or height of a shape or solid.

direct proportion
A5.5, A5.6

Two quantities are in direct proportion if one quantity increases at the same rate as the other.

distance-time graph
A3.6

A distance-time graph is a graph of distance travelled against time taken.
Time is plotted on the horizontal axis.

distribution
D3.4

A distribution is a set of observations of a variable.

divisible, divisibility
NA1.3

A whole number is divisible by another if there is no remainder after division.

divisor

The divisor is the number that does the dividing.
For example, in 14 ÷ 2 = 7 the divisor is 2.

edge (of solid)
S4.1

An edge is a line along which two faces of a solid meet.

edge

elevation
S4.2

An elevation is an accurate drawing of the side or front of a solid.

enlargement
33.5, S3.6

An enlargement is a transformation that multiplies all the sides of a shape by the same scale factor.

equation
A2.6

An equation is a statement showing that two expressions have the same value.

equation (of a graph)
A3.3

An equation is a statement showing the relationship between the variables on the axes.

equidistant
S4.7, S4.8

Equidistant means the same distance apart.

equivalent, equivalence
N3.5, N3.6

Two quantities, such as fractions which are equal, but are expressed differently, are equivalent.

estimate
S2.1, N3.2, N3.5, N3.6,
N4.4, N4.5, N4.6

An estimate is an approximate answer.

evaluate
P1.1, A4.5

Evaluate means find the value of an expression.

event
D1.1, D1.2

In probability an event is a trial or experiment.

expand
A2.4

To expand an expression you remove all the brackets.

Glossary

experiment
D4.2, D4.3

An experiment is a test or investigation to gather evidence for or against a theory.

experimental probability
D1.4, D1.5, D1.6

You can find the experimental probability of an event by conducting trials.

expression
A2.1, A2.4

An expression is a collection of terms linked with operations but with no equals sign.

exterior angle
S1.2, S1.3

An exterior angle is made by extending one side of a shape.

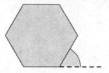

face
S4.1

A face is a flat surface of a solid.

face

factor
NA1.2, A2.4, N3.4

A factor is a number that divides exactly into another number. For example, 3 and 7 are factors of 21.

factorise
A2.4, A5.3

You factorise an expression by writing it with a common factor outside brackets.

formula, formulae
NA1.6, A2.4, S2.1, A4.5,
A4.6, A5.1

A formula is a statement that links variables.

frequency
D4.2

The frequency is the number of times an event occurs.

frequency diagram
D3.3

A frequency diagram uses bars to display data.
The height of the bars corresponds to the frequencies.

function, linear function
A5.5, A5.7

A function is a rule.
The graph of a linear function is a straight line.

generalise
A5.1, A5.2, P1.2, P1.3

Generalise means find a statement or rule that applies to all cases.

general term
NA1.6

The general term in a sequence allows you to evaluate unknown terms.

gradient
A3.1, A3.2, A3.3, A3.4,
A3.6, A5.5, A5.7

Gradient is a measure of the steepness of a line.

graph
A3.1

A graph is a diagram that shows a relationship between variables.

greater than or equal to (⩾)
A4.4

The symbol ⩾ means that the term on the left-hand side is greater than or equal to the term on the right-hand side.

hectare
S2.1, N4.6

A hectare is a unit of area equal to 10 000 (100 × 100) square metres.

hexagon
S1.3

A hexagon has six sides.

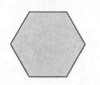

highest common factor (HCF)
NA1.3, A2.5, A5.3

The highest common factor is the largest factor that is common to two or more numbers.
For example the HCF of 12 and 8 is 4.

horizontal
A3.1, S4.9

A horizontal line is parallel to the ground.

hypotenuse
S1.6, S4.9

The hypotenuse is the side opposite the right angle in a right-angled triangle.

hypothesis
D3.1

A hypothesis is a statement used as a starting point for a statistical investigation.

identically equal to (≡)
A5.3

One expression is identically equal to another if they are mathematically equivalent.

identity
A5.3

An identity is an equation which is true for all possible values.
For example $3x + 6 \equiv 3(x + 2)$ for all values of x.

image
S3.6

An image is an object after it has been transformed.

implicit
A5.8

An equation in x and y is in implicit form if y is not the subject of the equation.

improper fraction
A4.2

In an improper fraction the numerator is bigger than the denominator.

index, indices
A2.1, A2.2

The index tells you how many of a quantity must be multiplied together. For example x^3 means $x \times x \times x$.

index laws
NA1.4, A2.3

To multiply powers of the same base add the indices, for example $2^5 \times 2^3 = 2^8$.
To divide powers of the same base subtract the indices.
For example $5^6 \div 5^3 = 5^2$.

index notation
NA1.4, A2.2

A number written as a power of a base number is expressed in index notation, for example $\frac{1}{1000} = 10^{-3}$.

inequality
NA1.1, A4.4, A5.4

An inequality is a relationship between two numbers or terms that are comparable but not equal.
For example, $7 > 4$.

infer
D2.3

Infer means to conclude from evidence.

inscribe, inscribed
S1.4

An inscribed polygon has every vertex lying on the perimeter of a shape, such as a circle.

Glossary

integer
NA1.1, N2.3, N3.5, N3.6,
A4.4, N4.1, N4.5

An integer is a positive or negative whole number (including zero).
The integers are: ..., ⁻3, ⁻2, ⁻1, 0, 1, 2, 3, ...

intercept
A3.1, A3.3, A5.7

The intercept is the point at which a graph crosses an axis.

interior angle
S1.2, S1.3

An interior angle is inside a shape, between two
adjacent sides.

interpret
D2.3

You interpret data whenever you make sense of it.

intersection
S1.6

The intersection of two lines is the point where they cross.

inverse function
A3.1

An inverse function acts in reverse to a
specified function.

function
input   output
x y
inverse
function

justify
NA1.5, A2.5, N3.5, N3.6,
N4.3, N4.4, A5.2

You justify a solution of a formula by explaining why it is correct.

less than or equal to (⩽)
A4.4

The symbol ⩽ means that the term on the left-hand side is less than
or equal to the term on the right-hand side.

like terms
A4.2

Like terms are terms with the same letter parts, for example $3x^2$ and
$-5x^2$ are like terms.

line graph
D2.6

On a line graph points are joined with straight lines.

line of best fit
A5.7

A line of best fit passes through the points on a scatter graph,
leaving roughly as many above the line as below it.

line segment
S4.6

A line segment is the part of a line between two points.

linear equation, linear graph
A4.1, A4.3, A5.7

A linear equation contains no squared or higher terms.
The graph of a linear equation is a straight line.

linear expression
A2.3

A linear expression contains no square or higher terms, for example
$3x + 5$ is a linear expression.

linear sequence
NA1.5

The terms of a linear sequence increase by the same amount each
time.

locus, loci
S4.7, S4.8

A locus is a set of points (a line, a curve or a region) that satisfies certain conditions.

lowest common multiple (LCM)
NA1.3, N2.2

The lowest common multiple is the smallest multiple that is common to two or more numbers, for example the LCM of 4 and 6 is 12.

mapping
A3.1

A mapping is a rule that can be applied to a set of numbers to give another set of numbers.

mass
S2.2

The mass of an object is a measure of the quantity of matter in it.

mean
D2.3, D2.5, D3.3

The mean is the average value found by adding the data and dividing by the number of data items.

median
D2.3, D2.5, D3.3

The median is the average which is the middle value when the data is arranged in order of size.

metric system
N4.6, P1.6

In the metric system, units of measurement are related by multiples of ten.

mid-point
S4.6

The mid-point of a line segment is the point that is halfway along.

mirror line
S3.1

A mirror line is a line or axis of symmetry.

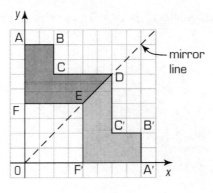

modal class
D2.3

The modal class is the most commonly occurring class when the data is grouped.
It is the class with the highest frequency.

mode
D2.3, D3.3

The mode is an average.
It is the value that occurs most often.

multiple
NA1.3

A multiple of an integer is the product of that integer and any other.
For example 12, 18 and 30 are multiples of 6.

multiple bar chart
D2.6

A multiple bar chart is a bar chart with two or more sets of bars.
It is used to compare two or more data sets.

mutually exclusive
D1.2

Two events are mutually exclusive if they cannot occur at the same time.

negative
NA1.1, A2.2

A negative number is a number less than zero.

net
S2.6, S4.2

A net is a 2-D shape that can be folded to make a 3-D solid.

nth term
NA1.6

The *n*th term is the general term of a sequence.

numerator
A4.2

The numerator is the top number in a fraction.
It tells you how many parts of the whole you have.

object, image
S3.1

The object is the original shape before a transformation.
An image is the shape after a transformation.

operation
N4.3

An operation is a rule for processing numbers.
The basic operations are addition, subtraction, multiplication and division.

order of operations
N4.3

The conventional order of operations is:
brackets first, then indices,
then division and multiplication,
then addition and subtraction.

order of rotational symmetry
S3.4

The order of rotation symmetry is the number of times that a shape will fit on to itself during a full turn.

origin

The origin is the point where the *x*- and *y*-axes cross, that is (0, 0).

outcome
D1.1, D1.2, D1.5

In probability an outcome is the result of a trial.

parallel
S1.1, A3.1,
A3.2, A3.4

Parallel lines are always the same distance apart.

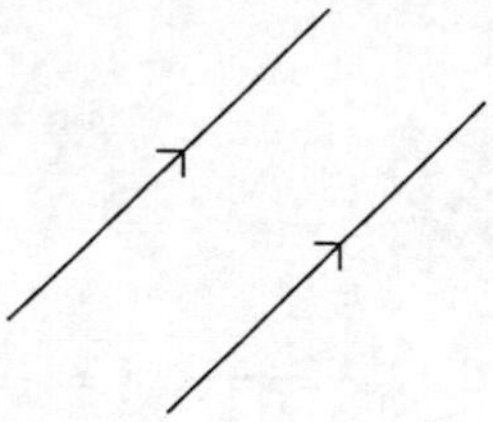

partitioning
N3.3, N3.4

Partitioning means splitting a number into smaller parts.

perimeter
A4.3, S3.6

The perimeter is the distance round the edge of a shape.

perpendicular
S1.6, A3.4

A line or plane is perpendicular to another line or plane if they meet at a right angle.

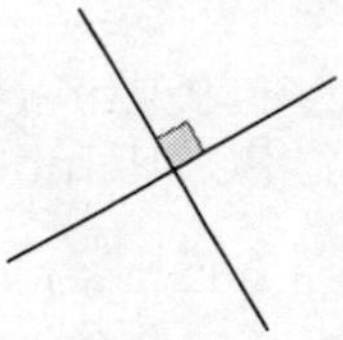

perpendicular bisector
S4.7

The perpendicular bisector of a line is the line that divides it into two equal parts and is at right angles to it.

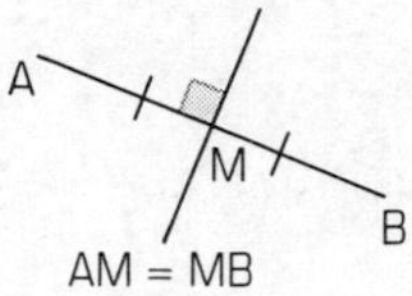

AM = MB

pi (π)
S2.3

The ratio $\frac{circumference}{diameter}$ is the same for all circles.
This ratio is denoted by the Greek letter π.

pie chart
D2.2

A pie chart is a circular diagram used to display data.
The angle in each sector is proportional to the frequency.

place value
N3.1

The place value is the value of a digit in a decimal number.
For example in 3.65 the digit 6 has a value of $\frac{6}{10}$.

plan, plan view
S4.2

The plan or plan view of a solid is an accurate drawing of the view from directly above.

plane
S4.1

A plane is a flat surface.

plane of symmetry
S3.4, S4.1

A plane of symmetry divides a solid into two halves.

polygon
S1.3

A polygon is a shape with three or more straight sides.

population
D4.2, D4.3

The population is the complete set of individuals from which a sample is drawn.

position-to-term rule
NA1.5

A position-to-term rule tells you how to calculate the value of a term if you know its position in the sequence.

positive
NA1.1, A2.2

A positive number is greater than zero.

power
NA1.4, NA1.6, A2.5, N3.1, N4.3

The power of a number or a term tells you how many of the number must be multiplied together.
For example 10 to the power 4 is 10 000.

primary data, primary source
D2.1, D3.1

Primary data is data you have collected yourself.

prime
NA1.3

A prime number is a number that has exactly two different factors.

prime factor
NA1.3

A prime factor is a factor that is a prime number.

prime factor decomposition
NA1.3

Prime factor decomposition means splitting a number into its prime factors.

prism
S2.5

A prism is a solid with a uniform cross-section.

product
NA1.2, N3.4

The product is the result of a multiplication.
For example, the product of 3 and 4 is 12.

proportion
A5.5, A5.6, N2.1, N2.6, N3.8, P1.4, P1.5, D4.3

A proportion compares the size of a part to the size of the whole.

Glossary

proportional to (∝)
N3.9

When two quantities are in direct proportion one quantity is proportional to the other.

prove, proof
S1.1

You prove a statement is true by arguing from known facts.

quadratic
A3.5

A quadratic expression contains a square term.

quadratic sequence
NA1.6

In a quadratic sequence the second difference is constant.

quadrilateral
S1.3

A quadrilateral is a polygon with four sides.

rectangle

All angles are right angles. Opposite sides equal.

parallelogram

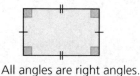

Two pairs of parallel sides.

kite

Two pairs of adjacent sides equal. No interior angle greater than 180°.

rhombus

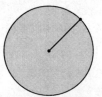

All sides the same length. Opposite angles equal.

square

All sides and angles equal.

trapezium

One pair of parallel sides.

quotient
NA1.2

A quotient is the result of a division.
For example, the quotient of $12 \div 5$ is $2\frac{2}{5}$, or 2.4.

radius
S1.4, S2.3, S2.4

The radius is the distance from the centre to the circumference of a circle.

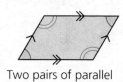

random process
D1.1

The outcome of a random process cannot be predicted.

range
D2.3, D2.5, D3.3

The range is the difference between the largest and smallest values in a set of data.

ratio
N3.7, N3.8, N3.9, S3.6, A5.6, P1.4, P1.5, P1.6, S4.4

A ratio compares the size of one part with the size of another part.

raw data
D2.5

Raw data is data before it has been processed.

reciprocal
A2.2, A3.4, N4.2, A5.6

The reciprocal of a quantity k is $1 \div k$.
For example the reciprocal of 5 is $\frac{1}{5}$ or 0.2; the reciprocal of x^2 is $\frac{1}{x^2}$.

recurring
N2.1

A recurring decimal has a repeating pattern of digits after the decimal point, for example 0.33333 ...

reflect, reflection
S3.1

A reflection is a transformation in which corresponding points in the object and the image are the same distance from the mirror line.

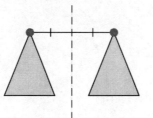

reflection symmetry
S3.4

A shape has reflection symmetry if it has a line of symmetry.

regular
S1.3

A regular polygon has equal sides and equal angles.

rotate, rotation
S3.1

A rotation is a transformation in which every point in the object turns through the same angle relative to a fixed point.

rotation symmetry
S3.4

A shape has rotation symmetry if when turned it fits onto itself more than once during a full turn.

right prism
S2.5

Apart from its two end faces, the faces of a right prism are all rectangles.

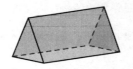

rounding
N3.2, N3.3, N3.5

You round a number by expressing it to a given degree of accuracy.

sample
D1.5, D4.1

A sample is a set of individuals or items drawn from a population.

sample space, sample space diagram
D1.1, D1.2

In probability the set of all possible outcomes in an experiment is called the sample space.
A sample space diagram is a diagram recording all the outcomes.

scale
S4.4

A scale gives the ratio between the size of an object and its diagram.

scale drawing
S4.5

A scale drawing is an accurate drawing of a shape to a given scale.

scale factor
N3.7, N3.9, S3.5, S3.6

A scale factor is a multiplier.

scatter graph
D2.4, D3.4

Pairs of variables, for example age and height, can be plotted on a scatter graph.

Glossary

secondary data, secondary source
D2.1, D3.1

Secondary data is data that someone else has collected. Common secondary sources include books, magazines and the Internet.

sector
S1.4, S2.4

A sector is part of a circle bounded by an arc and two radii.

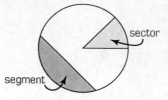

segment
S1.4

A segment is part of a circle bounded by an arc and a chord.

sequence
NA1.5, NA1.6

A sequence is a set of numbers, objects or terms that follow a rule.

similar, similarity
S2.3, A5.6

Similar shapes have the same shape but are different sizes.

simulation
D4.2

A simulation is an experiment designed to model a real-life situation.

simultaneous equations
A5.8

Simultaneous equations are two or more equations whose unknowns have the same values.

slope
A3.1

The slope of a line is measured by the angle it makes with the x-axis.

solid
S4.1, S4.2, S4.3

A solid is a shape formed in three-dimensional space.

cube

six square faces

cuboid

six rectangular faces

prism

the end faces are constant

pyramid

the faces meet at a common vertex

tetrahedron

all the faces are equilateral triangles

square-based pyramid

the base is a square

solution, solve
NA1.1, A2.6, A4.4, P1.1

The solution of an equation is the value that makes it true.

speed
A3.6

Speed is a measure of the rate at which distance is covered. It is often measured in miles per hour or metres per second.

sphere
S4.8

A sphere is a 3-D shape in which every point on its surface is equidistant from the centre.

square root
NA1.4

A square root is a number that when multiplied by itself is equal to a given number.
For example $\sqrt{25} = 5$, because $5 \times 5 = 25$.

steepness
A3.1

The steepness of a line depends on the angle the line makes with the x-axis.

stem-and-leaf diagram
D2.2, D2.3

A stem-and-leaf diagram is used to display raw data in numerical order.

straight-line graph
A3.1, A3.2, A3.3, A3.4, A5.8

A straight-line graph is the graph of a linear equation.

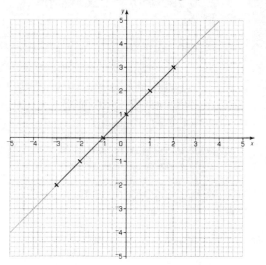

subject
A3.3, A4.5, A4.6, A5.8

The subject of an equation or formula is the term on its own in front of the equals sign. For example, the subject of $v = u + at$ is v.

substitute
A4.5, A5.4

To substitute is to replace a variable with a numerical value.

sum
N3.4

The sum is the total of an addition.

supplementary
S1.1

You can form a pair of supplementary angles on a straight line.
Supplementary angles add up to 180°.

$a + b = 180°$

Glossary

surface area
S2.6, S4.3

The surface area of a solid is the total area of its faces.

symmetry, symmetrical
S4.1

A shape is symmetrical if it is unchanged after a rotation or reflection.

T(n)
NA1.6

T(n) stands for the general term in a sequence.

term
A2.3

A term is a number or object in a sequence.
It is also part of an expression.

terminating

A terminating decimal has a limited number of digits after the decimal point.

tessellation
S3.2

A tessellation is a tiling pattern with no gaps.

theoretical probability
D1.1, D1.3, D1.6, D4.1

The theoretical probability of an event
$$= \frac{\text{number of favourable outcomes}}{\text{total possible number of outcomes}}$$

tonne
S2.2

The tonne is a unit of mass, equal to 1000 kg.

transform
A2.5

You transform an expression by taking out single-term common factors.

transformation
S3

A transformation moves a shape from one place to another.

translate, translation
S3.1

A translation is a transformation in which every point in an object moves the same distance and direction.
It is a sliding movement.

tree diagram
D1.2, D1.3

A tree diagram shows the possible outcomes of a probability experiment on branches.

trend
D2.4, D3.4

A trend is a general tendency.

trial
D1.1, D1.5, D4.1

In probability a trial is an experiment.

trial and improvement
NA1.4, A5.4

Square roots, cube roots and solutions to equations can be estimated by the method of trial and improvement.
An estimated solution is tried in the expression and refined by a better estimate until the required degree of accuracy is achieved.

triangle
S1.6, S4.9

A triangle is a polygon with three sides.

equilateral

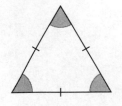

three equal sides

isosceles

two equal sides

scalene

no equal sides

right-angled

one angle is 90°

triangular number
NA1.6

A triangular number is the number of
dots in a triangular pattern:
The numbers form the sequence
1, 3, 6, 10, 15, 21, 28 ...

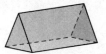

triangular prism
S2.6

A triangular prism is a prism with a triangular
cross-section.

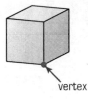

unit fraction
N2.3

A unit fraction has a numerator of 1.
For example, $\frac{1}{3}$ and $\frac{1}{7}$ are unit fractions.

unitary method
N2.5, N3.9

In the unitary method you calculate the value of one item or 1%
first.

variable
A4.4, D2.4

A variable is a quantity that can have a range of values.

vector
S3.1

A vector describes a translation by giving the x- and y-components
of the translation.

vertex, vertices
S4.1

A vertex of a shape is a point at which two or
more edges meet.

vertex

vertical
A3.1

A vertical line is at right angles to the horizontal.

vertically opposite angles
S1.1, S1.2

When two straight lines cross they form two
pairs of equal angles called vertically opposite
angles.

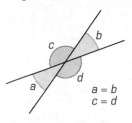

$a = b$
$c = d$

view
S4.2

A view of a solid is an accurate drawing of the appearance of the
solid above, in front or from the side.

volume
S2.2, S4.3

Volume is a measure of the space occupied by a 3-D shape.
Cubic millimetres, cubic centimetres and cubic metres are all units of
volume.

Glossary

x-axis, y-axis
A3.1

On a coordinate grid, the x-axis is the horizontal axis and the y-axis is the vertical axis.

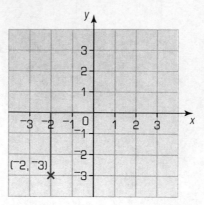

x-coordinate, y-coordinate
A3.1

The x-coordinate is the distance along the x-axis.
The y-coordinate is the distance along the y-axis.
For example, (⁻2, ⁻3) is ⁻2 along the x-axis and ⁻3 along the y-axis.

zero
N3.1

Zero is nought or nothing.
A zero place holder is used to show the place value of other digits in a number.
For example, in 1056 the 0 allows the 1 to stand for 1 thousand. If it wasn't there the number would be 156 and the 1 would stand for 1 hundred.

Thousands	Hundreds	Tens	Units
1	0	5	6

Answers

NA1 Check in

1 a 111   **b** 49   **c** 20
2 a 8   **b** ⁻8   **c** 36
   d 3   **e** 61   **f** 13

NA1 Check out

1 a $300 \times 1500 = 450\,000$
   b $20 \times 9 \times 5 = 900$ m
2 2.5 m by 2.5 m (square)
3 a i 10, 13, 16, 19, 22
     ii 3, 6, 11, 18, 27
     iii 5, 3, 1, ⁻1, ⁻3
     iv 3, 9, 19, 33, 51
   b i Linear   **ii** Quadratic   **iii** Linear
     iv Quadratic
   c $T(n) = \frac{n^3}{2}$ and 500, $T(n) = 10 - n^2$ and ⁻90,
     $T(n) = 2^n$ and 1024, $T(n) = (\frac{n}{2} + 3)^2$ and 64,
     $T(n) = (n + 1)(n + 4)$ and 154
4 a i $T(n) = 5n - 2$ **ii** $T(n) = n^2 - 1$
     iii $T(n) = 33 - 3n$ **iv** $T(n) = 3n^2$
     v $T(n) = \frac{(6n + 2)}{(n^2 + 2)}$
   b i 9, 16, 25, 36, 49
     ii $T(n) = (n + 5)^2$

S1 Check in

1 $a = 60°$, $b = 50°$, $c = 80°$
2 A Quadrilateral   **B** Isosceles trapezium
   C Regular pentagon

S1 Check out

1 $a = 42°$, $b = 138°$, $c = 42°$
   a is the corresponding angle to 42°.
   b is on a straight line with 42°
     $180° - 42 = 138°$.
   c is vertically opposite to *a*.

1 a $\frac{1}{10}$ **b** $\frac{1}{5}$

2 a $\frac{2}{3}$ **b** 0.76   **c** $\frac{3}{20}$

3

Pet	Tally	Frequency
Dog	⦀⦀ ⦀⦀ ⦀⦀	7
Cat	⦀⦀ ⦀⦀	5
Rabbit	⦀⦀	2

1 a $\frac{7}{8}$ **b** 2   **c** $\frac{5}{8}$

2 a $\frac{1}{16}, \frac{1}{8}, \frac{3}{16}, \frac{1}{4}, \frac{3}{16}, \frac{1}{8}, \frac{1}{16}$

b $\frac{3}{50}, \frac{7}{50}, \frac{4}{25}, \frac{6}{25}, \frac{9}{50}, \frac{3}{20}, \frac{2}{25}$

c When you convert the fractions to decimals they are similar.
The dice seem fair.

1 a $36 = 2^2 \times 3^2$, $48 = 2^4 \times 3$

b i 12   **ii** 144

2 a 0.45, 45%   **b** 14.5%, $\frac{29}{200}$ **c** 0.26, $\frac{13}{50}$

3 a i 93 mg   **ii** £160

b i 9.2   **ii** £42

1 a $\frac{-19}{55}$ **b** $1\frac{79}{136}$ **c** $1\frac{37}{60}$

d $21\frac{7}{33}$ **e** $\frac{4}{21}$ **f** $\frac{16}{63}$

g $7\frac{37}{45}$ **h** $1\frac{47}{100}$

2 a i 14 500   **ii** 5365

b 75 km

3 a i $66\frac{2}{3}\%$ **ii** $10\frac{1}{3}\%$

b £4480

4 Students $\approx 5 \times 140 = 700$, number of school days $\approx 40 \times 5 = 200$

So amount of water
$\approx 4 \times 700 \times 200 = 560\ 000$ litres

A2 Check in

1 $3 \times 4 + 3$, $3 + 2^2 \times 3$, $3^3 - 4 \times 3$
2 $\frac{5}{10}$ and $\frac{11}{22}$, $\frac{4}{14}$ and $\frac{10}{35}$, $\frac{6}{9}$ and $\frac{14}{21}$, $\frac{3}{13}$ and $\frac{12}{52}$
3 **a i** 1, 2, 3, 4, 6, 12
    ii 1, 2, 3, 5, 6, 10, 15, 30
    iii 1, 2, 3, 4, 6, 8, 12, 16, 24, 48
  b i 6
    ii 12
    iii 6
    iv 6

A2 Check out

1 **a i** 32   **ii** $^-27$ **iii** $\frac{1}{16}$
    iv 1   **v** $\frac{1}{25}$ **vi** 10 (or $^-10$)
  b i x^{14} **ii** y^8 **iii** z^6
    iv x^{-5} **v** $4x^6$
  c False, it equals a^{-1}.
2 **a** $7x^2 + 7x$ **b** $7ab$ **c** $30xyz$
  d $3p$ **e** $20x^7$ **f** $9a - 10b$
  g $7x$ **h** $x + 14$ **i** $24x^3$
3 **a** $9(x + 2)$ **b** $5(5y - 3)$
  c $3b(1 + 3a)$ **d** $5x(2y + x - 3)$
4 **a i** $x = 1\frac{6}{7}$ **ii** $x = 3\frac{1}{7}$
  b $x = 5$

S2 Check in

1 **a** 10 mm   **b** 0.56 m
2 **a** 21 m^2 **b** 31.5 m^2 **c** 2.88 m
  d 22.5 m^2
3 4.488 m^3
4 **a** 42.4   **b** 27.9   **c** 4.26
  d 116.0

S2 Check out

1 **a** 26.4 cm   **b** 55.4 cm^2
2 42.72 m^2
3 28.2 m^3

1 $abc = {}^-30$, $b^3 = {}^-8$, $a - 2c = {}^-7$, $\frac{(10 - b^2)}{a} = 2$, $c - 3b = 11$, $2a + 3b + 4c = 20$, $c^a = 125$

2 b The points form an arrow, which is a heptagon (7 sides).

3 a $y = 10 - x$

b $y = 2x + 4$

c $y = 3x + 3\frac{1}{2}$

d $y = \frac{1}{2}x + 7$

e $y = 4 - 1\frac{2}{3}x$

1 a i 3   **ii** $1\frac{1}{3}$ **iii** 1

    b i $-\frac{1}{3}$ **ii** $-\frac{3}{4}$ **iii** $^-1$

2 a i The car starts at a steady speed (A), continues at a slower steady speed (B), stops for a while (C), then returns to the starting point at a steady speed (D).

    b i Cup is full. Person takes a sip then waits a while, perhaps because it is too hot. Person drinks most of the tea, waits then finishes it or throws away the last bit.

    ii The water is heating slowly to start with, then gradually quicker, then tends to a steady temperature.

    iii The car accelerates, travels at a steady speed, then decelerates.

1 a 3   **b** 0   **c** 101

    d 7

2 a 13 717   **b** 35 441   **c** $107\frac{2}{9}$

    d 154.5

3

Fracton	Decimal	Percentage
$\frac{7}{12}$	$0.58\dot{3}$	$58.\dot{3}\%$
$3\frac{1}{4}$	3.25	325%
$2\frac{7}{10}$	2.7	270%
$\frac{1}{100}$	0.01	1%
$\frac{7}{20}$	0.35	35%
$\frac{33}{200}$	0.165	16.5%

1 a £16.42, 27.8%   **b** 9.5 m

2 a $(3600 + 400) \div 5 = 800$ **b** $\frac{(200 \div 40)}{9} = \frac{5}{9}$

S3 Check in

1

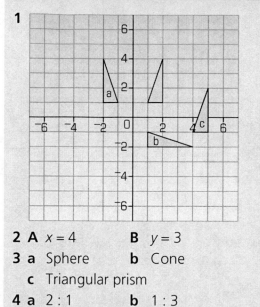

2 A $x = 4$ **B** $y = 3$

3 a Sphere   **b** Cone

  c Triangular prism

4 a $2 : 1$ **b** $1 : 3$

  c $20 : 1$ **d** $4 : 7$

S3 Check out

1 a B, C, E

  b i Translation $\binom{3}{0}$

    ii Translation $\binom{-2}{-5}$

    iii Rotation through 90° clockwise about (1.5, 0.5)

    iv Translation $\binom{-5}{-5}$

2 a 3.8 cm, 25.2 cm

  b 96 cm   **c** 16 cm

  d $1 : 6$ **e** $1 : 36$

A4 Check in

1 a $x = 17$ **b** $y = 4$ **c** $z = 3$

2 a $p = 3r + 2t$ **b** 5

3 a i $\frac{14}{45}$ **ii** $\frac{11}{21}$ **iii** $5\frac{7}{10}$

  b $4\frac{1}{9}$ cm

A4 Check out

1 a i $x = {}^-11$ **ii** $x = \frac{1}{3}$ **iii** $x = 2$

    iv $x = 1\frac{1}{4}$ **v** $x = {}^-13\frac{2}{3}$

  b i $x = 3$ **ii** $x = 66$ **iii** $x = 3$

  c i You can write an inequality, $3x + 6 > 42$.

    ii It is greater than 12.

2 a i $A = \pi r^2$ **ii** $r = \sqrt{(\frac{A}{\pi})}$

  b i $L = 10 - 2x$ **ii** $x = \frac{(10 - L)}{2}$

D2 Check in

1

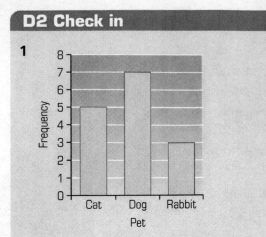

2 Mean = 5.14, Median = 5, Mode = 8, Range = 7

D2 Check out

1 a

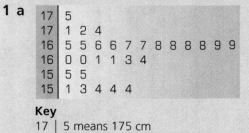

Key
17 | 5 means 175 cm

b Mean = 163.6 cm
Median = 165 cm
Range = 24 cm

c The modal class is 165–169 cm.
The median is a more representative average than the mean.

N4 Check in

1 a 33.64   **b** 219.76

2 a 27   **b** 70   **c** 11.5
    d 6570

3 a 30 366   **b** 11 696   **c** 10.5
    d 36.62

N4 Check out

2 a $^-5\frac{1}{4}$ **b** $2\frac{2}{3}$ **c** $5\frac{1}{24}$
    d $^-33$ **e** $\frac{11}{20}$ **f** $1\frac{59}{81}$

3 225.4 m

1 a 20, 23; T(n) = 3n + 2
  b 23, 27; T(n) = 4n − 1
  c 43, 53; T(n) = 10n − 7
  d 10, 8; T(n) = 22 − 2n
2 a 3x + 12   **b** 8x + 24
  c $x^2 - 6x$ **d** $3x^2 + 7x - 2xy$

1 a s = 3n + 1
  b i f = 4n + 2
    ii Middle cubes have 4 faces visible; both end cubes have an extra face visible.
2 a i y = 2.5x + 10
    ii He is paid £10 plus £2.50 per boat.
  b At birth your heart rate is 58 beats per minute. The rate increases by 3 beats per minute every 5 years.
3 a 8 camels   **b** 29 camels
4 2.7 cm

1 a i 25, 36   **ii** n^2
  b i 50, 72   **ii** $2n^2$
  c i 26, 37   **ii** $n^2 + 1$
2

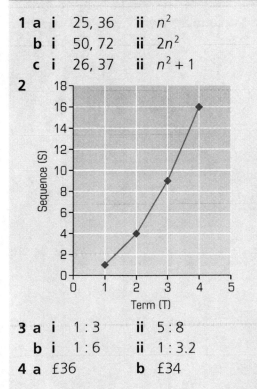

3 a i 1 : 3   **ii** 5 : 8
  b i 1 : 6   **ii** 1 : 3.2
4 a £36   **b** £34

1 64
2 Klearglass

1 5

2 a 42 cm³   **b** 8 mm²

3 a 40 mm   **b** 4.2 cm   **c** 0.52 m

  d 1 kg   **e** 3500 ml   **f** 2 g

4

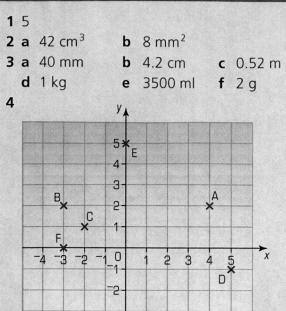

1 60 km

2 Perpendicular bisect of AB.

3 c 45°

  d The angle would be smaller.

1 Range = 7, Mean = 5, Median = 4.5, Mode = 4

2 a 2005   **b** 1.5

3

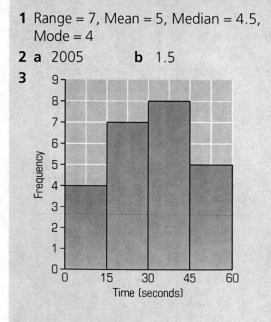

There are no unique answers.

D4 Check in

1 HH, HT, TH, TT
2 32
3 0.73
4 a $x = 210$ **b** $x = 100$ **c** $31\frac{1}{2}$

D4 Check out

1 b Hanif should expect to cycle home in five minutes 2 or 3 times a week.

Index

Index